Lecture Notes in Computer Science   13627

More information about this series at https://link.springer.com/bookseries/558

Marjan Mernik · Tome Eftimov ·
Matej Črepinšek (Eds.)

Bioinspired Optimization Methods and Their Applications

10th International Conference, BIOMA 2022
Maribor, Slovenia, November 17–18, 2022
Proceedings

 Springer

Editors
Marjan Mernik ⓘ
University of Maribor
Maribor, Slovenia

Tome Eftimov ⓘ
Jožef Stefan Institute
Ljubljana, Slovenia

Matej Črepinšek ⓘ
University of Maribor
Maribor, Slovenia

ISSN 0302-9743   ISSN 1611-3349 (electronic)
Lecture Notes in Computer Science
ISBN 978-3-031-21093-8   ISBN 978-3-031-21094-5 (eBook)
https://doi.org/10.1007/978-3-031-21094-5

This Springer imprint is published by the registered company Springer Nature Switzerland AG
The registered company address is: Gewerbestrasse 11, 6330 Cham, Switzerland

Preface

We live in a time of rapid development, high pace, and accelerated climate change. The primary focus of everyday life is the short-term and not the long-term benefits of the individual, where the exploitation of natural resources is at the forefront. At the same time, there is increasing awareness of the importance and effectiveness of biological systems that have thrived for billions of years. A good balance between exploitation and exploration of resources is crucial for living organisms' survival. The experiment of survival and development of living organisms on earth has been carried out for billions of years, with billions of participants in a dynamic and rapidly changing environment. The result of this evolution is a rich set of survival mechanisms that have enabled the survival of the most adapted species. Algorithms inspired by nature are those that try to transfer these proven mechanisms and their "survival wisdom" to the field of computer optimization while at the same time enabling a better understanding of it.

This volume contains recent theoretical and empirical contributions related to bioinspired optimization and its connection with machine learning presented at the Tenth International Conference on Bioinspired Optimization Methods and Their Applications (BIOMA 2022), held in Maribor, Slovenia, during 17–18 November, 2022. The BIOMA conference has been organized since 2004, with the primary purpose of bringing together a community for discussing recent advances in bioinspired optimization and existing challenges in transferring academic knowledge into real-world applications.

BIOMA 2022 received 23 submissions. The reviews were performed by 36 members of the international Program Committee. Each paper received three reviews. Based on these 19 papers were accepted, which is 82% of the submissions. The proceedings contain 19 papers by 55 (co)authors from 17 countries.

The conference invited two contributions to be presented as keynote lectures. The first keynote talk was provided by Carola Doerr from LIP6, Sorbonne Université and CNRS, Paris, France. The second keynote talk was provided by Shih-Hsi Liu from California State University, Fresno, USA.

The contributions presented at the conference covered genetic algorithms, swarm algorithms, genetic programming, surrogate-based optimization, reinforcement learning, improvements in differential evolution, sensitivity analysis for performance assessment, and explainable machine learning techniques for explaining bioinspired optimization algorithms' behavior. The applications of algorithms came from domains such as COVID-19 pandemic analysis and car rental problems.

Finally, we believe that the papers presented at BIOMA 2022 will provide readers with recent advances in bioinspired optimization and their applications and highlight possible gaps for future research. We are grateful to the conference sponsors, members of the Program and Organizing Committees, keynote speakers, authors, and other participants for being part of the conference.

November 2022

Marjan Mernik
Tome Eftimov
Matej Črepinšek

Organization

General Chair

Marjan Mernik   University of Maribor, Slovenia

Program Committee Chairs

Matej Črepinšek   University of Maribor, Slovenia
Tome Eftimov   Jožef Stefan Institute, Slovenia

Organizing Committee

Tomaž Kosar (Chair)   University of Maribor, Slovenia
Mario Casar   University of Maribor, Slovenia
Dragana Ostojić   University of Maribor, Slovenia

Steering Committee

Bogdan Filipič   Jožef Stefan Institute, Slovenia
Edmondo Minisci   University of Strathclyde, UK
Gregor Papa   Jožef Stefan Institute, Slovenia
Jurij Šilc   Jožef Stefan Institute, Slovenia
El-Ghazali Talbi   University of Lille, France

Program Committee

Aleš Zamuda   University of Maribor, Slovenia
Amer Draa   Université Constantine 2, Algeria
Bogdan Filipič   Jožef Stefan Institute, Slovenia
Boris Naujoks   Cologne University of Applied Sciences, Germany
Borko Bošković   University of Maribor, Slovenia
Carlos Coello Coello   CINVESTAV-IPN, Mexico
Carlos Cotta   University of Málaga, Spain
Carola Doerr   CNRS and Sorbonne University, France
Edmondo Minisci   University of Strathclyde, UK
Erik Dovgan   Jožef Stefan Institute, Slovenia
Fabio D'Andreagiovanni   CNRS, Sorbonne University - UTC, France

Contents

An Agent-Based Model to Investigate Different Behaviours in a Crowd Simulation

Carolina Crespi⬛, Georgia Fargetta⬛, Mario Pavone$^{(\boxtimes)}$⬛,
and Rocco A. Scollo⬛

Department of Mathematics and Computer Science, University of Catania,
Viale A. Doria 6, 95125 Catania, Italy
{carolina.crespi,georgia.fargetta,rocco.scollo}@phd.unict.it,
mpavone@dmi.unict.it

Abstract. This paper presents an agent-based model to evaluate the effects of different behaviours in a crowd simulation. Two different behaviours of agents were considered: collaborative, acting attentively and collaboratively, and defector who, on the other hand, acts individually and recklessly. Many experimental simulations on different complexity scenarios were performed and each outcome indicates how the presence of a percentage of defector agents helps and motivates the collaborative ones to be better and more fruitful. This investigation was carried out considering the (*i*) number of agents evacuated, (*ii*) exit times and (*iii*) path costs as evaluation metrics.

Keywords: Metaheuristics · Ant colony optimization · Swam intelligence · Optimization · Game theory

1 Introduction

Modeling and understanding a crowds' behaviour has become one of the most engaging and challenging topics of the last decades in different disciplines, such as establishing evacuation plans after an emergency [15, 19, 29], optimal architectural design [21–23], or even for entertainment purposes [12, 28]. All these fields are based on one key point: the comprehension of human behaviour. It is well known that how people act and react to different situations may positively or negatively affect the overall outcome, especially in emergencies and evacuations, where understanding the human behaviour is a primary issue to optimize evacuation plans [24]. However, the common problem encountered in these studies is the lack of human and social behavioural data [18]. For this reason, modeling and studying human behaviour is became one of the main purposes for many research areas [24]. Indeed, different models of crowd behaviour exist nowadays, each of which however focuses on different aspects of the problem depending on the framework used. Evacuation models can be classified into three main categories: (1) *macroscopic models* [4,6,9,10], which consider the crowd's dynamics as a flow; (2) *microscopic models* [8,14,17], which consider individual behaviour; and (3) *hybrid models* [1,2,7], which are a combination of both. Macroscopic models are mostly used to evaluate evacuation flow but cannot describe emergent crowd behaviour. Microscopic models, on the other hand, are used to investigate how small changes in the individual's characteristics affect the whole behaviour but are inefficient in large scenarios due to their computational cost. The main advantage in using hybrid models is that by combining macroscopic and microscopic methods, or even methods from different areas, one can exploit the best aspects of both or different approaches. Following the guidelines proposed in [24], in this paper an agent-based model is presented to evaluate whether and how different agent behaviours affect the collective behaviour of the whole group in a crowd simulation. In particular, a hybrid model has been developed in which agent-based models, which are one of the most powerful techniques to model individual-decision making and social behaviour, are combined to the features and dynamics of swarm intelligence methods. It consists of a set of agents that must reach a specific location, named *exit*, starting from a chosen point, adopting two different behavioural strategies: (*i*) the collaborative one, that is share information about the paths and/or repair destroyed paths; and (*ii*) defector that, on the other hand, doesn't share any information, can destroy some paths and/or nodes, but in any case exploits the help of the collaborative agents. Regardless of their behaviour, the goal of each agent is to reach the exit point. In this context, swarm intelligence algorithms, are useful not only for optimization purposes [13] but also to model the dynamics of the crowd [3,11,25,30], since they are capable to show the collective behaviours of the system under investigation. In the presented model, the Ant Colony Optimization (ACO) algorithm's principles and dynamics are considered to simulate the agents' behaviour and the environment setup. In particular, the agents are equipped with movement and decision rules that take inspiration from the ones used in ACO. The aim is to understand whether and how their behaviours, collaboratives, and defectors, affect the whole behaviour of the crowd. The investigation has been conducted by comparing simultaneously three evaluation metrics: (*i*) number of agents that have reached the exit; (*ii*) exit times, and (*iii*) cost of the paths to reach the exit. Using these metrics, the best expected performances are, therefore, the ones for which the number of outgoing agents is the highest possible, while the exit time and the path cost are the lowest possible.

2   The Mathematical Model

The idea of taking inspiration from the Ant Colony Optimization algorithm (ACO) to model the agents' behaviour comes from the observation that people in a crowd and ants seem to share some characteristics. Both of them, indeed, seem to behave following unwritten social rules. For instance, ants are able to find the shortest route from their anthill to a source of food and share it with the rest of the colony using some chemical signs called pheromones. This kind of communication is undirected because ants do not directly communicate with each other but release their pheromones along their path, and these pheromones act as roads to follow for the rest of the colony. The colony is able to find the best route thanks to this kind of communication and the finding itself is an example of emergent behaviour that is something indescribable if looking at the colony as a sum of single elements, the ants. On the other hand, in some contexts, for instance, in exiting and evacuation processes, people manifest the same local interactions, like ants do, since they take decisions following what their neighbours do, and they do it in absence of centralized decisions. Think for instance about how many times in a social context people find the exit of a place just by following the crowd. There is no one that guides the crowd, and the success of the process depends on how people are able to share information, directly by communicating with each other or indirectly by seeing what others do. In our model, the agents' behaviours take inspiration from the ones of the ants. The agents may be able to find promising routes in an unknown environment by communicating with each other in an indirect way. In addition, they may adopt an alternative behaviour of not sharing information about the path and destroying a part of it. The environment in which the agents move is represented as a weighted undirected graph $G = (V, E, w)$, where V is the set of vertices, $E \subseteq V \times V$ is the set of edges and $w: V \times V \to \mathbb{R}^+$ is a *weighted* function that assigns to each edge of the graph a positive cost. The *weighted* function highlights how hard is crossing an edge. Let define $A_i = \{j \in V : (i, j) \in E\}$ as the set of vertices adjacent to vertex i and $\pi^k(t) = (\pi_1, \pi_2, \ldots, \pi_t)$ as a non-empty sequence of vertices, with repetitions, visited by an agent k at the timestep t, where $(\pi_i, \pi_{i+1}) \in E$ for $i = 1, \ldots, t - 1$. Starting from a prefixed point, a population of N agents explore the environment trying to reach a destination point as quick as possible, through a path that has a lower cost. This population of agents is divided in Γ groups, each of which begins its exploration at regular intervals. For instance, it can be considered a simplified version of a delayed evacuation strategy, as the authors in [24] mention in their survey. Indeed, it is supposed that in some contexts, people do not evacuate all at the same time but organize themselves to evacuate in the ordered possible manner, for instance, in schools, public offices, and especially in recent pandemic plans to avoid Covid-19 diffusion. We have modeled this situation by establishing that each group starts its tour after a fixed time. At a specific time t, an agent k placed on a vertex i chooses as destination one of its neighbour vertices j, with a probability $p_{ij}^k(t)$ defined as the proportional transition rule defined in [5]:

$$p_{ij}^k(t) = \begin{cases} \frac{\tau_{ij}(t)^\alpha \cdot \eta_{ij}(t)^\beta}{\sum_{l \in J_i^k} \tau_{il}(t)^\alpha \cdot \eta_{il}(t)^\beta} & \text{if } j \in J_i^k \\ 0 & \text{otherwise,} \end{cases} \tag{1}$$

where $J_i^k = A_i \setminus \{\pi_{t-1}^k\}$ are all the possible displacements of the agent k from vertex i, $\tau_{ij}(t)$ is the trace intensity on the edge (i, j) and $\eta_{ij}(t)$ is the desirability of the edge (i, j) at a given time t, while α and β are two parameters that determine the importance of trace intensity with respect to the desirability of an edge. The trace intensity τ_{ij} on the edge (i, j) is a data that manifest how many times an edge is crossed by the agents and can help new agents to make a decision based on the actions of other agents. It is the equivalent of the pheromone in ACO algorithm. This value is a passive information, because the agents leave it unintentionally and after each movement the trace $\tau_{ij}(t)$ is increased by a constant quantity K, that is:

$$\tau_{ij}(t + 1) = \tau_{ij}(t) + K, \tag{2}$$

where K is a user-defined parameter. Equation 2 is the equivalent of the reinforcement rule of the ACO algorithm. This rule is so called because at each step, the amount of pheromone on a path (i, j) is augmented by the ants of a quantity that may be constant or not. In other words, every agent leaves a constant trace after crossing an edge (i, j). On the other hand, every T ticks[1] the amount of trace on the edges decays according to the global updating rule, which is also in this case the same present in ACO procedure. In ACO the algorithm, the global updating rule states that the amount of pheromone present in the environment is not fixed but it decays in time:

$$\tau_{ij}(t + 1) = (1 - \rho)\tau_{ij}(t), \tag{3}$$

where ρ is the evaporation decay parameter.

The desirability $\eta_{ij}(t)$ in Eq. 1, at a given time t, establish how much an edge (i, j) is promising. This information is not known a priory and it is released intentionally by an agent on a vertex after crossing an edge. In particular, $\eta_{ij}(t)$ is related to the discovered information by the agent k after crossing the edge (j, i). Its value depends on the inverse of the weight of the edge (i, j), that is:

$$\eta_{ij}(t) = 1/w(i, j). \tag{4}$$

It is important to note that the desirability is asymmetric because this information is present on the vertices, that is $\eta_{ij}(t) \leq \eta_{ji}(t)$ at a given time t. Lower the cost to cross an edge is, greater the desirability and the probability to follow a promising path is; vice versa, higher is the cost to cross an edge, lower is the desirability.

The agents are divided in two categories, each with its specific behaviour:

- **collaborators** C: they *leave* an information $\eta_{ij}(t)$ after crossing an edge (j, i) to help other agents during the escape and may *repair* a destroyed edge and/or a destroyed vertex before performing his movement, with a probability P_e^C and P_v^C respectively.
- **defectors** D: they *not leave* any information after crossing an edge and may *destroy* an edge and/or a vertex after performing his movement, with probability P_e^D and P_v^D respectively. A node or an edge destroyed is one no more traversable by other agents.

---

[1] The time unit used that corresponds to a single movement of all agents.

In other words, collaborators mainly perform actions that somehow help all the other agents to reach the rescue point as quick as possible. We can assume that they cross an edge or vertex in a cautious way, taking care not only to not destroy it, but also engaging themselves to repairing it if destroyed by a defector. Moreover, they leave an information $\eta_{ij}(t)$ about how hard is to cross a particular edge, so that the other agents can exploit in their own strategies. This increases the possibility to discover a more promising path toward the rescue point. To focus on a real situation, one can imagine leaving information to the other agent as, for example, a written message, a color mark, or a simple indication. On the other hand, the defectors mainly act in a hasty way, carrying out actions that can destroy the surrounding environment. Indeed, after crossing a node or an edge, and with a certain probability, they may destroy it decreasing the possibility of the other agents exploring the environment. This action may influence not only the cooperatives but also themselves especially if the destroyed path is an important one that is crucial to reach the location. The defectors' behaviour can be seen as a consequence of a stress and panic situation in which the agents like real people, due to this, are unable to be aware of their actions. They only try to find a good path by following the others and not informing the rest of the group about what they have found.

Table 1. Table of the variables and constants used.

Variable	Description
$w(i,j)$	Weight of an edge
$p_{ij}^k(t)$	Transition probability of the agents
$\tau_{ij}(t)$	Trace intensity on the edge
$\eta_{ij}(t)$	Desirability of an edge
$P_{e,v}$	Destruction/repair probability of a node and/or edge

To evaluate how these two different behaviour strategies influence the performance of all agents, the model takes into account as comparison metrics (i) the path cost, (ii) the exit time, and (iii) the number of agents that successfully reach the destination point. These three quantities have been considered together because the action of destroying/repair of nodes and/or edges makes the environment a dynamic environment. To clarify: once a simulation is launched, and if the population of agents is mixed with both kinds of agents, it may happen that one or more defectors cross an edge and/or a node and destroy it. Within the same simulation, it may also happen that one or more collaborator, approach the same nodes and/or edges destroyed by the defectors and may decide to repair them. Since a destroyed node and/or edge is no more traversable, it follows that these two actions change, from time to time, the structure of the environment, making the scenario dynamic. For this reason, considering just one of the three evaluation metrics mentioned above would have been incorrect because a promising path may not be the best in terms of cost, or just in terms of the success rate of the agents, or just in terms of exit time. A good path is one that minimizes its cost and exit time and, at the same time, maximizes the number of agents exiting that path.

Mathematically, the cost of a generic path $\pi(t) = (\pi_1, \pi_2, \ldots, \pi_t)$ is calculated as:

$$\sum_{i=1}^{t-1} w(\pi_i, \pi_{i+1}), \tag{5}$$

where π_1 and π_t are the starting and destination points, respectively. Since every tick all agents in the environment perform a single movement (from one vertex to another one), the exit time is calculated as the number of moves that an agent makes in its exploration and corresponds to the length of the path $\pi(t)$. All variables of interest are listed in Table 1.

3   NetLogo Model

The results are obtained using **NetLogo** [27], a multi-agent programmable modeling environment. As said in Sect. 2, the environments have been modelled as graphs with a topology similar to grid graphs, where each node can be connected with its 8-neighbours. The connectivity of a node with its neighbours is controlled by two parameters: $0 \leq p_1 \leq 1$, that represents the probability to create horizontal and vertical edges, and $0 \leq p_2 \leq 1$, that represents the probability to create oblique edges. The weight of each edge is a real value assigned with a uniform distribution in the range $[1, 100]$. Two different scenarios have been considered for the experiments:

– scenario **A** with $|V| = 100$ and $|E| = 213$, generated with $p_1 = 0.6$, $p_2 = 0.2$;
– scenario **B** with $|V| = 225$ and $|E| = 348$, generated with $p_1 = 0.6$, $p_2 = 0.0$.

They are both represented in Fig. 1.

4   Experimental Results

As the first step of this investigation, in both scenarios, have been considered $N = 1000$ agents divided into $\Gamma = 10$ groups. These values were chosen since 1000 agents represent a common number of people involved in a crowd, and 10 groups to better distribute the agents during the simulations. Depending on the value of the parameter $f \in [0, 1]$, user-defined and named *collaborative factor*, in each group there will be f collaborative agents, and $(1 - f)$ defectors. Therefore, in each group may be present both, or just one type of agent. In particular, if $f = 0.0$ groups with only defectors are considered; if $f = 0.5$ (for instance) each group is formed by half collaborative and half defectors; while for $f = 1.0$ only collaborative groups are generated.

Each group begins its exploration at different times, and precisely after a given time T_e from the group that precedes it, excepts the first group that

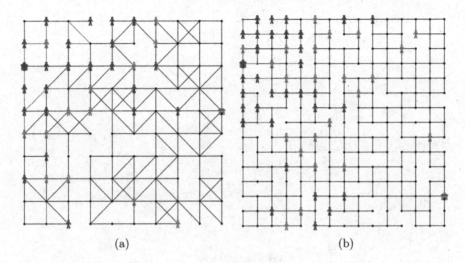

(a)   (b)

Fig. 1. Examples of networks used for the simulations in scenario **A** (Fig. 1a) and in scenario **B** (Fig. 1b). The starting point is represented by the house-shaped red node on the left of the network, while the exit point is the house-shaped green node on the right. Red nodes and edges represent the destroyed nodes by the defectors. The defectors themselves are represented by human-shaped blue agents, while the collaborators are of the same shape but in orange.

obviously starts at the time 0. In general, then, the i-th group will begin its exploration at the time $(T_e \times (i - 1))$. Note that the value to assign to T_e is related to the vertices number of the scenario considered ($T_e = |V|$). Therefore, at every T_e ticks, a new group starts its journey, having however a maximum time within which the agents must reach the exit. Let T_{max} the overall maximum time allowed to reach the exit, given by:

$$T_{max} = 2 \times \varGamma \times T_e, \tag{6}$$

where $\varGamma$ is the number of the groups, and 2 is a fixed parameter. It follows therefore that the time window within which each agent must reach the exit is from the begins of its exploration to the overall maximum time T_{max}, that is:

$$T_{max} - (T_e \times (i - 1)), \tag{7}$$

where i is the agent belonging group.

Analysing the Eqs. 6 and 7 one can note that the first groups have more time to explore the environment compared to the others. This is due because the groups begin their exploration even if in the environment are still present agents belonging to the previous groups. This means also that agents belonging to the same group can exit at different times (always within their time window) and those belonging to the first groups benefit more time to find the exit. It is important to highlight that the trace left by the collaborators along their path degrades over time with an evaporation interval fixed at $T_d = 50$. This means,

Average exit time per group - A

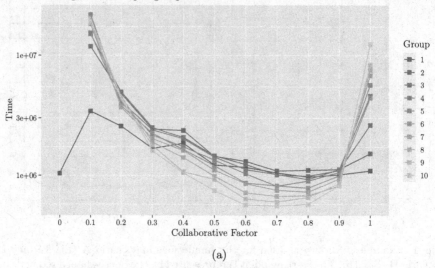

(a)

Average exit time per group - B

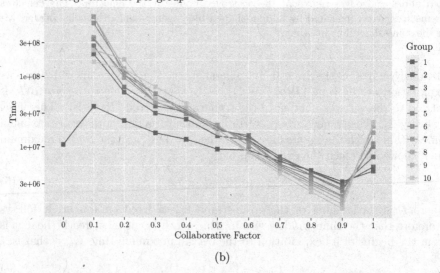

(b)

Fig. 2. The exit time for (a) scenario **A** and (b) scenario **B**.

then, that to every T_d ticks the global updating rule defined in Eq. 3 is applied with evaporation rate $\rho = 0.10$. Note that initially the trace on all edges is set to $\tau_{ij}(0) = 1.0$. Moreover, the parameters that regulate the importance of the trace and desirability in Eq. 1, i.e., α and β, are both set to 1.0. The destruction-repair probabilities on a vertex and an edge are $P_e^C = P_e^D = 0.02$ and $P_v^C = P_v^D = 0.02$, respectively and they are the same for both kinds of agents.

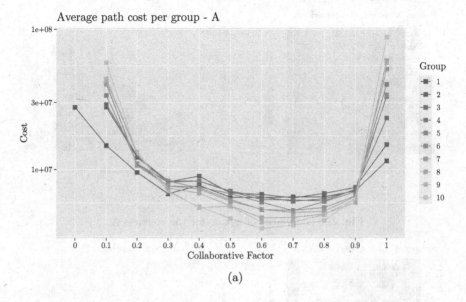

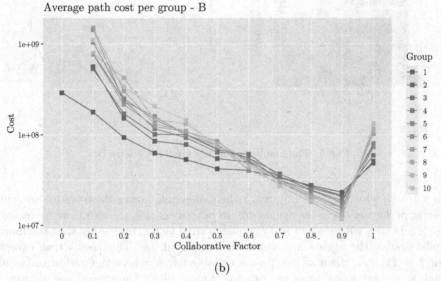

Fig. 3. The path cost for (a) scenario **A** and (b) scenario **B**.

Finally, to evaluate the effects of the two behaviours, collaboratives, and defectors, we have carried out experiments varying the collaborative factor f, that is the percentage of collaborators among the population of agents. For each value of f, from 0.0 to 1.0 with step of 0.1, we have performed 100 independent simulations. The exit time (Fig. 2) and path cost (Fig. 3) plots, have been normalized with respect to the group success rate, that is the percentage of agents

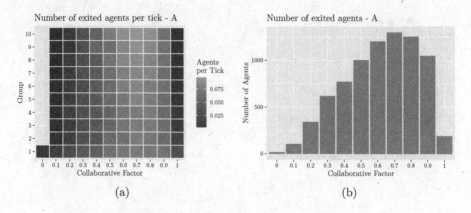

(a)

(b)

Fig. 4. The number of agents for the scenario **A**.

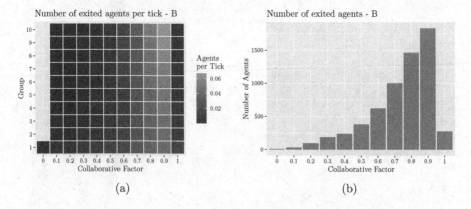

(a)

(b)

Fig. 5. The number of agents for the scenario **B**.

in a group, which successfully reach the exit point. Lower these values are, the better performances of the agents are. In both scenarios, the exit time decreases, so gets better, with respect to the collaborative factor, indicating that the more collaborative the agents are, faster their exit will be. This seems true except for $f = 1.0$, i.e., when all agents are collaboratives, where the performances of each group are worse than the previous values of f. The groups are indicated by different colored lines, and it also seems that the exit time decreases with respect to the group number, indicating that the groups that evacuate later, even if they have less allowed time, in some way, exploit the information left by those who have previously evacuated. In fact, looking for instance at group 1, in both scenarios, one can see how it has worse performances for low values of f, and better performances for high values of f. It means that the agents of this group can exploit better the information about the path especially when the crowd is composed mainly of collaborative agents. Do not be confused by the fact that the same group has good performance even for the lowest value of the collaborative factor f. Indeed, for that value, as it will be shown later, few agents find the exit from the environment and so, considering both metrics (the exit time and the number of agents exited) it can be concluded that it is not a significant result. On the other hand, by looking at group 10, in both scenarios, its performances improve with f, except for $f = 1.0$. This indicates that the last group can exploit better the information about the path when the crowd is composed mainly, but not totally, of collaborative agents. Same conclusions can be drawn for the path cost in Fig. 3, that decreases with respect to the collaborative factor and the group number. This is true for every value of f except for $f = 1.0$. This indicates, as above, that the more collaborative the agents are, the better path they will find, but if the collaboration is absolute, it seems to not work.

The heat maps in Figs. 4a and 5a represent the number of agents that reached the exit, and they are normalized with respect to the exit time available for each group. They represent how many agents have been evacuated in one unit of time (that is how many agents have been evacuated at each tick). The higher this value is, the better the performances of the agents are. The same trend is present, but opposite in value, observed for the exit time and the path cost: it seems that the number of exited agents increases with the collaborative factor except, also in this case, for $f = 1.0$, value for which few agents reach the exit. The performances of the agents are not better when all of them are collaborative but, oddly, when some of them act in a different way as defectors. The same quantity seems to increase with respect to the group number, indicating that the last groups benefit from the first ones, especially in scenario **A**.

Figures 4b and 5b represent the total number of exited agents for the **A** scenario and for the **B** scenario. Even without considering the group number, one can come to the same conclusions as above: the number of agents, that reaches the exit, increases with respect to the collaborative factor and one can better observe the collective behaviour of the simulated crowd. The maximum number of exited agents is obtained for $f = 0.7$ for scenario **A** and for $f = 0.9$ for scenario **B**. Considering the overall trend of the metrics used (the number of agents that reaches the exit, the path cost, and the exit time) it is possible to observe that the best performances of the agents are not when the entire group is composed only of collaborative agents, but when some of them are defectors. In other words, the crowd seems to perform better when some agents act differently and, in general, when there is a condition of mixed strategy among the agents.

5   Conclusions and Future Works

In this work, we proposed an agent-based model to evaluate the effects of two different behaviours in a crowd simulation: a collaborative and a defector one. Each strategy corresponds to different actions performed by the agents. The metrics used to evaluate the strategies are the number of exited agents, the path cost, and the exit time. From the results presented we can conclude that:

– a completely collaborative crowd has, in general, bad performances because it exits spending more time, by a more expensive path and does not maximize the number of agents that reach the exit point. This is since the destroy action performed by the defectors may help the rest of the group in pruning undesirable paths;
– a mixed crowd, in which are present both behaviours, is more efficient not only in obtaining the best values of the metrics used but also in the transmission of the information from one group to another;
– the results are confirmed for two scenarios with different characteristics, indicating that they may be generalized to more complex ones.

Since the agents' behaviour and environment setup follow the ACO rules, the results obtained can be used to improve the algorithm itself. In the literature, there exist several studies which demonstrate that a hybrid ant colony, or an ant colony with hybrid strategies, has, in general, better performances in solving different kinds of optimization problems [16,20,26]. However, it seems that no or few studies have been made about how these performances vary with respect to the composition of the colony with a user-defined parameter (in our case, the collaborative factor f). This aspect may open interesting opportunities in the optimization field because it can make it easier to make a step-by-step study and control the percentage of heterogeneity within a colony. The proposed model is still under investigation and surely needs more validation, both qualitative and quantitative. However, the abstraction used for the agents' communications seems to be reasonable if compared to what common sense suggests. In a situation in which we have to choose a direction, we mediate the information by considering not only what other people do (if we see some people that choose a path we are more confident in choosing the same path) but also what we objectively know about that path (if we know that a path has been chosen by a lot of people but we also know that it is not a good path, we may not choose it). This behaviour, represented by the product between the trace parameter and the desirability parameter, be adapted to different situations like, for instance when we have to choose a new mobile phone, a new job position, or a new car. We often mediate between what we see other people do and what we objectively know is true or good for us.

Future works include simulations with more complex scenarios, more starting points, and end points and, a sensitivity analysis of the parameters used.

References

1. Battegazzorre, E., Bottino, A., Domaneschi, M., Cimellaro, G.P.: Idealcity: a hybrid approach to seismic evacuation modeling. Adv. Eng. Softw. **153**, 102956 (2021). https://doi.org/10.1016/j.advengsoft.2020.102956
2. Chang, D., Cui, L., Huang, Z.: A cellular-automaton agent-hybrid model for emergency evacuation of people in public places. IEEE Access **8**, 79541–79551 (2020). https://doi.org/10.1109/ACCESS.2020.2986012

3. Crespi, C., Fargetta, G., Pavone, M., Scollo, R.A., Scrimali, L.: A game theory approach for crowd evacuation modelling. In: Filipič, B., Minisci, E., Vasile, M. (eds.) BIOMA 2020. LNCS, vol. 12438, pp. 228–239. Springer, Cham (2020). https://doi.org/10.1007/978-3-030-63710-1_18

4. Dogbe, C.: On the modelling of crowd dynamics by generalized kinetic models. J. Math. Anal. Appl. **387**(2), 512–532 (2012). https://doi.org/10.1016/j.jmaa.2011.09.007

5. Dorigo, M., Gambardella, L.: Ant colony system: a cooperative learning approach to the traveling salesman problem. IEEE Trans. Evol. Comput. **1**(1), 53–66 (1997). https://doi.org/10.1109/4235.585892

6. Fargetta, G., Scrimali, L.: Optimal emergency evacuation with uncertainty. In: Parasidis, I.N., Providas, E., Rassias, T.M. (eds.) Mathematical Analysis in Interdisciplinary Research, vol. 179, pp. 261–279. Springer, Cham (2021). https://doi.org/10.1007/978-3-030-84721-0_14

7. Gu, T., Wang, C., He, G.: A VR-based, hybrid modeling approach to fire evacuation simulation. In: Proceedings of the 16th ACM SIGGRAPH International Conference on Virtual-Reality Continuum and Its Applications in Industry. VRCAI 2018, Association for Computing Machinery, New York (2018). https://doi.org/10.1145/3284398.3284409

8. Han, Y., Liu, H.: Modified social force model based on information transmission toward crowd evacuation simulation. Phys. Stat. Mech. Appl. **469**, 499–509 (2017). https://doi.org/10.1016/j.physa.2016.11.014

9. Han, Y., Liu, H., Moore, P.: Extended route choice model based on available evacuation route set and its application in crowd evacuation simulation. Simul. Model. Pract. Theor. **75**, 1–16 (2017). https://doi.org/10.1016/j.simpat.2017.03.010

10. Hu, J., Li, Z., You, L., Zhang, H., Wei, J., Li, M.: Simulation of queuing time in crowd evacuation by discrete time loss queuing method. Int. J. Mod. Phys. C **30**(08), 1950057 (2019). https://doi.org/10.1142/S0129183119500578

11. Huang, Z.M., Chen, W.N., Li, Q., Luo, X.N., Yuan, H.Q., Zhang, J.: Ant colony evacuation planner: An ant colony system with incremental flow assignment for multipath crowd evacuation. IEEE Trans. Cybern. **51**(11), 5559–5572 (2021). https://doi.org/10.1109/TCYB.2020.3013271

12. Ijaz, K., Sohail, S., Hashish, S.: A survey of latest approaches for crowd simulation and modeling using hybrid techniques. In: Proceedings of the 2015 17th UKSIM-AMSS International Conference on Modelling and Simulation, UKSIM 2015, pp. 111–116. IEEE Computer Society (2015). https://doi.org/10.5555/2867552.2868182

13. Khamis, N., Selamat, H., Ismail, F.S., Lutfy, O.F., Haniff, M.F., Nordin, I.N.A.M.: Optimized exit door locations for a safer emergency evacuation using crowd evacuation model and artificial bee colony optimization. Chaos Solitons Fractals **131**, 109505 (2020). https://doi.org/10.1016/j.chaos.2019.109505

14. Li, Y., Chen, M., Dou, Z., Zheng, X., Cheng, Y., Mebarki, A.: A review of cellular automata models for crowd evacuation. Phys. A: Stat. Mech. Appl. **526**, 120752 (2019). https://doi.org/10.1016/j.physa.2019.03.117

15. Liu, H., Xu, B., Lu, D., Zhang, G.: A path planning approach for crowd evacuation in buildings based on improved artificial bee colony algorithm. Appl. Soft Comput. **68**, 360–376 (2018). https://doi.org/10.1016/j.asoc.2018.04.015

16. Meng, L., You, X., Liu, S.: Multi-colony collaborative ant optimization algorithm based on cooperative game mechanism. IEEE Access **8**, 154153–154165 (2020). https://doi.org/10.1109/ACCESS.2020.3011936

17. Mohd Ibrahim, A., Venkat, I., De Wilde, P.: The impact of potential crowd behaviours on emergency evacuation: an evolutionary game-theoretic approach. J. Artif. Soc. Soc. Simul. **22**(1) (2019). https://doi.org/10.18564/jasss.3837
18. Pan, X., Han, C.S., Dauber, K., Law, K.H.: A multi-agent based framework for the simulation of human and social behaviors during emergency evacuations. AI & Soc. **22**(2), 113–132 (2007). https://doi.org/10.1007/s00146-007-0126-1
19. Peng, Y., Li, S.W., Hu, Z.Z.: A self-learning dynamic path planning method for evacuation in large public buildings based on neural networks. Neurocomputing **365**, 71–85 (2019). https://doi.org/10.1016/j.neucom.2019.06.099
20. Randall, M.: Competitive ant colony optimisation. In: Okuno, H.G., Ali, M. (eds.) IEA/AIE 2007. LNCS (LNAI), vol. 4570, pp. 974–983. Springer, Heidelberg (2007). https://doi.org/10.1007/978-3-540-73325-6_97
21. Shahhoseini, Z., Sarvi, M.: Traffic flow of merging pedestrian crowds: how architectural design affects collective movement efficiency. Transp. Res. Rec. **2672**(20), 121–132 (2018). https://doi.org/10.1177/0361198118796714
22. Shi, X., Ye, Z., Shiwakoti, N., Tang, D., Lin, J.: Examining effect of architectural adjustment on pedestrian crowd flow at bottleneck. Phys. A Stat. Mech. Appl. **522**, 350–364 (2019). https://doi.org/10.1016/j.physa.2019.01.086
23. Shiwakoti, N., Shi, X., Ye, Z.: A review on the performance of an obstacle near an exit on pedestrian crowd evacuation. Saf. Sci. **113**, 54–67 (2019). https://doi.org/10.1016/j.ssci.2018.11.016
24. Siyam, N., Alqaryouti, O., Abdallah, S.: Research issues in agent-based simulation for pedestrians evacuation. IEEE Access **8**, 134435–134455 (2020). https://doi.org/10.1109/ACCESS.2019.2956880
25. Wang, S., Liu, H., Gao, K., Zhang, J.: A multi-species artificial bee colony algorithm and its application for crowd simulation. IEEE Access **7**, 2549–2558 (2019). https://doi.org/10.1109/ACCESS.2018.2886629
26. Wickramage, C., Ranasinghe, D.N.: Modelling altruistic and selfish behavioural properties of ant colony optimisation. In: 2014 14th International Conference on Advances in ICT for Emerging Regions (ICTer), pp. 85–90 (2014). https://doi.org/10.1109/ICTER.2014.7083884
27. Wilensky, U.: Netlogo: center for connected learning and computer-based modeling, Northwestern University, Evanston, IL (1999). http://ccl.northwestern.edu/netlogo/
28. Yücel, F., Sürer, E.: Implementation of a generic framework on crowd simulation: a new environment to model crowd behavior and design video games. Mugla J. Sci. Technol. **6**, 69–78 (2020). https://doi.org/10.22531/muglajsci.706841
29. Zhang, L., Liu, M., Wu, X., AbouRizk, S.M.: Simulation-based route planning for pedestrian evacuation in metro stations: A case study. Autom. Constr. **71**, 430–442 (2016). https://doi.org/10.1016/j.autcon.2016.08.031
30. Zong, X., Yi, J., Wang, C., Ye, Z., Xiong, N.: An artificial fish swarm scheme based on heterogeneous pheromone for emergency evacuation in social networks. Electronics **11**(4) (2022). https://doi.org/10.3390/electronics11040649

Accelerating Evolutionary Neural Architecture Search for Remaining Useful Life Prediction

Hyunho Mo[iD] and Giovanni Iacca[(✉)][iD]

Department of Information Engineering and Computer Science,
University of Trento, Trento, Italy
giovanni.iacca@unitn.it

Abstract. Deep neural networks (DNNs) obtained remarkable achievements in remaining useful life (RUL) prediction of industrial components. The architectures of these DNNs are usually determined empirically, usually with the goal of minimizing prediction error without considering the time needed for training. However, such a design process is time-consuming as it is essentially based on trial-and-error. Moreover, this process may be inappropriate in those industrial applications where the DNN model should take into account not only the prediction accuracy but also the training computational cost. To address this challenge, we present a neural architecture search (NAS) technique based on an evolutionary algorithm (EA) that explores the combinatorial parameter space of a one-dimensional convolutional neural network (1-D CNN) to search for the best architectures in terms of a trade-off between RUL prediction error and number of trainable parameters. In particular, a novel way to accelerate the NAS is introduced in this paper. We successfully shorten the lengthy training process by making use of two techniques, namely architecture score without training and extrapolation of learning curves. We test our method on a recent benchmark dataset, the N-CMAPSS, on which we search for trade-off solutions (in terms of prediction error vs. number of trainable parameters) using NAS. The results show that our method considerably reduces the training time (and, as a consequence, the total time of the evolutionary search), yet successfully discovers architectures compromising the two objectives.

Keywords: Evolutionary algorithm · Multi-objective optimization · Convolutional neural network · Remaining useful life · N-CMAPSS

1 Introduction

Predictive maintenance (PdM) is one of the key enabling technologies for Industry 4.0. It develops a maintenance policy using predictions of future failures of industrial components. Considering that this can be realized by estimating remaining useful life (RUL) of the target components, the RUL prediction has attracted considerable research interest, and much attention also from industry stakeholders.

M. Mernik et al. (Eds.): BIOMA 2022, LNCS 13627, pp. 15–30, 2022.
https://doi.org/10.1007/978-3-031-21094-5_2

Today, data-driven approaches using various deep learning (DL) models have gained increasing attention for developing RUL prediction tools. However, these models are usually handcrafted and their performance depends on the network architecture, usually set empirically. Such a design process can be time-consuming and computationally expensive because of the needed trial-and-error. Neural architecture search (NAS), a technique that enables to design the architectures automatically, can be a reasonable solution for this problem. Particularly, the realization of NAS through an evolutionary algorithm (EA), the so called evolutionary NAS, has attracted considerable attention.

In the field of RUL prediction, a recent work [1] applied evolutionary NAS to design the architecture of a data-driven DL model automatically. The authors use a genetic algorithm (GA) to optimize the architecture of a complex DL architecture that was manually designed in their previous work [2], aimed at improving RUL prediction accuracy. Solving such an optimization problem for better prediction accuracy can be formalized as follows:

$$w^*(a) = \arg\min_w \mathcal{L}_{train}(w, a) \tag{1}$$

$$a^* = \arg\min_a \mathcal{L}_{val}(w^*(a), a) \tag{2}$$

where Eq. (1) describes an inner evaluation loop that aims to find the optimal weights w^* for a given architecture (described by its parameters a) w.r.t. the training loss $\mathcal{L}_{train}$, while the outer loop defined by Eq. (2) searches for the optimal architecture (i.e., the one described by the parameters a^*) w.r.t. the validation loss $\mathcal{L}_{val}$.

There are two problems in the above optimization task. As shown in Eq. (2), the algorithm searches for an optimal architecture w.r.t. the prediction accuracy, regardless of the size of the network. Although this aspect has not been thoroughly discussed so far in the existing literature, limiting the size of the network determined by the number of trainable parameters is an important objective in industrial contexts that normally seek to save cost by minimizing access to expensive computing infrastructures. To solve this problem, in this paper we propose to evolve a one-dimensional convolutional neural network (1-D CNN) simultaneously subject to the two objectives of reducing the RUL prediction error and minimizing the number of trainable parameters. For this multi-objective optimization (MOO) task, we use the well-known non-dominated sorting genetic algorithm II (NSGA-II) [3], which has already been applied successfully to NAS tasks [4,5].

Another challenge of the aforementioned task is that it typically requires a lengthy and rather expensive training process. As shown in Eq. (1), evolutionary NAS is computationally expensive because each individual (i.e., candidate network architecture) should tune its parameters iteratively with gradient-based computations until convergence, before being evaluated on the validation data. To address this issue in our MOO approach, we propose a method for speeding up the training formulated in Eq. (1) by combining two techniques: architecture score without training [6] and extrapolation of learning curves.

The idea behind the architecture score is to predict the performance of a trained network from its initial state. This score measures the overlap of activations in an untrained network between different inputs from a mini-batch of data, so that a higher score at initialization implies better performance in terms of prediction error after training. Based on our preliminary experiments (not reported here for brevity), we found that this score is distinctive for networks with less than a certain number of trainable parameters. For those networks, we replace the expensive training step with the architecture score. For the networks with a larger number of trainable parameters, we instead apply extrapolation of learning curves. This technique prevents the need for full training (as done in the existing literature, where training is typically continued until a given maximum epoch, set large enough to allow convergence, before computing the validation loss), by training the network for a smaller number epochs, and observing the validation root mean square error (RMSE) after each training epoch. The observations are then used for estimating the validation RMSE at the maximum epoch. Specifically, we derive a learning curve based on the observations, and extrapolate it to take the predicted validation RMSE at the maximum epoch.

To test the proposed method, we have used the new commercial modular aero-propulsion system simulation (N-CMAPSS) dataset provided by NASA [7], which is a well-established benchmark in the area of RUL prediction. On this dataset, we search for optimal CNN architectures compromising the RUL prediction error and the number of trainable parameters. The experimental results verify that speeding up the evolutionary search causes the reduction of the hypervolume (HV) of just around 3% (compared to the NAS without acceleration), but the proposed method provides a considerable overall runtime reduction of approximately 75% in terms of GPU hours.

To summarize, the main contributions of this work can be identified in the following elements:

- The proposed method significantly shortens the evaluation time of the evolutionary NAS process.
- The networks discovered by the architecture search process represent successful trade-off between two conflicting objectives, namely the RUL prediction error and the number of trainable parameters.

The rest of the paper is organized as follows: in the next section Sect. 2, the general concepts on RUL prediction are introduced. The details of the proposed methods are presented in Sect. 3. Then, Sect. 4 describes the specifications of our experiments, while Sect. 5 presents the numerical results of the experiments. Finally, Sect. 6 discusses the conclusions of this work.

2   Background

Recently, various RUL prediction methods have been proposed, which can be mainly categorized into two approaches [8]: physics-based approaches and data-driven approaches. The former require extensive knowledge to analytically model the physical degradation process [9]. In practice, these implementations have been limited because the physics underlying degradation is well-understood only for relatively simple components, despite the huge amount of efforts [10]. On the other hand, data-driven approaches do not suffer from the above problems as they assume that the information relevant to the health and lifetime of components can be learned from past monitoring data [11].

Due to their ability to learn degradation patterns directly from historical data without knowing the underlying physics, data-driven approaches have gained increasing attention. Especially, black-box models based on deep learning have been widely used for prediction [12]. Figure 1 illustrates the flowchart of a data-driven RUL prediction task with a black-box model. The object of the RUL prediction is a target component. The sensors installed on the target collect the health monitoring data, usually recorded in the form of multivariate time series. The data are then fed into a black-box model that derives a RUL prediction as its output. This model is trained on historical data collected by run-to-failure operations. The training loss is then defined based on the difference between the predicted RUL and the actual RUL. After training, the model can directly provide the predicted RUL w.r.t. current sensor measurements. However, determining an appropriate black-box model is a key issue for developing successful data-driven RUL prediction tools.

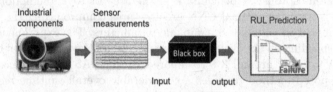

Fig. 1. Flow chart of a data-driven RUL prediction task.

Over the past decade, extensive research on data-driven approaches for RUL prediction using neural networks has been performed. One of the earliest works, introduced in [13], propose to use a multi-layer perceptron (MLP) for predicting the RUL of aircraft engines. The authors also propose to employ a convolutional neural network (CNN). Instead, the authors of [14] propose to a recurrent neural networks (RNN), in particular a long short term memory (LSTM), to recognize the temporal patterns in the time series. Considering the advantages of both CNNs and LSTMs, a combination of them has been used to predict RUL in [2]. As an alternative to use back propagation neural networks (BPNNs), Yang et al. [15] employ an extreme learning machine (ELM), a model originally introduced in [16], that achieves a much faster training compared to BPNNs. Recently, an autoencoder (AE) has been combined with RNNs [17] to obtain unsupervised learning. In [18], attention mechanism has been applied to a DL-based framework. Finally, deeper CNNs have been proposed in [10,19], showing comparable performances to the aforementioned combined architectures.

3   Method

We present now the details of the proposed method: Sect. 3.1 describes the individual encoding and the optimization algorithm we used, while Sect. 3.2 explains how we defined and applied the two techniques for speeding up the evaluation.

3.1 Multi-objective Optimization

Individual Encoding. Deep CNN architectures have provided outstanding performances on multivariate time series processing [20], also including RUL prediction [10,19]. Therefore, we adopt a 1-D CNN as our backbone network, whose architecture should be optimized. This network consists of a set of 1-D convolution layers and one fully connected layer: the n_l stacked convolution layers aim to extract high-level feature representations, while the following fully connected layer uses all the extracted features for regression.

Figure 2 visualizes a 1-D convolution layer in the network. Each of n_f filters of length l_f slides over its input features to apply convolution in the temporal direction. Each convolution layer is followed by an activation layer applying the rectified linear unit (ReLU) activation function. The feature map of the last convolution layer is flattened and fed into the fully connected layer comprising $n_{f.c.}$ neurons to predict the RUL.

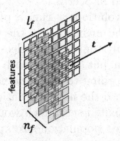

Fig. 2. Illustration of 1-D convolution layer with n_f filters of length l_f.

The number of convolution layers n_l and the two hyper-parameters regarding the convolution filter, n_f and l_f, contribute to the feature extraction. The number of neurons in the fully connected layer, $n_{f.c.}$, works on the regression task based on the extracted feature. All the four hyper-parameters largely affect the prediction error and determine the total number of trainable parameters in the network.

Based on the above description, we consider the optimization of the following architecture parameters:

- n_l, number of convolution layers;
- n_f, number of filters in each convolution layer;
- l_f, length of convolution filters;
- $n_{f.c.}$, number of neurons in the fully connected layer.

Considering that the architecture parameters are all integers, the encoding of solutions consists of four integers. The lower and upper bounds for each parameter considered in our evolutionary search are set as follows: [3, 8] for n_l, [5, 25] for both n_f and l_f, and [5, 15] (multiplied by a fixed value of 10) for $n_{f.c.}$. These values have been chosen empirically. In particular, the smaller networks, which have too few trainable parameters, cannot decrease the training loss (i.e., they underfit), while the larger networks, containing too many parameters, may overfit the training data. Taking these two aspects into account, we set the bounds so to explore a parameter space of approximately 30,000 1-D CNN configurations. One additional note is that n_f is not valid for the last convolution layer. The number of filters in the layer is set to 5, to prevent the fully connected layer from receiving a too long flattened feature.

Optimization Algorithm. In order to optimize the architecture of the CNN described in Sect. 3.1, we use the well-known NSGA-II algorithm [3], to look explicitly for the best trade-off solutions in terms of RUL prediction error and number of trainable parameters. In the evaluation step of our evolutionary search, the fitness of each individual is calculated by generating a CNN (the phenotype) associated to the corresponding genotype, i.e., a vector containing the four parameters introduced in Sect. 3.1.

At the beginning of the evolutionary run, a population of n_{pop} individuals is initialized at random. In the main loop of the GA, an offspring population of the same size is generated by tournament selection, crossover and mutation. The new individuals are then put together with the parents. The combined population is then sorted according to non-domination. Finally, the best non-dominated sets are inserted into the new population until no more sets can be taken. For the next non-dominated set, which would make the size of the new population larger than the fixed population size n_{pop}, only the individuals that have the largest crowding distance values are inserted into the remaining slots in the new population. Subsequently, the next generation starts with the new population by creating its offspring population. We stop this loop after a fixed number of generations n_{gen}.

Regarding the genetic operators, we consider 1-point crossover with crossover probability p_{cx} set to 0.5 and uniform mutation with mutation probability p_{mut} set to 0.5. The probabilities have been chosen such that, in most cases, individuals are produced by either mutation or crossover (exclusively), so to avoid disruptive effects due to the combination of mutation and crossover that may lead to bad individuals. The expected number of mutations per individual is determined by the probability p_{gene}, set to 0.4. It indicates the probability of applying the mutation operator to a single gene. This means that we have, on average, 1.6 mutated genes out of 4, which allows us not only to have a relatively faster architecture search process, but also to avoid disruptive mutations.

Finally, we set n_{pop} and n_{gen} to 20 and 10 respectively. We have empirically found that these values allow enough evaluations to observe an improvement on the HV spanned by the discovered solutions. After 10 generations, the algorithm returns a Pareto front, that is defined as the set of trade-off solutions at the top dominance level.

3.2   Speeding up Evaluation

Architecture Score Without Training. In almost all evolutionary NAS methods, the evaluation is the most time-consuming stage, because these methods typically evaluate a number of candidate networks on the validation data after the computationally expensive training [21]. In detail, a training set D_{train} is set to include all the available training data. This set is split into training purpose data, E_{train}, and validation purpose data, E_{val} (i.e., $D_{train} = E_{train} \cup E_{val}$). Then, E_{train} is used for the training process defined in Eq. (1) while the architecture search process, defined in Eq. (2), is based on E_{val}.

To reduce the time needed for the evolutionary search, we employ a speed-up technique called architecture score without training [6]. This method predicts the performance of a trained network based on its ability to discriminate between the different inputs of the network upon initialization, instead of training it.

Given that we use ReLU as the activation function in the networks, the output activation of each unit can indicate whether the unit is active or inactive; if the activation value is non-zero positive, the unit is active; otherwise, it is inactive. This is encoded as a binary bit, representing the former case as 1 and the latter as 0, i.e., we set the output activation of the non-zero positive case to 1. Given a data mini-batch $X = \{x_i\}_{i=1}^M$, we feed an input sample x_i into a network containing N_{ReLU} activation units, and gather all the binary bits. Then, we obtain a binary code $c_i \in \{0,1\}^{N_{ReLU}}$ for each sample x_i, thus in total we have M binary codes for the mini-batch.

The underlying intuition for the binary activation codes is that the similarity of two binary codes from two different inputs reveals how difficult it is to separate them for the network. For instance, if two different inputs have the same binary code, they lie within the same linear region defined by the activation function and therefore they are particularly difficult to distinguish [6].

The similarity between two different binary codes for x_i and x_j can be measured by the Hamming distance $d_H(c_i, c_j)$, and the correspondence between binary codes for X can be computed by the kernel matrix $\boldsymbol{K}_H$:

$$\boldsymbol{K}_H = \begin{bmatrix} N_{ReLU} - d_H(c_1, c_1) & \cdots & N_{ReLU} - d_H(c_1, c_M) \\ \vdots & \ddots & \vdots \\ N_{ReLU} - d_H(c_M, c_1) & \cdots & N_{ReLU} - d_H(c_M, c_M)f \end{bmatrix}. \tag{3}$$

Based on $\boldsymbol{K}_H$, the architecture score is then defined as:

$$s = \frac{c}{\ln |\boldsymbol{K}_H|}. \tag{4}$$

Following Eq. (4), the determinant of the kernel matrix $|\boldsymbol{K}_H|$ is higher for the kernel closest to the diagonal, and large distances between two different codes mean that those can be well-separated by the neural network. Thus, a lower score for the same input batch at initialization implies a better prediction accuracy after training. We set the value of the constant c to 10^4, so that the score values in our work range approximately from 1 to 20.

Extrapolation of Learning Curves. The evaluation step in evolutionary NAS typically requires training each network for a given number of epochs, with a relatively small learning rate. Following our previous work [1], we know that using a large learning rate enables to reduce the number of epochs, but it makes the validation curve fluctuate, thus providing unreliable evaluations caused by overfitting. Early stopping policy has been widely used for saving a few epochs, but its result is largely affected by how we define the performance improvement and the amount of patience. Moreover, the early stopping policy can lead to inaccurate performance estimations [21]. To mitigate this problem, we propose an extrapolation of learning curves that allows to save half of the training time w.r.t. a predetermined number of epochs. Note that here the learning curve is based on the validation RMSE across epochs.

Our basic approach is to derive the learning curve based on a set of functions $f(x)$, which are combined after fitting each of them to the observations. Specifically, we terminate the training at n_t epochs, that is half of the maximum epoch n_m planned for convergence, and collect all the validation RMSE for each of the n_t epochs. The observations are then used to fit each function defined in Table 1 by non-linear least squares minimization:

$$\text{minimize} \sum_{j=1}^{n_t} (y_j^o - f(x_j))^2$$

where y_j^o indicates the observed validation RMSE at x_j. The obtained function is denoted by f^*. The algorithm to solve the least squares problem is the Levenberg-Marquardt algorithm [22].

Table 1. Functions $f(x)$ used for extrapolation of learning curves. We chose a set of functions from the literature [23], whose shape coincides with our prior knowledge about the trend of the validation RMSE.

Name	Formula
MMF	$\alpha - \frac{\alpha - \beta}{1 + \gamma x^\delta}$
Janoschek	$\alpha - (\alpha - \beta)e^{-\gamma x^\delta}$
Weibull	$\alpha - (\alpha - \beta)e^{-(\gamma x)^\delta}$
Gompertz	$\alpha + (\beta - \alpha)(1 - e^{-e^{-\gamma(x-\delta)}})$
Hill custom	$\alpha + \frac{\beta - \alpha}{1 + 10^{(x-\gamma)^\delta}}$

As shown in Fig. 3, each curve drawn by f^* is close to the observation curve, but no single function can sufficiently describe the learning curve. Therefore, we combine all the obtained functions by solving a linear regression:

$$\text{minimize} \| \boldsymbol{F}^* \boldsymbol{a} - \boldsymbol{y}^o \|_2^2$$

where $\boldsymbol{F}^* \in \mathbb{R}^{n_t \times k}$ contains all the function values from k functions f^* (in our experiments, $k = 5$) for n_t epochs, and $\boldsymbol{y}^o \in \mathbb{R}^{n_t}$ is a vector of observations.

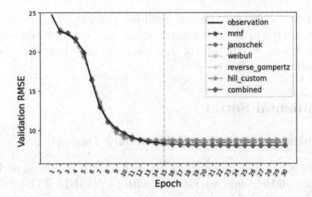

Fig. 3. An example of how the learning curve is derived from the $k = 5$ functions and the observations for 15 epochs. The red colored curve called "combined" represents the obtained learning curve. We take its value at 30 epochs and use it as the predicted validation RMSE. (Color figure online)

The optimal $a \in \mathbb{R}^k$, obtained by solving the linear problem problem, can be written as $a^* = [a_1^*, \cdots, a_k^*]$. Our target value is the predicted validation RMSE at x_{n_m}. For that, first we take the function value at x_{n_m} for each function f^*, i.e., $f^*(x_{n_m}) = [f_1^*(x_{n_m}), \cdots, f_k^*(x_{n_m})]$. The linear combination of these values with the weights a^* is then the target value $y_{n_m}^p$:

$$y_{n_m}^p = f^*(x_{n_m}) \cdot a^*. \tag{5}$$

If the validation RMSE has not converged yet, then our defined curve sufficiently decreases with x and the minimum observed value, $\min(y^o)$, is greater than $y_{n_m}^p$. This decay can be defined as $d = \min(y^o) - y_{n_m}^p$, where we take $y_{n_m}^p$ as the fitness value if the decay d is greater than 0.

During the evolutionary NAS process, many networks appear and each network converges at a different speed. Based on preliminary observations, we found that $y_{n_m}^p$ cannot be directly used as the fitness for some networks for which the learning curve decreases rapidly in the first few epochs and then reaches a plateau. In this case, our derived curve does not decrease with x and the decay d can be negative, while the actual validation RMSE may decrease even a little if we keep training the network. We compensate for this scenario by subtracting the absolute value of the decay to the minimum observed value. This way, we can assign a lower fitness value to the network that shows a validation RMSE trend that converges very quickly. Overall, the predicted fitness in terms of validation RMSE is then defined as:

$$fitness_{RMSE} = \begin{cases} y_{n_m}^p, & d > 0 \\ \min(y^o) - |d|, & d \leq 0. \end{cases} \tag{6}$$

In our experiments, we set the maximum epoch n_m to 30, based on our previous works [1,2,5]. If we terminate the training too early (i.e., after less than half n_m), then the predicted value may be too small because the learning curve shows no sign of convergence. On the other hand, using too many training epochs (close to n_m) would reduce any benefit of this speed-up technique. For these reasons, n_t is set to half n_m, i.e., 15.

4   Experimental Setup

4.1   Computational Setup and Benchmark Dataset

The 1-D CNNs are implemented using TensorFlow 2.4. All the experiments have been conducted on the same workstation with an NVIDIA TITAN Xp GPU, so that we can have a reliable comparison of the GA runtime in terms of GPU hours. To get reproducible results, we use the tensorflow-determinism library[1], which allows the DNNs implemented by TensorFlow to provide deterministic outputs when running on the GPU. The GA is implemented using the DEAP library[2]. Our code is available online[3].

To test the proposed method, we use the N-CMAPSS dataset [7] that consists of the run-to-failure degradation trajectories of nine turbofan engines with unknown and different initial conditions. The trajectories were generated with the CMAPSS dynamic model implemented in MATLAB, employing real flight conditions recorded on board of a commercial jet. Among the nine engines, we use 6 units (u_2, u_5, u_{10}, u_{16}, u_{18} and u_{20}) for the training set D_{train}, and the remaining 3 units (u_{11}, u_{14} and u_{15}) for the test set D_{test}. We select and use 20 condition monitoring signals following the setup in [10].

4.2   Data Preparation and Training Details

As shown in Fig. 2, the DNNs used in our work require time-windowed data as an input to apply 1-D convolution in the temporal direction. To prepare the input samples for the networks, first each time series is normalized to $[-1, 1]$ by min-max normalization. Then, we apply a time window of length 50 and stride 50 so that the given multivariate time series consisting of the 20 signals is divided into input samples, with each sample of size 50×20. After slicing the time series into samples, we assign 80% randomly selected samples from D_{train} to E_{train}. The remaining samples in D_{train} are assigned to E_{val}, which is used for the fitness evaluation.

For training, we use stochastic gradient descent (SGD). In particular, AMS-grad [24] is used as optimizer after initializing weights with the $Xavier$ initializer. We set the initial learning rate to 10^{-4} and divide it by 10 after 20 epochs, following our previous observations on the effect of learning rate decay [1]. The size of the mini-batch for the SGD is set to 512. This size is also used for defining a mini-batch for the architecture score. We randomly choose 512 samples from

---

[1] https://github.com/NVIDIA/framework-determinism.

[2] https://github.com/DEAP/deap.

[3] https://github.com/mohyunho/ACC_NAS.

E_{val}, and use it as the mini-batch X. On this regard, the ablation study in [6] verified that the choice of the mini-batch has little impact on the score trend over different network architectures.

5 Results

First, we generate 20 individuals (i.e., 1-D CNNs) randomly. Then, the multi-objective evolutionary process starts from the initial population. To perform a comparative analysis, we consider 5 different configurations w.r.t. the way of defining the fitness, denoted by $fitness_{RMSE}$: 1) using the architecture score, without training any networks; 2) using the validation RMSE, after training for 30 epochs; 3) using the validation RMSE, but training only for 15 epochs; 4) using the predicted validation RMSE at 30 epochs based on learning curve extrapolation, after training for 15 epochs; 5) using the architecture score if the network contains less than 5×10^4 trainable parameters, and the predicted validation RMSE with learning curve extrapolation otherwise.

The last configuration corresponds to our proposed method. We determine the decision threshold value to be 5×10^4 by analyzing the correlation between the number of trainable parameters and the architecture score. In Fig. 4, we can observe a negative correlation below the decision threshold. The difference in the architecture score for the range between 4 and 5 ($\times 10^4$) on the horizontal axis is trivial, but we take a large threshold value so that we can apply the architecture score based evaluation to as many networks as possible, because our major concern is to speed up the evolutionary search.

Here, we should note that the two different proposed surrogate mechanisms, i.e., the architecture score and the validation RMSE predicted by means of learning curve extrapolation, provide evaluation metrics that are obviously in different

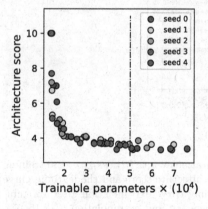

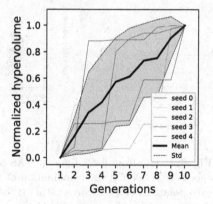

Fig. 4. Architecture score vs. number of trainable parameters on 100 randomly generated networks (20 for each seed). The dash-dotted line indicates the decision threshold.

Fig. 5. Normalized validation HV across generations (mean ± standard deviation across 5 independent runs) for the proposed NAS approach.

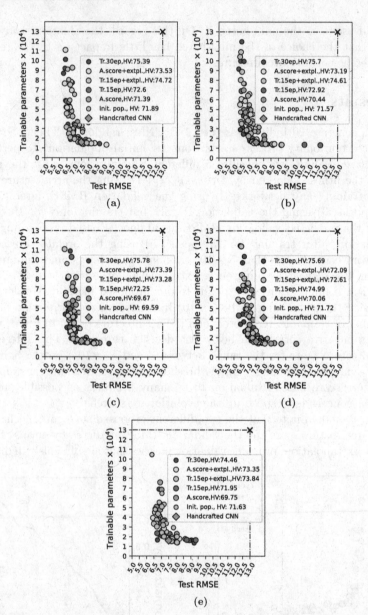

Fig. 6. Pareto front for 5 different NAS configurations w.r.t. $fitness_{RMSE}$: 30 training epochs ("Tr.30ep"); the combination of the architecture score and the learning curve extrapolation for ("A.score+extpl."); the learning curve extrapolation after 15 training epochs ("Tr.15ep+extpl."); 15 training epochs without extrapolation ("Tr.15ep"); merely using the architecture score ("A.score"). Each HV is calculated on the space shown in the figure which is defined by the test RMSE and the number of trainable parameters, and its value indicates the size of the space covered by the solutions of the corresponding configuration, with reference point (13, 13). Each figure shows the solutions found in 5 independent runs. The results of the handcrafted CNN, used as baseline, are taken from [10].

Table 2. Summary of the comparative analysis for 5 different NAS configurations w.r.t. $fitness_{RMSE}$. The HV is an avg.±std. of the values in Fig. 6 that are based on the test RMSE and the number of trainable parameters. The boldface indicates the proposed method, which includes both architecture score and learning curve extrapolation. It gives the shortest GA runtime of all the methods that achieve better results than randomly generated solutions.

Methods (w.r.t. $fitness_{RMSE}$)	Test HV (Avg. ± Std.)	GA runtime (GPU hours)
Initial population (without GA)	71.28 ± 0.95	–
Architecture score	70.26 ± 0.70	0.03 ± 0.01
Training 30 epochs	75.40 ± 0.55	4.96 ± 0.51
Training 15 epochs	72.94 ± 1.20	2.59 ± 0.30
Training 15 epochs + Extrapolation	73.81 ± 0.89	2.53 ± 0.15
Architecture score + Extrapolation	73.11 ± 0.58	1.23 ± 0.09

ranges. In order to use the score as fitness value (from the GA perspective), we proceed as follows. For all the individuals in the initial population, we calculate both the architecture score and the actual validation RMSE value. Then, we fit a cubic function to these values, by means of least squares minimization, as explained in Sect. 3.2. This fitted curve is then used to convert, for any new network, the architecture score to the corresponding best fit validation RMSE value. This mechanism is meant to prevent any potential bias in the relative comparison of architectures evaluated by means of different metrics.

We execute 5 independent runs with different random seeds to improve the reliability of the results. While searching for the solutions, we consider the validation HV, which is calculated on the fitness space defined by the validation RMSE and the number of trainable parameters; we collect the validation HV across 10 generations, and normalize it to $[0, 1]$ by min-max normalization. The monotonic increase of the mean of the normalized validation HV in Fig. 5 indicates that the GA keeps finding new non-dominated solutions across the generations.

After finding the solutions, our result analysis is based on the test RMSE, which is evaluated a posteriori. Therefore, the HV in the rest of this paper is calculated on the space defined by the test RMSE and the number of trainable parameters. Figure 6 shows the results of our experiments and Table 2 describes the summary of the comparative analysis. In the result analysis, we assess how the speed-up techniques affect the GA in terms of two metrics: 1) the quality of the solutions, represented by the HV, and 2) the GA runtime, in GPU hours.

It is obvious that the solutions based on the full training NAS are always the best in terms of HV, but it takes a rather long time (about 5 hours) to obtain them. When we merely use the architecture score without training, the NAS fails to find better solutions w.r.t. the initial population, because as said this approach alone cannot discriminate complex networks with a larger number of trainable parameters. If we terminate the training after 15 epochs, the obtained solutions are still better than the initial populations, but worse than the solutions obtained by training for 30 epochs. This implies that the learning curves of most of the networks appeared in our search converge later than 15 epochs. In this case, our extrapolation technique helps find better solutions for the same 15 epochs training time, i.e., it improves the HV without significantly increasing the GA runtime. Finally, the proposed method, which combines the two techniques, further decreases the runtime while the HV slightly decreases. Its HV is not comparable to the HV obtained when training for 30 epochs, but this method allows to save a considerable amount of search time. Compared to the case of training for 15 epochs, the proposed method not only achieves better HV, but it saves more than 50% of GPU hours.

6   Conclusions

In this work, we presented a multi-objective evolutionary NAS approach that uses a custom GA to optimize the architecture parameters of a 1-D CNN specialized to make RUL predictions. The multi-objective optimization is based on NSGA-II and aims to achieve a trade-off between two competing objectives: the RUL prediction error and the number of trainable parameters. To improve the efficiency of evaluations in the NAS process, we introduced two acceleration methods for evaluating networks with either training for a reduced number of epochs or no training at all. The experimental results on the benchmark show that the speed-up techniques save about 75% of the GA runtime, while the solutions are slightly worse but still much better than randomly generated networks.

The most important limitation of this work is that the learning curve cannot fully simulate the actual learning trend for some networks. In future work, we can consider a variety functions (i.e., more than the 5 functions considered in this work), to enforce the learning curve decay for all the networks.

References

1. Mo, H., Custode, L., Iacca, G.: Evolutionary neural architecture search for remaining useful life prediction. Appl. Soft Comput. **108**, 107474 (2021)
2. Mo, H., Lucca, F., Malacarne, J., Iacca, G.: Multi-head CNN-LSTM with prediction error analysis for remaining useful life prediction. In: 2020 27th Conference of Open Innovations Association (FRUCT), pp. 164–171. IEEE (2020)
3. Deb, K., Pratap, A., Agarwal, S., Meyarivan, T.: A fast and elitist multiobjective genetic algorithm: NSGA-II. IEEE Trans. Evol. Comput. **6**(2), 182–197 (2002)
4. Lu, Z., et al.: NSGA-Net: neural architecture search using multi-objective genetic algorithm. In: Genetic and Evolutionary Computation Conference (GECCO), pp. 419–427 (2019)
5. Mo, H., Iacca, G.: Multi-objective optimization of extreme learning machine for remaining useful life prediction. In: Jiménez Laredo, J.L., Hidalgo, J.I., Babaagba, K.O. (eds.) EvoApplications 2022. LNCS, vol. 13224, pp. 191–206. Springer, Cham (2022). https://doi.org/10.1007/978-3-031-02462-7_13

6. Mellor, J., Turner, J., Storkey, A., Crowley, E.J.: Neural architecture search without training. In: International Conference on Machine Learning, pp. 7588–7598. PMLR (2021)
7. Arias Chao, M., Kulkarni, C., Goebel, K., Fink, O.: Aircraft engine run-to-failure dataset under real flight conditions for prognostics and diagnostics. Data **6**, 5 (2021)
8. Atamuradov, V., Medjaher, K., Dersin, P., Lamoureux, B., Zerhouni, N.: Prognostics and health management for maintenance practitioners-review, implementation and tools evaluation. Int. J. Prognostics Health Manage. **8**(3), 1–31 (2017)
9. Bolander, N., Qiu, H., Eklund, N., Hindle, E., Rosenfeld, T.: Physics-based remaining useful life prediction for aircraft engine bearing prognosis. In: Annual Conference of the PHM Society, vol. 1 (2009)
10. Arias Chao, M., Kulkarni, C., Goebel, K., Fink, O.: Fusing physics-based and deep learning models for prognostics. Reliab. Eng. Syst. Saf. **217**, 107961 (2022)
11. Schwabacher, M., Goebel, K.: A survey of artificial intelligence for prognostics. In: AAAI Fall Symposium: Artificial Intelligence for Prognostics, Arlington, VA, pp. 108–115 (2007)
12. Khan, S., Yairi, T.: A review on the application of deep learning in system health management. Mech. Syst. Sig. Process. **107**, 241–265 (2018)
13. Sateesh Babu, G., Zhao, P., Li, X.-L.: Deep convolutional neural network based regression approach for estimation of remaining useful life. In: Navathe, S.B., Wu, W., Shekhar, S., Du, X., Wang, X.S., Xiong, H. (eds.) DASFAA 2016. LNCS, vol. 9642, pp. 214–228. Springer, Cham (2016). https://doi.org/10.1007/978-3-319-32025-0_14
14. Zheng, S., Ristovski, K., Farahat, A., Gupta, C.: Long short-term memory network for remaining useful life estimation. In: International Conference on Prognostics and Health Management (ICPHM), pp. 88–95. IEEE (2017)
15. Yang, Z., Baraldi, P., Zio, E.: A comparison between extreme learning machine and artificial neural network for remaining useful life prediction. In: Prognostics and System Health Management Conference (PHM), pp. 1–7 (2016)
16. Huang, G.B., Zhu, Q.Y., Siew, C.K.: Extreme learning machine: a new learning scheme of feedforward neural networks. In: International Joint Conference on Neural Networks (IJCNN), vol. 2, pp. 985–990. IEEE (2004)
17. Ye, Z., Yu, J.: Health condition monitoring of machines based on long short-term memory convolutional autoencoder. Appl. Soft Comput. **107**, 107379 (2021)
18. Chen, Z., Wu, M., Zhao, R., Guretno, F., Yan, R., Li, X.: Machine remaining useful life prediction via an attention-based deep learning approach. IEEE Trans. Ind. Electron. **68**(3), 2521–2531 (2021)
19. Li, X., Ding, Q., Sun, J.Q.: Remaining useful life estimation in prognostics using deep convolution neural networks. Reliab. Eng. Syst. Saf. **172**, 1–11 (2018)
20. Kiranyaz, S., Ince, T., Abdeljaber, O., Avci, O., Gabbouj, M.: 1-d convolutional neural networks for signal processing applications. In: ICASSP 2019–2019 IEEE International Conference on Acoustics, Speech and Signal Processing (ICASSP), pp. 8360–8364. IEEE (2019)
21. Liu, Y., Sun, Y., Xue, B., Zhang, M., Yen, G.G., Tan, K.C.: A survey on evolutionary neural architecture search. IEEE Trans. Neural Netw. Learn. Syst. (2021)

22. Moré, J.J.: The Levenberg-Marquardt algorithm: implementation and theory. In: Watson, G.A. (ed.) Numerical Analysis. LNM, vol. 630, pp. 105–116. Springer, Heidelberg (1978). https://doi.org/10.1007/BFb0067700

23. Domhan, T., Springenberg, J.T., Hutter, F.: Speeding up automatic hyperparameter optimization of deep neural networks by extrapolation of learning curves. In: Twenty-Fourth International Joint Conference on Artificial Intelligence (2015)

24. Kingma, D., Ba, J.: Adam: a method for stochastic optimization. In: International Conference on Learning Representations (ICLR) (2014)

ACOCaRS: Ant Colony Optimization Algorithm for Traveling Car Renter Problem

Elvis Popović[1,2]($\boxtimes$) (iD), Nikola Ivković[1] (iD), and Matej Črepinšek[2] (iD)

[1] Faculty of Organization and Informatics, University of Zagreb, Pavlinska 2, 42000 Varaždin, Croatia
elvpopovi@foi.hr

[2] Faculty of Electrical Engineering and Computer Science, University of Maribor, Koroška cesta 46, 2000 Maribor, Slovenia

Abstract. The Traveling Car Renter Salesman (CaRS) is a combinatorial optimization problem that is NP-hard and thus evolutionary and swarm computation metaheuristics are natural choices for designing a new practical algorithm. Considering that Ant Colony Optimization (ACO) is well suited for other routing type of problems - in this paper we propose ACOCaRS - an algorithm for solving CaRS based on ACO. The proposed algorithm was investigated experimentally and compared with other published algorithms for CaRS. The first results are encouraging since the proposed algorithm was significantly better for smaller problem instances than all the other published algorithms. However, for problem instances of size 100 and larger, ACOCaRS was the second best algorithm, and was outperformed significantly by a Transgenetic Algorithm. These results are based on the average performance of the algorithm and ranks, taking into account the number of wins and average ranks for the algorithms. A Friedman test confirmed that the results are statistically significant. In addition to average performance, data for assessing the peak performance of ACOCaRS are reported, along with a few new best known solutions for CaRS obtained in this research.

Keywords: Ant colony optimization · Algorithm · Combinatorial optimization · Car rental

1 Introduction

The Traveling Car Renter Salesman (CaRS) is a combinatorial optimization problem [1,2] that is related to the Traveling Salesmen problem and Vehicle Routing Problem. The problem concerns finding the optimal closed tour with the possibility of changing cars on route. Considering that CaRS is NP-hard [1],

---

This work has been fully supported by the Croatian Science Foundation under the project IP-2019-04-4864.

M. Mernik et al. (Eds.): BIOMA 2022, LNCS 13627, pp. 31–45, 2022.
https://doi.org/10.1007/978-3-031-21094-5_3

exact algorithms are viable only for solving small instances of the problem, and hence evolutionary and swarm computation metaheuristics are a natural choice for designing new practical algorithms. One extension of CaRS is the Traveling Car Renter Salesman With Passengers (CaRSP) where, in addition to multiple cars, there are a set of passengers who can share the traveling cost, and want to travel from one city to another. Several methods of solving these problems are published, such as evolutionary algorithms or optimization solvers. The motivation for implementing these optimizations can be found in the growing trends of reducing pollution in cities, as well as in the reduction of traffic jams and parking capacity problems. For such projects to come to life in practice, it is necessary to make the price of renting a car accessible to a wide range of customers, which ultimately means efficient use of the available cars for transporting passengers. It is also important to enable informatics systems to find the optimal routes and strategies for picking up cars on route quickly, so that the real situation can be monitored and managed in such organized traffic [3].

In this paper, we propose a new method for solving the CaRS problem with the Ant Colony Optimization (ACO) algorithm, labeled as ACOCaRS. In order to evaluate ACOCaRS' performance, a new method is compared experimentally with other published algorithms on available problem instances. The result suggests that ACOCaRS is currently the state-of-the-art for smaller problem instances, while it was the second best algorithm for larger problem instances.

The rest of the paper is structured as follows. Section 2 lists papers dealing with solving methods for CaRS problems and other publications that are relevant to this research. In Sect. 3, we describe the combinatorial problem we want to solve (CaRS), and we mention the symmetry of the problem concerning the starting node. In Sect. 4 we present the proposed algorithm ACOCaRS, that is based on Ant Colony Optimization metaheuristics with a theoretical background. Section 5 contains the experimental procedure, used benchmarks and method parameters. There are also comparative Tables of results. In Sect. 6, we discuss the results obtained and research limitations. In Sect. 7 we give a conclusion, and propose further guidelines for the research and development of the algorithm, including an upgrade to the Car Rental Salesman With Passengers (CaRSP) problem.

2 Related Work

The CaRS problem is defined by the authors of the article [1] as described in the next section. Goldbarg et al. have published several similar papers, but it is important that they define the problem and provide Tables with the results. Unfortunately, the Tables give only the Mean and best values and not the Standard Deviations, percentiles and other useful data needed to make a more exhaustive statistical comparison. In the paper, the authors describe the application of the Greedy Randomized Adaptive Search (GRASP) and Variable Neighborhood Descent (VND) procedures.

Since CaRS and CaRSP are new problems that have not been addressed by the ACO at all so far, we have compared a methodology applied to sufficiently similar problems, such as the classic Traveling Salesman Problem (TSP) [5]. On the other hand, we consider solutions to the same problem where the authors used methods different but comparable to ACO.

The proposal of Stützle and Hoos related to the MAX-MIN method of convergence regulation had the greatest impact on the development of our algorithm. The problems they solved were the classical Traveling Salesman Problem and the Quadratic Assignment Problem [6]. According to the results of their research, the authors confirm that the MAX-MIN ant system is among the best algorithms for solving such problems.

In paper [7], the authors deal with the problem of waste collection vehicle routing using the Ant Colony System (ACS) implemented in MATLAB (http://comopt.ifi.uni-heidelberg.de/software/TSPLIB95) [7]. The Capacitated Vehicle Routing Problem (CVRP) and the Capacitated Arc Routing Problem (CARP) have similarities with our CaRS problem. The authors use five stages of problem-solving. In the first three parts, the authors solve the Fleet Size Problem. Households from which waste needs to be collected are grouped into zones and clusters. The ACO is then applied to find the optimal route among the clusters. In the last step, the limits on the maximum permissible mass and volume of waste taken over for a particular vehicle are taken into account.

A Distributed Multilevel Ant Colonies Approach proposed by Taškova, Korošec, and šilc in their paper [8], uses artificial ants distributed in k colonies. The way artificial ants look for food is represented by mapping the graph to a particular grid in the first phase. The graph contraction procedure (coarsening) and the graph expansion procedure (refinement) were applied in the second phase.

In paper [1], Goldbarg, Asconavieta et al. describe the CaRS problem and give results for MA1 and MA2.

Another approach to solving CaRS problems is given in their next paper [2], where they propose the Transgenetic Algorithm (TA), evolutionary computing techniques inspired on the mutualistic intracellular endosymbiotic evolution. This algorithm surpassed all previous ones in terms of results.

In these papers, their authors have not given all the artifacts that would serve us in elaborating our method (e.g. Standard Deviations, medians and percentiles), but they provide results as usual.

Two articles of Sabry, G., Goldbarg, M. et al. [9,10] deal with a CaRSP problem. In the first paper, the authors propose a method of linearization as a form or problem relaxation to apply existing solvers [9]. In the second paper, the authors use an evolutionary approach to solve the same problem [10].

3   Problem Description

The CaRS problem is a generalization of the traveling salesman problem, in such a way that the salesman drives rented cars that can be exchanged during the route. In some problem variants one car can carry several passengers at the same time who share the costs (CaRSP). If the car is replaced in a city where it has not been rented, the car return cost is charged. In the CaRSP variant, the return cost is not shared between passengers, and is covered fully by the traveling salesman. The goal is to find a route, i.e. a sequence of cities and a sequence of rented cars that minimize the salesman's travel costs. Once a car is rented and returned, it cannot be rented again. In this way, the number of possible rental cars is reduced while traveling on a particular route where the vertices of the graph correspond to the problem nodes, i.e. cities.

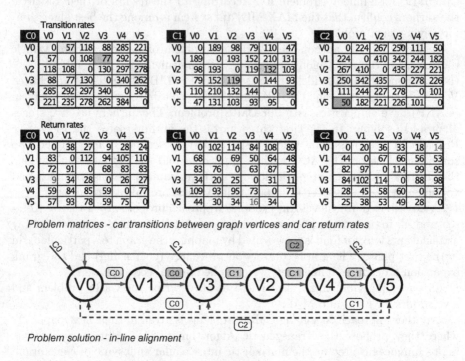

Transition rates

C0	V0	V1	V2	V3	V4	V5
V0	0	57	118	88	285	221
V1	57	0	108	77	292	235
V2	118	108	0	130	297	278
V3	88	77	130	0	340	262
V4	285	292	297	340	0	384
V5	221	235	278	262	384	0

C1	V0	V1	V2	V3	V4	V5
V0	0	189	98	79	110	47
V1	189	0	193	152	210	131
V2	98	193	0	119	132	103
V3	79	152	119	0	144	93
V4	110	210	132	144	0	95
V5	47	131	103	93	95	0

C2	V0	V1	V2	V3	V4	V5
V0	0	224	267	250	111	50
V1	224	0	410	342	244	182
V2	267	410	0	435	227	221
V3	250	342	435	0	278	226
V4	111	244	227	278	0	101
V5	50	182	221	226	101	0

Return rates

C0	V0	V1	V2	V3	V4	V5
V0	0	38	27	9	28	24
V1	83	0	112	94	105	110
V2	72	91	0	68	83	83
V3	9	34	28	0	26	27
V4	59	84	85	59	0	77
V5	57	93	78	59	75	0

C1	V0	V1	V2	V3	V4	V5
V0	0	102	114	84	108	89
V1	68	0	69	50	64	48
V2	83	76	0	63	87	58
V3	34	20	25	0	31	11
V4	109	93	95	73	0	71
V5	44	30	34	16	34	0

C2	V0	V1	V2	V3	V4	V5
V0	0	20	36	33	18	14
V1	44	0	67	66	56	53
V2	82	97	0	114	99	95
V3	84	102	114	0	88	98
V4	28	45	58	60	0	37
V5	25	38	53	49	28	0

Problem matrices - car transitions between graph vertices and car return rates

Problem solution - in-line alignment

Fig. 1. Example of a CaRS problem for 6 nodes i.e. cities and 3 cars. The problem matrices, node transitions for a particular car, and car return rates are shown on the top. In the bottom is the in-line arranged solution tour.

Solving the problem can be divided into two parts: Calculating the cost of a closed path on a graph (tour) and finding a path for which the price will be closest to the optimum. The costs of transitions between cities are different for each car, as are the costs of returning cars. Therefore, there are two groups of matrices: Transition rate matrices and car return rate matrices, as shown in Fig. 1 and proposed in article [1]. When the car is returned to the same city where it was rented, there is no charge for returning. It is also visible in the return rate matrices, where we have zeros on their diagonals. The only possibility of such a scenario is if one car covers the whole tour. If we move the starting city cyclically to any city where the cars are changed, the tour cost remains the same. It is the symmetry of the CaRS problem, so this is why we can have more than one optimal tour. In the example in Fig. 2, the tour cost remains unchanged if we set V0 or V3 or V5 as the starting city, but it changes if we use any other node as the starting point.

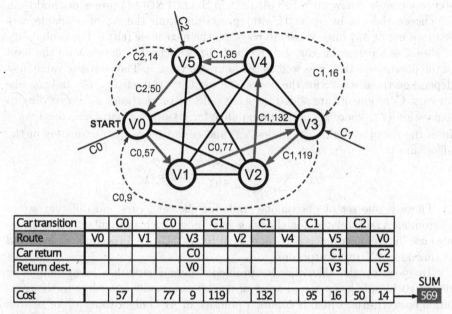

Car transition		C0		C0		C1		C1		C1		C2	
Route	V0		V1		V3		V2		V4		V5		V0
Car return					C0						C1		C2
Return dest.					V0						V3		V5

													SUM
Cost		57		77	9	119		132		95	16	50	14 → 569

Solution cost calculation - uses node transitions and car return rates

Fig. 2. Example of a CaRS problem for 6 nodes i.e. cities and 3 cars. Graphical presentation of the procedure for calculating the cost of the solution tour and problem graph with solution tour in red. The solution transitions, as well as the car return costs, are marked with the appropriate colors. (Color figure online)

4 ACOCaRS Algorithm

To address the problem we propose the algorithm using the meta-heuristics techniques of Ant Colony Optimization (ACO). "These are non-deterministic techniques, meaning there is no guarantee that the result will be the same for every run." [11]. ACO is a general framework with two main procedures: the solution construction procedure and the pheromone trails update procedure. The pseudocode of the ACOCaRS algorithm is presented in Algorithm 1, and its solution construction procedure in Algorithm 2. The entire source code written in language C is available on https://github.com/ElvisPopovic1/FERI.

The ACO uses a constructive procedure to build solution, starting with an empty solution to which solution components are added. In Algorithm 1 this procedure is denoted as CONSTRUCTSOLUTION() and explained in Algorithm 2. Both procedures use the arguments *Prob* which denotes the data loaded from the problem file, and *Param* which denotes the algorithm's parameters. The starting node (labeled with 0) is defined by the problem, and the first car is chosen according to the pheromone values in SELECTCAR() procedure. In each step, we choose a new node to visit next in SELECTNODE() procedure and than we choose the car in SELECTCAR() procedure from the set of available cars (current car or any unused car) to travel to the next node (city). The probability of choosing a particular car and the next city we travel to depends on the level of the pheromone trail τ as well as the heuristic value η. The heuristic value may depend on the distance for the nodes, the capacity, the return cost, and the like for cars. Components are added by using random-proportional rule according to expression (1), where p_{ij}^k is the probability for choosing a particular neighbor j from the set of available neighbors $\mathcal{N}_i^k$, and α and β are the parameters of the algorithm.

$$p_{ij}^k = \frac{[\tau_{ij}]^\alpha [\eta_{ij}]^\beta}{\sum_{l \in \mathcal{N}_i^k} [\tau_{il}]^\alpha [\eta_{il}]^\beta}, \text{if } j \in \mathcal{N}_i^k \tag{1}$$

There is one set of pheromone trails for choosing cars and different set of pheromone trails related for choosing next node with particular car, thus we also use different parameters α_c and β_c in SELECTCAR() procedure and α_n and β_n in SELECTNODE() procedure.

To speed up the algorithm for problems' instances with a large number of nodes, each node has a sorted array of neighboring nodes according to the transition cost criteria, which is common practice in ACO algorithms [6] for related (Symmetrical) Travelling Salesmen Problem and Asymmetrical and Travelling Salesmen Problem. Although speed is not the goal of the algorithm, this approach is necessary so that larger problems can be solved in a reasonable amount of time. When choosing the next node, we use only a limited set of nearest neighbors $\mathcal{N}_i^l, l < k$ and the cardinal number of this set as the algorithm parameter l. In the case that all neighbors in that set have already been used, the set is expanded.

The procedure is conducted until the solution is complete. The solution is complete when all the directed edges connecting the cities on the closed path are determined, and when the cars used to travel between cities are determined. It is not necessary to use all available cars in the solution. The solution is invalid if it does not connect all cities in one closed path.

More ants are searching for solutions independently, and the one whose solution is the best, leaves a pheromone trail for the components used in that solution. The contribution of the pheromone trail left by the elitist ant is inversely proportional to the cost of the solution as described in expression (2),

$$\Delta \tau_c^{(best)} = \begin{cases} \frac{1}{f(s^{best})}, & \text{if } c \in s^{best} \\ 0 & \text{else} \end{cases} \tag{2}$$

where s^{best} is the "best solution" according to some trail reinforcement strategy. In the iteration best strategy, the best obtained solution in the current iteration is used, while, in the global best strategy, the best obtained solution is used from the beginning of the algorithm [6]. More general strategies, like κ-best and max-κ-best, can be used to fine tune the behavior of the algorithm [12]. The strategy choice is a categorical (non-numerical) parameter. In addition to the deposition of pheromones, here also occurs evaporation of the existing pheromone trails, according to expressions (3) and (4) for arcs and cars, respectively.

$$\tau_{ij}^n \leftarrow (1 - \rho_n)\tau_{ij}^n, \forall(i,j) \in L \qquad (3)$$
$$\tau_{ik}^c \leftarrow (1 - \rho_c)\tau_{ik}^c, \forall ik \in C \qquad (4)$$

where ρ_n and ρ_c are parameters of the algorithm, and L and C are sets of all node arcs or node cars, respectively.

In general, the pheromone trail on the best ant path is updated by a combination of evaporation and deposition according to expression (5)

$$\tau_c = (1 - \rho) \cdot \tau_c + \Delta\tau_c^{(best)} . \qquad (5)$$

Normally, after some iterations, the pheromone separation should occur Because of pheromone update [13]. The pheromone trail of good components from previous best solutions will be much larger then the pheromone trail of other components. The pheromone values of components that are not part of a best solution for many iterations decrease exponentially according to expression (3) and expression (4). This can lead to algorithmic stagnation, because the probability of choosing some new component can become extremely small, or even become zero because of the limited precision of floating point number representation in computers. To avoid algorithmic stagnation the pheromone trails are limited between τ_{max} and τ_{min} as in the MAX-MIN Ant System [6]. The upper limit is determined by expression (6)

$$\tau_{max} = \frac{1}{\rho \cdot f(s^*)}, \qquad (6)$$

where $f(s^*)$ is the cost of the optimal solution s^*. For the lower limit there are analytical expressions for the Traveling Salesmen Problem, the Asymmetrical Traveling Salesmen Problem and the Quadratic assignment problem [13,14], but, generally for each problem type and algorithm variant, this expression would be different. For ACOCaRS, as with most other ACO algorithms, the analytical expression is not known. In these situations the lower limit τ_{min} is calculated by multiplying the upper limit with algorithmic parameter ϑ, as in expression (7). The value of parameter ϑ must be greater than 0 and less than 1. Generally, the appropriate value for ϑ decreases with the problem size and increases with parameter α [13,14].

$$\tau_{min} = \vartheta \cdot \tau_{max} \qquad (7)$$

Algorithm 1. ACOCaRS

Input: $(Problem, Param)$ ▷ $Param$ has all the parameters from Table 1
Output: $(bestSolution)$
1: $(\tau, \eta, sol, bestSolution) \leftarrow$ INIT(Problem, Param)
2: **for** $iteration = 0$ to $Param.n_{ants}$ **do**
3:      **for** $ant = 0$ to $Param.M_{iter}$ **do**
4:         $sol[ant] \leftarrow$ CONSTRUCTSOLUTION(Problem, Param, τ, η)
5:      **end for**
6:      $\tau \leftarrow$ UPDATEPHEROMONES(τ, Param, sol)
7:      $bestSolution \leftarrow$ FINDBESTSOL(sol)
8: **end for**
9: **return** $bestSolution$

Algorithm 2. ConstructSolution for ACOCaRS

Input: $(Prob, Param, \tau, \eta)$
Output: $(AntSol)$
1: $AntSol \leftarrow \emptyset$ ▷ $AntSol$ starts as empty partial solution
2: $currNode \leftarrow 0$ ▷ Each route start with node 0
3: $currCar \leftarrow$ SELECTCAR(Prob, Param, currNode, currCar)
4: $AntSol \leftarrow$ ADDTOSOL(AntSol, 0, currentCar, currentNode)
5: **for** $i = 1$ to $Prob.numOfNodes$ **do**
6:      $currtNode \leftarrow$ SELECTNODE(τ,η, Prob, Param, currtNode, currCar)
7:      $currCar \leftarrow$ SELECTCAR(τ,η, Prob, Param, currNode, currCar)
8:      $AntSol \leftarrow$ ADDTOSOL(AntSol, i, currCar, currNode)
9: **end for**
10: **return** $AntSol$

---

5   Experiment

We based our experimental work on the following research hypothesis: The ACO optimization approach, ACOCaRS is suitable for optimizing CaRS problems, and it is comparable with the state-of-the-art algorithms.

5.1   Testbed

To test the hypothesis that the developed ACOCaRS algorithm is comparable to the state-of-the-art algorithms in solving CaRS, we conducted experiments on available test data and compared our results with known results [1]. We used a group of non-Euclidean data since we prepared an appropriate parser for them. Test problems can be divided into two groups: Those with up to 50 nodes (cities), and those with 100 or more than 100 nodes. All problems have between 2 and 5 types of cars. To speed up the construction solution procedure of the algorithm we used parameter limited lists of potential nodes for problem instances with 100 nodes or larger. To make this possible, the set of neighbor nodes must be an n-tuple sorted by distance, from smallest to largest. We applied this technique only for problems with 100 or more than 100 nodes (cities), while, for smaller problems, the total number of nodes was considered, which is equivalent to setting it equal to the size of the problem. In the pheromone update procedure we used upper and lower pheromone limits like in the MAX-MIN Ant System. The upper pheromone limit was set according to the mathematical expression (6), and the lower limit τ_{min} was calculated by expression (7), where parameter ϑ was obtained experimentally. For a pheromone reinforcement strategy we used iteration best solution for all problem instances.

Before performing experimental research on problem instances, we adjusted the algorithm parameters by using the R studio and iRace package, "an extension of the Iterated F-race method for the automatic configuration of optimization algorithms, that is, (offline) tuning their parameters by finding the most appropriate settings given a set of instances of an optimization problem" [15]. The parameters of the algorithm were tuned for smaller problem instances and for some larger instances. For other larger problem instances the parameters were set manually to be similar to those tuned for other problems. The resulting parameter settings, as shown in Table 1, were used in the experimental research, aimed to validate the algorithmic performance of ACOCaRS. We have used standard symbols for algorithmic parameters and their role is specified in mathematical expressions (1) to (7) for some and in the pseudocode of Algorithm 1 and 2 for the others. The parameters with subscript "c" are related to cars and parameters with subscript "n' are related to nodes.

Table 1. Parameters for the calculated problems

Problem	ρ_n	α_n	β_n	ρ_c	α_c	β_c	n_{ants}	M_{iter}	ϑ_n	ϑ_c	n_{fv}	stall
BrasilRJ14n	0.15	0.7	1.2	0.17	0.7	1.2	550	10000	0.002	0.005	14	100
BrasilRN16n	0.15	1.6	2	0.17	1	2.2	900	10000	0.05	0.001	16	50
BrasilPR25n	0.1	1	2.5	0.07	1	1.2	850	10000	0.01	0.001	25	225
BrasilAM26n	0.1	1	2.5	0.07	1	1.2	850	10000	0.01	0.001	26	225
BrasilMG30n	0.16	0.68	3.3	0.12	0.73	1.19	1117	5000	0.0008	0.0003	30	206
BrasilRS32n	0.07	0.6	2.5	0.08	0.6	1.1	970	10000	0.006	0.004	32	60
BrasilSP32n	0.03	0.63	2.8	0.05	0.64	0.33	1242	5000	0.0009	0.0004	32	782
BrasilCO40n	0.07	0.62	3.29	0.12	0.63	1.93	1294	5000	0.0003	0.0002	40	800
BrasilNO45	0.03	0.7	2.37	0.14	0.64	1.85	1500	5000	0.0004	0.0004	45	600
BrasilNE50	0.05	0.7	2.5	0.12	0.7	2	800	6000	0.0001	0.0004	50	100
Londrina100n	0.05	1.05	2	0.1	1.05	2	500	10000	0.0002	0.0004	100	600
rd100nB	0.05	1.05	2	0.1	1.05	2	500	10000	0.00002	0.00005	65	600
kroB150n	0.12	1.14	1.46	0.067	1.92	2.07	952	7000	0.0005	0.0002	68	274
d198n	0.14	1.3	1.07	0.16	0.88	2.1	734	5000	0.0005	0.0004	68	277
Teresina200n	0.05	1.05	2	0.1	1.05	2	600	7000	0.00002	0.00004	68	1200
Curitiba300n	0.14	1.3	1.07	0.16	0.88	2.1	734	5000	0.0005	0.00004	68	277

5.2　Results

While the conducted experiment aimed to evaluate the performance of ACO-CaRS, for each problem instance, we ran the algorithm $n_{exec} = 51$ times, and reported the lowest price, 20th percentile, median, 80th percentile, the highest price, mean price, and Standard Deviation. The ACOCaRS improved solutions with the number of iterations, which can be attributed to the learning properties of the algorithm stored in the pheromone trails. Figure 3 shows an example of the observed behavior of ACOCaRS on the problem BrasilNO45n. The ordinate axis measures the cost of the solution, while the abscissa measures the number of iterations. The red discrete marks stand for the iteration costs obtained by one particular execution of the algorithm, and the rest are statistically calculated values for 51 executions. The graph is truncated so that the abscissa range from 0 to 1300 is reduced to the range from 540 to 940.

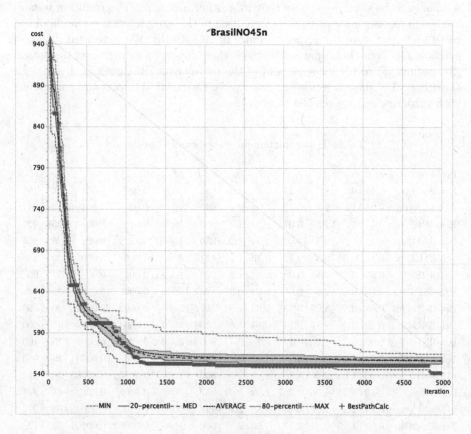

Fig. 3. An example of algorithm runs for the BrasilNO45n instance

The region between the 20th and 80th percentiles is filled with light blue. Obviously the median is in between the 20th and 80th percentiles, and the arithmetic mean in this case is very close to the median. For some other problem instances the arithmetic mean and median have observably different curves. Everything is contained between the lines that represent the minimum (best) and the maximum (worst) cost solutions.

We compared the obtained values for each problem with the available results of other algorithms from paper [1] and paper [2]. Exact stopping criteria for GVND1, GVND2, MA1, MA2, and TA were not published, only running times in range from 1 s to multiple hours for problems like Curitiba300n on personal computer. ACOCaRS found solutions in much less time, even for Curitiba300n it took less than 10 min on personal computer. Comparing our results with it, it turns out that ACOCaRS managed to find better solutions to problems of less than 100 nodes. For larger problems, the TA gave better results.

The results of the experiments are compared with other algorithms in Table 2 and Table 3. Table 2 shows the arithmetic mean solution obtained for a particular algorithm on each problem instance. Since the results for other published algorithms were rounded to integers, we reported our results in the same manner. Along with the arithmetic mean, in the brackets are reported ranks calculated for the Friedman test. The best results for each problem instance are set to bold font. It is observed that ACOCaRS had the best arithmetic mean for all problems instances with up to 50 nodes. For all larger instances the TA had the best arithmetic mean, while ACOCaRS was the second best, and sometimes the third best algorithm. Similarly, ACOCaRS had the best average rank for smaller instances, and the second best average rank, after TA, for larger instances. The average ranks are reported in separate rows in Table 2 as $\bar{r}_{.j}$. Friedman test confirmed statistically different performance with significance $\alpha = 0.05$ for both groups of problems. Since TA was already established as the best algorithm in previous publications, we have compared newly proposed ACOCaRS with TA and performed Wilcoxon signed-rank test. For instances with up to 50 nodes ACOCaRS had smaller mean solution for 7 problem instances, 3 ties and it was never worse than TA. This gives seven non-tie cases ($N = 7$) and statistic value $W = 0$, so the Wilcoxon signed-rank test confirms that ACOCaRS obtained better mean results with statistical significance $\alpha = 0.05$ (critical value for W is 2). For instances with at least 100 nodes, TA obtained smaller mean solutions than ACOCaRS in all six cases ($N = 6$) which gives value $W = 0$ and so the Wilcoxon signed-rank test confirmed that TA was better than ACOCaRS with statistical significance $\alpha = 0.05$ (critical value for W is 0).

Additional data about the experiments are contained in Table 3, with the best known solutions for particular problem instances and the algorithms that obtained them. As is shown, the ACOCaRS obtained the new best known solutions for some problem instances. The Table also contains other relevant statistics for ACOCaRS: The best solution, 20th percentile, which is relevant for assessing the peak performance of the algorithm, median, which is more suited for average performance, and arithmetic mean with Standard Deviation.

Table 2. Mean solutions with corresponding ranks of different algorithms on CaRS

Problem	ACOCaRS	GVND1	GVND2	MA1	MA2	TA
BrasilRJ14n	**167 (2.5)**	171 (5)	217 (6)	**167 (2.5)**	**167 (2.5)**	**167 (2.5)**
BrasilRN16n	**188 (1)**	203 (5)	260 (6)	195 (4)	191 (3)	189 (2)
BrasilPR25n	**227 (1.5)**	311 (5)	325 (6)	256 (4)	241 (3)	**227 (1.5)**
BrasilAM26n	**202 (1.5)**	242 (6)	236 (5)	212 (3)	213 (4)	**202 (1.5)**
BrasilMG30n	**274 (1)**	375 (6)	332 (5)	328 (4)	299 (3)	277 (2)
BrasilRS32n	**269 (1)**	372 (6)	346 (5)	340 (4)	311 (3)	271 (2)
BrasilSP32n	**257 (1)**	336 (6)	292 (4)	296 (5)	284 (3)	259 (2)
BrasilCO40n	**580 (1)**	826 (5)	845 (6)	743 (4)	660 (3)	583 (2)
BrasilNO45	**557 (1)**	889 (5)	922 (6)	764 (4)	667 (3)	561 (2)
BrasilNE50	**618 (1)**	1044 (5)	1071 (6)	861 (4)	736 (3)	629 (2)
$\overline{r}_{.j}$	**1.25**	5.4	5.5	3.85	3.05	1.95
Londrina100n	1410 (3)	1783 (5)	1785 (6)	1522 (4)	1369 (2)	**1201 (1)**
rd100nB	1789 (2)	2953 (6)	2844 (5)	2271 (4)	1832 (3)	**1442 (1)**
kroB150n	3738 (3)	5368 (6)	5300 (5)	4259 (4)	3675 (2)	**3018 (1)**
d198n	4518 (2)	7138 (5)	7289 (6)	5449 (4)	4665 (3)	**3264 (1)**
Teresina200n	1940 (2)	3793 (6)	3772 (5)	2551 (4)	2241 (3)	**1467 (1)**
Curitiba300n	3541 (2)	6125 (6)	6082 (5)	4076 (4)	3726 (3)	**2272 (1)**
$\overline{r}_{.j}$	2.33	5.67	5.33	4.00	2.67	**1.00**

Table 3. Best known solutions for CaRS and ACOCaRS statistics

| Problem | Best known solution | | ACOCaRS | | | | |
	Cost	Found by algorithm	Best	20th perc	Median	Mean	SD
BrasilRJ14n	167	ACOCaRS, MA1&2, TA	**167**	167	167	167.00	0.00
BrasilRN16n	188	ACOCaRS,MA2,TA	**188**	188	188	188.00	0.00
BrasilPR25n	226	ACOCaRS,TA	**226**	226	228	227.35	0.91
BrasilAM26n	202	ACOCaRS,TA	**202**	202	202	202.14	0.35
BrasilMG30n	271	ACOCaRS	**271**	272	274	273.67	1.23
BrasilRS32n	269	ACOCaRS,TA	**269**	269	269	269.12	0.33
BrasilSP32n	254	TA	257	257	257	257.18	0.99
BrasilCO40n	574	ACOCaRS	**574**	579	580	579.80	1.82
BrasilNO45	542	ACOCaRS	**542**	553	557	556.59	4.89
BrasilNE50	611	ACOCaRS	**611**	617	617	617.53	1.78
Londrina100n	1186	TA	1264	1360	1402	1409.65	61.27
rd100nB	1412	TA	1635	1749	1789	1788.73	58.95
kroB150n	2966	TA	3622	3674	3732	3737.86	98.51
d198n	3188	TA	4128	4338	4499	4517.59	236.73
Teresina200n	1410	TA	1786	1868	1919	1939.61	100.09
Curitiba300n	2222	TA	3281	3418	3512	3540.78	141.63

6   Discussion

In the process of creating a successful new algorithm for a combinatorial optimization problem based on the Ant Colony Optimization framework, the key factors are knowledge about the problem, experience with the ACO algorithm for other problems, and intuition. Applications of this algorithm can be outside of transport problems, such as telecommunications with different transmission media or analysis of electronic circuits. The general framework like ACO, or more specific variants, like MMAS that we used, provide only abstract ideas that can be realized in many different ways. There are many design choices that are critical, so the resulting performance must be evaluated experimentally. In that process of creation we did change our algorithm more then a few times until we got the ACOCarRS that we used in this research. Although ACO algorithms often perform better when some sort of local optimization is used, like 2-opt, short runs of a tabu search, Lin-Kernighan [6,16]. In this study we first wanted to explore basic ACO capabilities without local optimization (which is optional in ACO). Like in the case of other evolutionary computation and swarm intelligence algorithms, the behavior of ACOCaRS also depends on parameter setting, so we did some partial parameter tuning. In our future work there are different possibilities that could be explored. One research direction should be about trying different local optimization techniques. Some other design choices regarding basic ACO procedures might lead to a better ACO algorithm for this problem. There are extended pheromone trail reinforcement strategies like κ-best and max-κ-best that might provide finer control over algorithmic behavior [12]. Other ACO variants, like the Three Bound Ant System [4], showed some interesting theoretical and practical properties, so we can try to improve performance by adopting these techniques. To make parameter tuning easier we can try to develop analytical expressions for appropriate lower pheromone trails.

In this paper we compared ACOCaRS with other published algorithms for CaRS on non-Euclidean CaRS instances. Other papers use Euclidean and non-Euclidean instances, but unfortunately we could not decode the file format in which Euclidean instances were stored. We tried to contact the authors of these instances about their file format but we did not receive any reply. This is an obvious limitation of our research that we were unable to mitigate. In the future work we might try to generate a new set of Euclidean instances that will be well documented and available for public research.

Experimental research on other algorithms for CaRS published in the literature did not provide all the important information, but we did our best to make a fair comparison. Taking into account all the limitations, our experimental results show that, with high statistical significance, ACOCaRS outperformed the other algorithms for smaller problem instances, and went second best after the TA algorithm for larger problem instances. Although one of a few experimental studies are certainly not enough to (over)generalize conclusions about the performance of the algorithms for all imaginable instances, this study suggest that ACOCaRS is the state-of-the-art for small CarS instances, but not for larger problem instances, where the TA reported significantly better results.

In our results we have provided some additional information, like the 20th percentile as a measure of peak performance, and median as an appropriate measure of average performance, so that other independent researchers can compare their results with ours. Perhaps the 20th percentile is arguably the most interesting measure of performance, since, with arbitrarily high probability, one can obtain solutions of at least that quality if the algorithms is repeated more than once, e.g. with at least 89% after 10 repetitions or 99% after 20 repetitions of the algorithm [17].

7 Conclusion and Future Work

Solving CaRS and CaRSP classes of problems using ACO algorithms is a new research area. These problems have significant practical applications, and there is a lack of research on how to solve these problems with the Ant Colony Optimization method. There are various methods of solving these problems, such as the already mentioned MA1, MA2 and TA [1,2]. We built ACOCaRS - a complete algorithm to solve this class of problems using the ant colony optimization technique. As with other bioinspired optimization methods, the behavior of ACOCaRS is parameter-dependent, therefore, we tuned the parameters of the algorithm partially by the iRace framework. We compared our solutions with known solutions in the literature, where the authors used algorithms such as GVND, MA and TA. In the performed experiments, our algorithm outperformed GVNDs and the MA completely, while, at this point, it has outperformed the TA significantly only for smaller problem instances. For instances with 100 or more cities and 3 to 5 types of cars the TA found significantly better mean solutions than ACOCaRS.

In the future work we will try to improve ACOCaRS for larger problem instances by using various methods already mentioned in the Discussion (by adding local optimization, trying to adopt TBAS methods, using generalized pheromone reinforcement strategies, and possibly changing other algorithmic design choices). Our goal is also to develop an effective algorithm based on ACO for Traveling Car Renter with Passengers.

References

1. Goldbarg, M.C., Asconavieta, P.H., Goldbarg, E.F.G.: Memetic algorithm for the traveling car renter problem: an experimental investigation. Memet. Comput. **4**(2), 89–108 (2012)
2. Goldbarg, M.C., Goldbarg, E.F.G., Asconavieta, P.H., Menezes, M., Luna, H.P.L.: A transgenetic algorithm applied to the traveling car renter problem. Expert Syst. Appl. **40**(16), 6298–6310 (2013). https://doi.org/10.1016/j.eswa.2013.05.072
3. Bruglieri, M., Pezzella, F., Pisacane, O.: A two-phase optimization method for a multiobjective vehicle relocation problem in electric carsharing systems. J. Comb. Optim. **36**(1), 162–193 (2018). https://doi.org/10.1007/s10878-018-0295-5
4. Dorigo, M., et al. (eds.): ANTS 2014. LNCS, vol. 8667. Springer, Cham (2014). https://doi.org/10.1007/978-3-319-09952-1

5. Dorigo, M., Di Caro, G., Gambardella, LM.: Ant algorithms for discrete optimization. Artif Life **5**(2), 137–172 (1999) https://doi.org/10.1162/106454699568728
6. Stützle, T., Hoos, H.H.: MAX-MIN ant system. Future Gener. Comput. Syst. **16**(8), 889–914 (2000)
7. Reed, M., Evering, R.: An ant colony algorithm for recycling waste collection. In: BIOMA, pp. 221–230 (2012)
8. Taškova, K., Korošec, P., Šilc, J.: A distributed multilevel ant colonies approach. Informatica (Slovenia). **32**, 307–317 (2008)
9. Sabry, G.A., Goldbarg, M.C., Goldbarg, E.F.G., Menezes, M.S., Calheiros, Z.S.A.: Models and linearizations for the traveling car renter with passengers. Optim. Lett **15**(1), 59–81 (2020). https://doi.org/10.1007/s11590-020-01585-0
10. Sabry, G., Goldbarg, M., Goldbarg, E., Menezes, M. and J. Filho, G.: Evolutionary algorithms for the traveling car renter with passengers. In: IEEE Congress on Evolutionary Computation (CEC), pp. 1–8 (2020). https://doi.org/10.1109/CEC48606.2020.9185693
11. Eftimov, T., Korošec, P., Koroušić Seljak, B.: Data-driven preference-based deep statistical ranking for comparing multi-objective optimization algorithms. In: Korošec, P., Melab, N., Talbi, E.-G. (eds.) BIOMA 2018. LNCS, vol. 10835, pp. 138–150. Springer, Cham (2018). https://doi.org/10.1007/978-3-319-91641-5_12
12. Panigrahi, B.K., Suganthan, P.N., Das, S., Satapathy, S.C. (eds.): SEMCCO 2011. LNCS, vol. 7076. Springer, Heidelberg (2011). https://doi.org/10.1007/978-3-642-27172-4
13. Ivkovic, N., Golub, M., Malekovic, M.: A pheromone trails model for MAX-MIN ant system. In: Proceedings of 10th Biennal International Conference on Artificial Evolution, Angers, France, pp. 35–46 (2011)
14. Matthews, D.C.: Improved lower limits for pheromone trails in ant colony optimization. In: Rudolph, G., Jansen, T., Beume, N., Lucas, S., Poloni, C. (eds.) PPSN 2008. LNCS, vol. 5199, pp. 508–517. Springer, Heidelberg (2008). https://doi.org/10.1007/978-3-540-87700-4_51
15. López-Ibáñez, M., Dubois-Lacoste, J., Pérez Cáceres, L., Stützle, T., Birattari, M.: The Irace package: iterated racing for automatic algorithm configuration. Oper. Res. Perspect. **3**, 43–58 (2016). https://doi.org/10.1016/j.orp.2016.09.002
16. Dorigo, M., Stützle, T.: Ant Colony Optimization year. Bradford Company, USA (2004)
17. Ivkovic, N., Jakobovic, D., Golub, M.: Measuring performance of optimization algorithms in evolutionary computation. Int. J. Mach. Learn. Comput. **6**(3), 167–171 (2016). https://doi.org/10.18178/ijmlc.2016.6.3.593
18. Alfian, G., Farooq, U., Rhee, J.: Ant Colony Optimization for Relocation Problem in Carsharing (2013)
19. Eftimov, T., Korošec, P., Seljak, B.K.: Disadvantages of statistical comparison of stochastic optimization algorithms. In: Proceedings of the Bioinspired Optimizaiton Methods and their Applications, BIOMA, pp. 105–118 (2016)

A New Type of Anomaly Detection Problem in Dynamic Graphs: An Ant Colony Optimization Approach

Zoltán Tasnádi and Noémi Gaskó[(✉)]

Faculty of Mathematics and Computer Science, Babeş-Bolyai Unversity,
Cluj-Napoca, Romania
zoltan.tasnadi@stud.ubbcluj.ro, noemi.gasko@ubbcluj.ro

Abstract. Anomaly detection has gained great attention in complex network analysis. Every unusual behavior in a complex system can be viewed as an anomaly. In this article, we propose a new anomaly type in dynamic graphs, an existing community-based anomaly detection problem combined with the heaviest k-subgraph problem. Searching the heaviest subgraphs in dynamic graphs viewed as an anomaly problem can give new insights into the studied dynamic networks. An ant colony optimization algorithm is proposed for the heaviest k-subgraph problem and used for the community detection problem. Numerical experiments on real-world dynamic networks are conducted, and the results show the importance of the proposed problem and the potential of the solution method.

Keywords: Anomaly detection · ACO · Heaviest k-subgraph · Community detection

1 Introduction

Anomaly detection is an essential task with applications in biology [9], fraud detection [12], and social networks [16]. Since some problems can be represented as graphs, anomaly detection in graphs plays a key role in computational network problems. Generally, anomaly detection consists in finding an element of the graph (e.g., edge or node) that differs significantly from the rest of the graph [2].

In [1], anomalies are categorized into three main classes: 1) point anomaly, meaning that only one point deviates from the normal behavior, 2) contextual anomaly, meaning that data behave unusually in a certain context, and 3) collective anomaly, meaning that a group of data behave differently than the rest.

The survey [2] reviews the state of the art of anomaly detection in static and dynamic networks. [13] identifies and analyzes the types of anomalies in

This work was supported by a grant of the Ministry of Research, Innovation and Digitization, CNCS/CCDI - UEFISCDI, project number 194/2021 within PNCDI III.

dynamic networks: anomalous nodes, anomalous vertices, anomalous subgraphs, and event and change detection.

Although studies and algorithms exist regarding various anomalies, new perspectives and formulations of anomalies can capture new insights in network study. The goal of this paper is to propose a new anomaly type based on two properties. The first is community-based anomaly detection, which considers the different behaviors of the communities over time: splitting, shrinking, growing, or emerging new communities. The second is the heaviest k-subgraph problem [3], which consists in finding k connected nodes of a weighted graph, such that the total edge weight is maximized.

The rest of the paper is organized as follows: Sect. 2 presents the new anomaly detection problem in more detail, Sect. 3 presents the proposed ant colony-based approach, and Sect. 4 describes the numerical experiments. The article ends with a conclusion and further work possibilities.

2 Anomaly Detection Problem

In the next section, we introduce the two main problems, the community-based anomaly detection problem and the heaviest subgraph problem, and formalize the new anomaly type in dynamic graphs.

Although no formal definition for communities exists, we refer to them as disjoint groups that are more connected among themselves than to the rest of the graphs.

Definition 1. *Given a graph $G = (G_1, G_2, \dots)$ which varies in time, detection of communities in each step and observation if a change appears in the community structures (e.g., splitting or growing) means the community based anomaly detection.*

Definition 2. *The heaviest k-subgraph problem consists in finding $G' = (V', E'), V' \subset V, |V'| = k$ and $E' \subset E$ such that $\sum_{e \in E'} w(e)$ is maximal.*

Definition 3. *Given a graph $G = (G_1, G_2, \dots)$ which varies in time $(G_i = (V_i, E_i), i = 1, 2, \dots)$, the dynamic heaviest k-subgraph problem consists in finding $G'_i = (V'_i, E'_i), V'_i \subset V_i, |V'_i| = k$ and $E'_i \subset E_i$ such that $\sum_{e \in E'_i} w(e)$ is maximal, $i = 1, 2, \dots$.*

The heaviest k-subgraph and community based anomaly detection in dynamic graph can be defined as follows:

Definition 4. *Given a graph $G = (G_1, G_2, \dots)$ varying in time, the combined heaviest k-subgraph and community based anomaly consists in finding changes in the following way: a subgraph $G'_i = (V'_i, E'_i), V'_i \subset V_i, |V'_i| = k, E'_i \subset E_i$ such that $\sum_{e \in E'_i} w(e)$ is maximal and G_i differs from $G_{i+1}, i = 1, 2, \dots$ (G_i is a part of the community modification).*

3 Proposed Approach

Several studies exist concerning community detection methods (for a survey see [5]). Ant colony optimization [4] seems to be a powerful optimization tool in combinatorial optimization problems. As observed, ants communicate indirectly with pheromones, and based on this information they can find possible food sources.

The probability of choosing an edge between i and j is calculated with the following formula:

$$p(i,j) = \frac{(\tau_{ij})^\alpha (\eta_{ij})^\beta}{\sum_{v_q \in \mathcal{N}_i} (\tau_{iq})^\alpha (\eta_{iq})^\beta}, if v_j \in \mathcal{N}_i, \tag{1}$$

where τ_{ij} is the pheromone level between nodes i and j, η_{ij} is the a priori knowledge, and $\mathcal{N}_i$ is the set of the neighbors of i.

The ant starts from a node and chooses a next node, based on the probabilities. In one iteration, each ant generates a solution, for which the modularity of the communities will be calculated [11]:

$$Q = \frac{1}{2m} \sum_{v_i, v_j \in V} \left[A_{ij} - \frac{k_i k_j}{2m} \right] \delta(c_i, c_j), \tag{2}$$

where δ is the Kronecker delta function, m is the total number of edges, A_{ij} is 1 if there is an edge between node i and node j otherwise 0, k_i and k_j are the degrees and c_i and c_j is the communities of node i and node j, respectively.

The algorithm proposed in [6] uses Pearson correlation for the heuristic information determination. We modified this to another function, which is based on the common neighbors of the nodes:

$$C(i,j) = \frac{|\mathcal{N}_i \cap \mathcal{N}_j| + 1}{n} \tag{3}$$

The final heuristic function has the following form:

$$\eta_{ij} = \frac{1}{1 + e^{-C(i,j)}}. \tag{4}$$

The pheromone update is based on the following formula:

$$\tau_{ij}^{(t)} = (1 - \rho)\dot{\tau}_{ij}^{(t-1)} + Q(i,j), \tag{5}$$

where τ_{ij} is the pheromone level between nodes i and j at iteration t, ρ is the evaporation coefficient and $Q(i,j)$ is the quality of the solution. The pheromone level is limited in the interval $[\tau_{min}, \tau_{max}]$, where $\tau_{max} = Q(s_{gb})/(1 - \rho)$ $\tau_{min} = \vartheta\dot{\tau}_{max}$.

The main steps of the algorithm are outlined in Algorithm 1.

Algorithm 1. Modified ACO algorithm for community detection

Set parameters and initialize pheromone trails;
i=0
while $i < iter_{max}$ **do**
    $S \longleftarrow \emptyset$
    repeat
        if $i\%3 < 2$ **then**
            Construct a new solution s according to Eq. 3;
        else
            Construct a new solution s according to Eq. 3, but forbidding cycles made out
            of two points
        end if
        $S \longleftarrow S \cup s$
        $i \longleftarrow i+1$
    until $|S| = n_s$
    Calculate the iteration-best (s_{ib}), and the global-best (s_{gb})
    Compute pheromone trail limits (τ_{min} and τ_{max})
    Update pheromone trail
end while
return s_{gb}

For the heuristic information the following function is used to determine the
heaviest k-subgraph:

$$\eta_{ij} = \begin{cases} w_i + w_{ij} - min_w + 1 & \text{if } w_i \neq 0 \\ 0 & \text{if } w_i = 0 \end{cases}, \tag{6}$$

where w_i is the sum of the weights associated to node i.

Dealing with the Dynamic Environment. For each time step the algorithm is
applied. Firstly, ACO determines several types of community based anomalies:
separation and union of communities - a community can be divided several
smaller communities, and vice versa different communities can be unified to one
community, appearance of new communities - when new communities are born,
and unexpected change of communities - when a node changes the community
to which belongs. Secondly, ACO determines the k nodes, which are forming the
heaviest subgraph.

4   Numerical Experiments

4.1   Benchmarks

To test the parameters of the algorithm we used synthetic benchmarks. For the
community detection testing LFR benchmarks were used [7] with 128 nodes and
having different mixing parameters. For the heaviest k-subgraph problem we
generated graphs with 100 nodes, and for the edges between k nodes we set a
larger number than between the rest of the nodes ($k = 10$ in the parameter
setting tests).

4.2 Parameter Setting

Different values for α, β and ρ values were tested. As performance measure in the case of the community detection we used the NMI [8]. An NMI value of 1 indicates that a correct solution were found. For the heavies k-subgraph problem the proportion of correctly detected nodes in the subgraph is reported. Table 1 presents the parameter tuning of the values α, β and ρ.

Table 1. Parameter testing of α, β and ρ for the community detection problem and for the heaviest k-subgraph problem. Mean, standard deviation, minimum value and maximum value of the NMI over 20 independent runs is reported for a network with $p_{in} = 0.3$ and for the k-heaviest graph problem the proportion of the correctly found components over 20 independent runs is reported for a network with 100 nodes and $k = 10$.

α	β	ρ	Community detection			Heaviest k-subgraph		
			Mean ± St.dev.	Min	Max	Mean ± St.dev.	Min	Max
1	1	0.1	0.7232 ± 0.0463	0.6580	0.8243	0.9200 ± 0.0092	0.9000	0.9400
1.5	1	0.1	0.7318 ± 0.0491	0.6554	0.8522	0.9220 ± 0.0128	0.9000	0.9400
2	1	0.1	0.7301 ± 0.0355	0.6807	0.8139	0.9200 ± 0.0130	0.9000	0.9400
1	1.5	0.1	0.8703 ± 0.0338	0.7776	0.9242	0.9340 ± 0.0131	0.9200	0.9600
1.5	1.5	0.1	0.8586 ± 0.0305	0.7902	0.9073	0.9360 ± 0.0105	0.9200	0.9600
2	1.5	0.1	0.8671 ± 0.0327	0.8134	0.9242	0.9370 ± 0.0117	0.9200	0.9600
1	2	0.1	0.9578 ± 0.0200	0.9245	1.0000	0.9490 ± 0.0102	0.9400	0.9600
1.5	2	0.1	0.9639 ± 0.0231	0.9248	1.0000	0.9550 ± 0.0089	0.9400	0.9600
2	2	0.1	0.9604 ± 0.0277	0.8825	1.0000	0.9550 ± 0.0089	0.9400	0.9600
1	1	0.2	0.7384 ± 0.0510	0.6718	0.8630	0.9280 ± 0.0120	0.9000	0.9400
1.5	1	0.2	0.7129 ± 0.0423	0.6355	0.7740	0.9220 ± 0.0182	0.9000	0.9600
2	1	0.2	0.7185 ± 0.0472	0.6155	0.8084	0.9230 ± 0.0117	0.9000	0.9600
1	1.5	0.2	0.8637 ± 0.0412	0.7986	0.9497	0.9380 ± 0.0111	0.9200	0.9600
1.5	1.5	0.2	0.8625 ± 0.0347	0.7755	0.9078	0.9380 ± 0.0128	0.9200	0.9600
2	1.5	0.2	0.8739 ± 0.0317	0.8084	0.9496	0.9390 ± 0.0121	0.9200	0.9600
1	2	0.2	0.9512 ± 0.0233	0.8996	1.0000	0.9500 ± 0.0103	0.9400	0.9600
1.5	2	0.2	0.9446 ± 0.0204	0.8998	0.9748	0.9560 ± 0.0082	0.9400	0.9600
2	2	0.2	0.9534 ± 0.0181	0.9245	0.9748	0.9570 ± 0.0073	0.9400	0.9600
1	1	0.3	0.7384 ± 0.0402	0.6757	0.8351	0.9250 ± 0.0110	0.9000	0.9400
1.5	1	0.3	0.7140 ± 0.0492	0.6270	0.8080	0.9230 ± 0.0134	0.9000	0.9400
2	1	0.3	0.7451 ± 0.0620	0.6371	0.9325	0.9220 ± 0.0111	0.9000	0.9400
1	1.5	0.3	0.8697 ± 0.0358	0.7851	0.9176	0.9380 ± 0.0111	0.9200	0.9600
1.5	1.5	0.3	0.8722 ± 0.0385	0.8130	0.9245	0.9370 ± 0.0134	0.9200	0.9600
2	1.5	0.3	0.8456 ± 0.0319	0.7728	0.8932	0.9380 ± 0.0111	0.9200	0.9600
1	2	0.3	0.9543 ± 0.0206	0.9072	1.0000	0.9480 ± 0.0101	0.9400	0.9600
1.5	2	0.3	0.9493 ± 0.0301	0.8824	1.0000	0.9540 ± 0.0094	0.9400	0.9600
2	2	0.3	0.9638 ± 0.0221	0.9247	1.0000	0.9590 ± 0.0045	0.9400	0.9600
1	1	0.4	0.7269 ± 0.0384	0.6614	0.8138	0.9220 ± 0.0144	0.9000	0.9400
1.5	1	0.4	0.7174 ± 0.0414	0.6510	0.7792	0.9200 ± 0.0145	0.9000	0.9400
2	1	0.4	0.7253 ± 0.0503	0.6457	0.8352	0.9160 ± 0.0167	0.9000	0.9600
1	1.5	0.4	0.8700 ± 0.0354	0.7700	0.9182	0.9340 ± 0.0114	0.9200	0.9600
1.5	1.5	0.4	0.8587 ± 0.0382	0.7938	0.9246	0.9390 ± 0.0102	0.9200	0.9600
2	1.5	0.4	0.8802 ± 0.0333	0.8078	0.9497	0.9330 ± 0.0117	0.9200	0.9600
1	2	0.4	0.9522 ± 0.0262	0.8905	1.0000	0.9540 ± 0.0094	0.9400	0.9600
1.5	2	0.4	0.9648 ± 0.0189	0.9246	1.0000	0.9560 ± 0.0082	0.9400	0.9600
2	2	0.4	0.9534 ± 0.0225	0.9245	1.0000	0.9580 ± 0.0062	0.9400	0.9600

4.3   Anomaly Detection in Real-World Networks

To test the new anomaly type on dynamic networks three real-world weighted dynamic networks are used [15]: insecta-ant-colony1 [10], an animal interaction dataset with 113 nodes and 41 time steps, mammalia-raccoon proximity network [14] (another animal interaction network) with 24 nodes and 52 time steps, and ia-workplace-contacts network, a communication network with 92 nodes and 239 time steps.

Figure 1 presents a visualization of a run for mammalia-raccoon proximity network in 4 time steps. The dynamics of the anomaly change is represented with different colours. Figure 2 presents the changes detected in the same network over the 41 time steps. It is obvious, that the combined new anomaly type significantly reduces the observed changes, while we are seeking for nodes which changes heaviest k-subgraph component and the communities as well at the same time. For the other two real-world networks results are similar. This new type of anomaly can be useful if we want to be less sensitive for all minor changes, but to a more complex behavior of nodes and belonging edges.

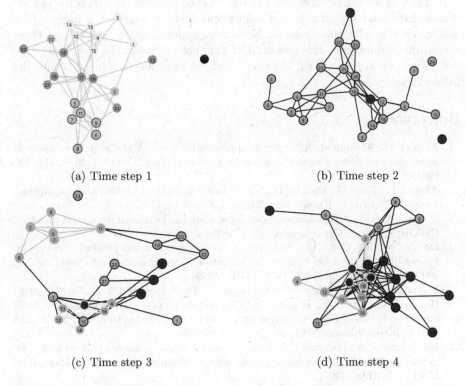

(a) Time step 1

(b) Time step 2

(c) Time step 3

(d) Time step 4

Fig. 1. Visualization of the mammalia-raccoon proximity dynamic network in four time steps. In the first figure detected communities are presented with different colors, red contoured nodes are the detected 6-subgraphs. In next figures blue nodes illustrates the community change, pink edges mean leaving the 6-heaviest subgraph, and yellow edges are the new joining edges and associated nodes of the 6-heaviest subgraph (Color figure online)

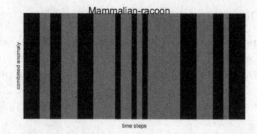

Fig. 2. Heatmap for each time step for mammalia-raccoon proximity dynamic network. Red color indicates change, black color indicates no change in the network. (Color figure online)

5   Conclusion and Further Work

We proposed a new anomaly type for dynamic graphs, where in one step we detect not only the modification of communities but the heaviest k-subgraphs as well. Thereby we can observe other changes in the network, based on the change in connected edge weights. An ant colony optimization approach is designed for the heaviest k-subgraph problem. Numerical experiments conducted on three real-world graphs show the potential of this new anomaly. In the same step, two properties are examined, through which we can gain new insights into the networks analyzed.

References

1. Ahmed, M., Mahmood, A.N.: Novel approach for network traffic pattern analysis using clustering-based collective anomaly detection. Ann. Data Sci. **2**(1), 111–130 (2015)
2. Akoglu, L., Tong, H., Koutra, D.: Graph based anomaly detection and description: a survey. Data Min. Knowl. Disc. **29**(3), 626–688 (2015)
3. Billionnet, A.: Different formulations for solving the heaviest k-subgraph problem. INFOR: Inf. Syst. Oper. Res. **43**(3), 171–186 (2005)
4. Dorigo, M., Di Caro, G.: Ant colony optimization: a new meta-heuristic. In: Proceedings of the 1999 congress on evolutionary computation-CEC99 (Cat. No. 99TH8406), vol. 2, pp. 1470–1477. IEEE (1999)
5. Fortunato, S.: Community detection in graphs. Phys. Rep. **486**(3–5), 75–174 (2010)
6. Honghao, C., Zuren, F., Zhigang, R.: Community detection using ant colony optimization. In: 2013 IEEE Congress on Evolutionary Computation, pp. 3072–3078 (2013). https://doi.org/10.1109/CEC.2013.6557944
7. Lancichinetti, A., Fortunato, S.: Benchmarks for testing community detection algorithms on directed and weighted graphs with overlapping communities. Phys. Rev. E **80**(1), 016118 (2009)
8. Lancichinetti, A., Fortunato, S., Kertész, J.: Detecting the overlapping and hierarchical community structure in complex networks. J. Phys. **11**(3), 033015 (2009)
9. Mall, R., Cerulo, L., Bensmail, H., Iavarone, A., Ceccarelli, M.: Detection of statistically significant network changes in complex biological networks. BMC Syst. Biol. **11**(1), 1–17 (2017)

10. Mersch, D.P., Crespi, A., Keller, L.: Tracking individuals shows spatial fidelity is a key regulator of ant social organization. Science **340**(6136), 1090–1093 (2013)
11. Newman, M.E.: Modularity and community structure in networks. Proc. Natl. Acad. Sci. **103**(23), 8577–8582 (2006)
12. Pourhabibi, T., Ong, K.L., Kam, B.H., Boo, Y.L.: Fraud detection: a systematic literature review of graph-based anomaly detection approaches. Decis. Support Syst. **133**, 113303 (2020)
13. Ranshous, S., Shen, S., Koutra, D., Harenberg, S., Faloutsos, C., Samatova, N.F.: Anomaly detection in dynamic networks: a survey. Wiley Interdisc. Rev. Comput. Stat. **7**(3), 223–247 (2015)
14. Reynolds, J.J., Hirsch, B.T., Gehrt, S.D., Craft, M.E.: Raccoon contact networks predict seasonal susceptibility to rabies outbreaks and limitations of vaccination. J. Anim. Ecol. **84**(6), 1720–1731 (2015)
15. Rossi, R.A., Ahmed, N.K.: The network data repository with interactive graph analytics and visualization. In: AAAI (2015). https://networkrepository.com
16. Savage, D., Zhang, X., Yu, X., Chou, P., Wang, Q.: Anomaly detection in online social networks. Soc. Netw. **39**, 62–70 (2014)

CSS–A Cheap-Surrogate-Based Selection Operator for Multi-objective Optimization

Lingping Kong[1] , Abhishek Kumar[2] , Václav Snášel[1](✉) ,
Swagatam Das[3](✉) , Pavel Krömer[1] , and Varun Ojha[4]

[1] Department of Computer Science, VSB-Technical University of Ostrava,
Ostrava, Czechia
{lingping.kong,vaclav.snasel,pavel.kromer}@vsb.cz
[2] Department of Artificial Intelligence, Kyungpook National University,
Daegu 701702, Republic of Korea
abhishek.kumar.eee13@iitbhu.ac.in
[3] Electronics and Communication Sciences Unit, Indian Statistical Institute,
203, B. T. Road, Kolkata 700108, India
swagatam.das@isical.ac.in
[4] Department of Computer Science, University of Reading, Reading, UK

Abstract. Due to the complex topology of the search space, expensive multi-objective evolutionary algorithms (EMOEAs) emphasize enhancing the exploration capability. Many algorithms use ensembles of surrogate models to boost the performance. Generally, the surrogate-based model either works out the solution's fitness by approximating the evaluation function or selects the solution by weighting the uncertainty degree of candidate solutions. This paper proposes a selection operator called Cheap surrogate selection (CSS) for multi-objective problems by utilizing the density probability on a k-dimensional tree. As opposed to the first type of surrogate models, which approximate the objective function, the proposed CSS only estimates the uncertainty of the candidate solutions. As a result, CSS does not require extensive sampling or training. Besides, CSS makes use of neighbors' density and builds the tree with low computational complexity, resulting in an accelerated surrogate process. Moreover, a new EMOEA is proposed by integrating spherical search as the core optimizer with the proposed selection scheme. Over a wide variety of benchmark problems, we show that the proposed method outperforms several state-of-the-art EMOEAs.

Keywords: Multi-objective optimization · Evolutionary algorithm · Cheap surrogate selection · Spherical search

1 Introduction

Consider a bound-constrained multi-objective optimization problem (MOP) $F(\mathbf{x})$ with M objectives to be optimized, where $F(\mathbf{x})$ is the vector of objective functions with $F(\mathbf{x}) = (f_1(\mathbf{x}), \ldots, f_M(\mathbf{x}))$ and $\mathbf{x} = (x_1, \ldots, x_D)$ is a D-dimensional variable vector in the decision space. Suppose $\min\{f(\mathbf{x})\}$ can be

M. Mernik et al. (Eds.): BIOMA 2022, LNCS 13627, pp. 54–68, 2022.
https://doi.org/10.1007/978-3-031-21094-5_5

denoted as a minimization problem, and $\max\{f(\mathbf{x})\}$ can be the opposite one, the maximization problem. In general, for the real number domain $\mathbb{R}^D$, these two kinds of problems can be correlated to each other as $\min\{f(\mathbf{x})\} \leftrightarrow \max\{-f(\mathbf{x})\}$ or $\min\{-f(\mathbf{x})\} \leftrightarrow \max\{f(\mathbf{x})\}$. Thus, without losing the generality, a MOP can be defined using Eq (1).

$$
\begin{cases}
\text{minimize } F(\mathbf{x}) = (f_1(\mathbf{x}), \ldots, f_M(\mathbf{x})), \\
\text{subject to : } \mathbf{x} \in \mathbb{R}^D \mid C(\mathbf{x}) \le 0,
\end{cases}
\tag{1}
$$

where, $C(\mathbf{x}) \le 0$ represents the box-constraints, e.g. the upper and lower bounds of the decision variables.

There are various kinds of development on MOEAs, such as _1)_ decomposition-based method [29], which decomposes a MOP into several scalar optimization subproblems and collaboratively optimizes them, such as MOEA/D [31] and NSGA-III [3]); _2)_ indicator-based methods that use performance indicators as a secondary selection criterion, like AR-MOEA [26], R2-IBEA [22], MaOEA/IGD [24], SMS-EMOA [7], and DLS-MOEA [11]; and _3)_ preference-based methods (dominance-based) that use the rank of the population members, which are determined by both the Pareto dominance and the preference information from the decision-maker, such as SPEA2 [35] and PESA-II [2], AGE-MOEA [21], NSGA-II [4] and Clustering-based adaptive MOEA (CA-MOEA) [12]. _4)_ Hybrid MOEAs combine multiple techniques to optimize the algorithms.

Except these, _Surrogate-model-based methods_ [15] have become popular. Generally, there are two types of surrogates for problems, _type 1)_ it approximates the expensive objective function to reduce the needs of fitness calculation, such as Kriging (Gaussian process regression), radial basis function, etc. _type 2)_ it evaluates the uncertainty degree of candidate solutions by weighting them on the distance to the known solutions, such as solution filter [34]. Here, the Surrogate-assisted evolutionary algorithms (SAEAs) work as an approximated-objective function, and also can be a solution classifier. The popular _type 1_ surrogate models are highly complex due to the prediction accuracy requirement. This accuracy partly depends on the massive sampling data, such as KTA2 [20], OSPNSDE [9], CSEA [20], etc., making those algorithms slow in the running process. The above-mentioned fact limits the surrogate models' application for non-expensive optimization problems. Further, KRVEA [1] needs a complex process to model function multiple times, which hinders its wide application. GMOEA [10] adopts a generative adversarial network, and it needs GPU supporting device to work efficiently.

The contribution of this work is summarized as follows.

1. A new cheap surrogate selection operator (CSS) is proposed based on the k-d tree (short for the k-dimensional tree) and density probability estimation. In the k-d tree, a tree node (leaf or internal node) represents a candidate of the population. Unlike the ordinary k-d tree building, where each node has M-dimensional fitness values, we first sort the solution set based on an objective value in each dimension and get a sorted index. We then use the

M-tuple sorted index of M objectives to denote the tree node. The role of translation operation from objective values to sorted index is twofold. *1)* The sorted index is all in the interval of 1 to the population size of N, which reduces the effect of numerical scale. And *2)* the operation of the sorted data guarantees $\mathcal{O}(kN \log N)$ complexity to build a k-d tree, which simplifies the tree building process.

2. We proposed the CSS-MOEA algorithm for multi-objective problems. In CSS-MOEA, a parent produces multiple offspring in one iteration. Then we pick one candidate among those offspring based on CSS.

3. We experiment with the proposed method on various test problems. Experimental results show that our proposed algorithm has promising performance on MOPs.

The rest of this paper is organized as follows. First, in Sect. 2 a brief introduction about the Spherical Search (SS), projection matrix, and single-objective cheap surrogate model is given. Then, the proposed CSS-MOEA algorithm is described in Sect. 3. Next, the experimental environment and results are presented in Sect. 4. Finally, the conclusion is given in Sect. 5.

2   Background

In this section, we first present a brief description of the SS algorithm proposed in [17]. Then the single-objective-specific surrogate model introduced in [8] based on density probability is introduced.

2.1   Spherical Search

SS is a swarm-intelligence-based optimization method that is simple yet highly effective for solving single-objective optimization problems [16,17]. On non-linear bound-constrained global optimization problems with or without constraints, this algorithm shows promising results compared to state-of-the-art algorithms. SS uses the vector space as a representation of the search space where every target's location is represented by a vector based on its location in the search space. In each iteration, a $(\hat{D} = D - 1)$ spherical boundary is created for a D-dimensional search space to produce candidate solutions for next iterations. The main axis of the hyperspherical space points towards a search direction, greatly influenced by either a random solution or the best solution ever found. Simultaneously, SS generates hyperspherical boundaries based on orthogonal projection matrices, which are composed of a rotation matrix and diagonal matrices. Meanwhile, the radius of the hyperspherical boundaries (dependent on the step-size parameter) is used to control the search range. Since the spherical boundary is wide when an individual is far away from the target, the individual tends to explore. On the other hand, when the spherical border is narrow, the exploitation of search becomes more dominant. After that, SS randomly picks a point on each hyperspherical boundary as a candidate solution.

Compared to the particle swarm optimization (PSO) algorithm, the SS algorithm adds an extra projection step where it puts a projection matrix to the updated trail solution. This projection matrix comprises an orthogonal and diagonal matrix controlling the explorations. In the SS algorithm, the search step is calculated using the following equation.

$$step^t(i) = c^t \mathbf{A}^t \mathbf{B}^t (\mathbf{A}^t)^T \mathbf{z}^t(i), \tag{2}$$

where the leading direction $\mathbf{z}^t$ generated by the targets (best solutions) and the current solution, which will be detailed later in Sect. 3. t is the current iteration index. c^t is the step-size parameter, and $\mathbf{A}^t \mathbf{B}^t (\mathbf{A}^t)^T = \mathbf{P}^t$ represents a projection matrix. Matrices $\mathbf{A}$ and $\mathbf{B}$ are random orthogonal and binary diagonal matrices, respectively. $step^t$ is the final step vector of solution, which will be added to the original position vector for constructing a new trial solution.

2.2 Cheap Surrogate Selection (CSS)

In some real-world applications, it is expensive to evaluate an individual. A major problem involves balancing several fitness values computations and the quality of a population. The surrogate model is a technique that models an approximate evaluation system that the system estimates the function values instead of the actual function computation. The surrogate model is built based on some of the obtained solutions $(\mathbf{x}, f(\mathbf{x}))$, and estimates some new solutions' values through this model. In this way, the original expensive computation is reduced. A surrogate model can be used in all stages of an optimization process, such as population initialization, and offspring selection. Some widely used models include linear models, support vector machine [14], Kriging (Gaussian) process regression [23,30].

A cheap surrogate model for solving single-objective optimization problem is introduced by [8] inspired by [34]. In the model given by [8], an assumption is made that the objective function $f(\mathbf{x})$ is continuous and not constant in the feasible region, which is reasonable for most cases. It is a *type 2* surrogate model, which estimates the quality of trial solutions by calculating their density value and selects the solution with the biggest density value. A density function calculates a trial solution's density value (or density estimation). This function qualifies the weighted Euclidean distance between this trial solution and reference solutions whose actual fitness values are known. Moreover, this density function comprises a kernel equation, which will be introduced in the next section.

3 Proposed Method

3.1 General Framework of CSS-MOEA

The proposed CSS-MOEA[1] has basic processes including Offspring production, Offspring pre-selection and Population updating, besides that SS and CSS

---

[1] code link: https://drive.google.com/drive/folders/14uif4ozlZCrJ8EAi4FrlUUk9GKr h72IR?usp=sharing.

operators are adopted. In addition, we build a cheap, multi-objective surrogate model for pre-selection offspring. CSS-MOEA starts with one population set and an archive set. Then, there are two kinds of offspring sets; an M-size memory set, which stores M offspring produced by one parent; an N-size memory set, which contains the final selected candidates from the first offspring set. Finally, an N-size k-d tree is built at each iteration, where a solution represents each tree node. If there is no specific noting in this paper, N is the population size, and M represents the objective size and also the generated offspring size by a parent. A simple flowchart of CSS-MOEA is shown as in Fig. 1.

Algorithm 1: Main Framework of the proposed CSS-MOEA

Input: N, population size; the maximal number of iteration t_{max}, iteration index, $t = 0$; M: problem objective number

Output: Archive $\mathbf{Q}^{t_{max}}$

1 $\mathcal{P}^0 \leftarrow$ createInitialPopulation(N);
2 $\mathcal{P}^0, \mathcal{Q}^0, \mathcal{H}^0 \leftarrow$ evaluatePopulation($\mathcal{P}^0$);
3 Nadir, ideal $\leftarrow \mathcal{P}^0$;
4 **while** $t < t_{max}$ **do**
5 $\quad$ $\mathcal{O}^t_{NM} \leftarrow$ offspring production($\mathcal{P}^t, \mathcal{Q}^t, \mathcal{H}^t$) ;
6 $\quad$ Nadir, ideal $\leftarrow \mathcal{O}^t_{NM}$;
7 $\quad$ $\mathcal{O}^{t+1}_N \leftarrow$ CSS pre-selection($\mathcal{O}^t_{NM}$);
8 $\quad$ $\mathcal{O}^{t+1}_N \leftarrow$ PolynomialMutation ;
9 $\quad$ $\mathcal{P}^{t+1} \leftarrow$ EnvironmentSelection($\mathcal{O}^{t+1}_N, \mathcal{P}^t$);
10 $\quad$ $\mathbf{H}^{t+1}, \mathbf{Q}^{t+1} \leftarrow$ evaluatePopulation($\mathcal{P}^{t+1}$);
11 $\quad$ $t \leftarrow t + 1$;
12 **end**

We use $\mathcal{P}, \mathcal{Q}, \mathcal{H}$ to present the population, archive, and the history best solution set. The solution i in population is denoted as $\mathcal{P}(i)$. Then the $\mathcal{Q}^t(i)$ is called a target which corresponds to $\mathcal{P}(i)$ and leads the searching direction. $\mathcal{O}^t_N$ store the offspring. The subscript marks the size of this storage at the t iteration in $\mathcal{O}^t_N$. $\mathcal{O}^t_{NM}$ stores all generated offspring of N parent, and each parent produces M offspring. $\mathcal{O}^{t+1}_N$ preserves the final pre-selected N offspring, which will pass to the environment selection process with $\mathcal{P}^t$.

Line 5 in Algorithm 1 is for producing offspring, where each solution produces M offspring with its target solution (a target is a solution in $\mathcal{Q}^t$). Then a k-d tree with a solution as a tree node is built, where the solution set is from $\mathcal{Q}^t$. Each node has M-tuple indices where it presents the sorted solution ranks instead of a solution's objective values. Then a final offspring will be selected from M offspring, where the selection follows the density estimation as in Line 7 of Algorithm 1. We use polynomial mutation (built-in function in jMetal) with mutation distribution index 20 and mutation probability as $\frac{1}{D}$, where D is the problem ~~variable size~~ dimension.

The pre-selection of offspring has the following steps. First, find the number of J nearest neighbors of a target $\mathcal{Q}^t(i)$ as $V_{\mathcal{Q}^t(i)}$. Second, calculate the density estimation for each offspring to $V_{Q^t(i)}$. Then compare the density values, select the offspring with the largest value as the candidate of $O^{t+1}(i)$.

The k-d tree builds once for one iteration by solution in Q^t. The R is $M \times J$ size matrix with the sorted index of neighbors at M objective dimension. R_{kj} is an element of R, representing the sorted j_{th} neighbor at k objective. The primary purpose of using the k-d tree is to find stored data points (solutions) efficiently and eliminate the various value scale effects of the Euclidean distance because we need to find the nearest neighbors in multi-dimension space.

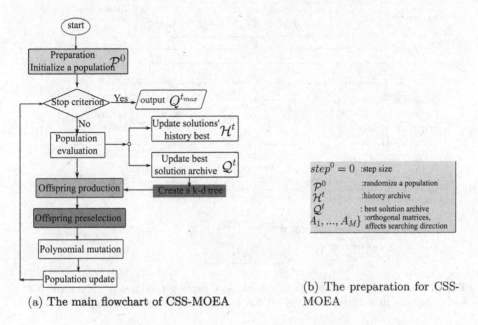

(a) The main flowchart of CSS-MOEA

(b) The preparation for CSS-MOEA

Fig. 1. The main flowchart of the proposed CSS-MOEA.

The density estimation function $\Theta(\mathbf{x})$ is defined as in Eq (3).

$$\Theta(\mathbf{x}) = \sum_{k=1}^{M} \sum_{j=1}^{J} \left(\frac{R_{kj}}{J} \frac{1}{\omega} \psi \left(\frac{\|\mathbf{x} - V_{Q^t(i)}^j\|_2}{\omega} \right) \right), \tag{3}$$

where $\|\mathbf{x}\|_2 = \sqrt{\sum_{j=1}^{M} x_j^2}$ denotes the L_2 norm; $\psi(u)$ is the kernel function. In this paper, we use the Epanechnikov Kernel [19] as in Eq (4) and ω is the window width calculated by Eq (5).

$$\psi(u) = \frac{3}{4}(1 - u^2), \; Support : (\|u\| \le 1), \tag{4}$$

$$\omega = \sqrt{\frac{1}{M} \sum_{j=1}^{M} (nadir_j - ideal_j)^2}, \tag{5}$$

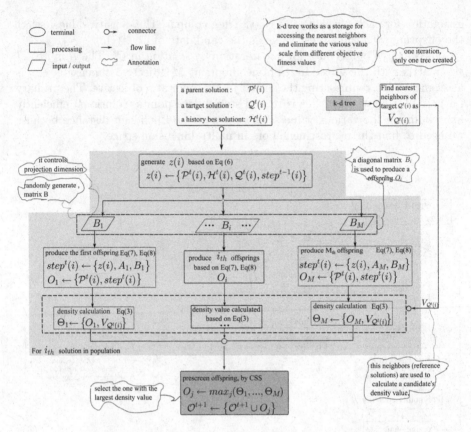

Fig. 2. The descriptive flowchart of sub-block 'Offspring production' (blue) and 'Offspring preselection' (green) of CSS-MOEA in Fig 1 (Color figure online).

where *nadir* is a virtual point which has the worst value in each objective dimension, the *ideal* is the opposite point which has the best value in each dimension (In this paper, the nadir and ideal points are calculated and updated based on population of the whole algorithm process, which is different with that updated based on one iteration or some consecutive iterations). Whenever a new solution is generated, these nadir and ideal points will be updated.

3.2   The Detailed Process of CSS-MOEA

Figure 2 shows the detailed process of CSS-MOEA. The inputs include the population size and problem size. $step^0$ is the step size that is initialized as 0. The SS algorithm updates a solution by the projection matrix, composed of an orthogonal matrix A and diagonal matrix B. CSS-MOEA requires $\{A_1, \ldots, A_M\}$ of M matrices for each objective to produce M direction offspring.

We built a k-d tree for finding the nearest neighbor fast. Moreover, the tree nodes of this tree are from Archive Q^t. Each tree node has a M-tuple ranked indices. The role of translation operation (from objective dimension values to dimension sorted index) is twofold: *1*) to guarantee the nearest neighbors chosen are not from some Euclidean distance measure (a particular objective dimension all objectives values varied small). *2*) the sorted index expressed tree node reduced the computation cost of building the tree, which will simplify the process. For the population with multi-objective problems, it is reasonable to consider that most of the solutions are non-dominated by each other. Therefore, the sorted index of the solution should proceed in each dimension since the candidate solutions are highly towards non-dominated status even in the early stage.

During the number of M offspring production, the number of M diagonal matrices $B_1, \ldots, B_M$ is randomly generated and used to update the current solution. First, we generate the uniform $z(i)$ for current solution $\mathcal{P}^t(i)$ as Eq (6).

$$\mathbf{z}(i) \leftarrow r_1 step^{t-1} + r_2 r_3 (\mathcal{Q}^t(i) - \mathcal{P}^t(i)) + r_4 r_5 (\mathcal{H}^t(i) - \mathcal{P}^t(i)), \qquad (6)$$

where r_1 to r_5 are random numbers. In our experiemnt, we set r_1 to 0.1, r_2, r_4, and r_3, r_5 are random values in the range of $[0, 1]$ and $[1.5, 2.5]$, respectively.

O_1 to O_M denote number of M offspring generated by solution $\mathcal{P}^t(i)$, where each offspring is calculated as Eq (7) and Eq (8).

$$step^t(i) \leftarrow c^t A_j B_j A_j^T z(i) \text{ as in Eq (2)}, \qquad (7)$$

$$\mathbf{O}(j) \leftarrow \mathcal{P}^t(i) + step^t(i) \qquad (8)$$

Then we measure the density estimation value on all M offspring from one parent and select the one with the largest one. At this step, the nearest neighbors $V_{\mathcal{Q}^t(i)}$ are repeatedly used for O_1 to O_M.

The CSS-MOEA will stop until the iteration t reach t_{max}.

4   Experiment Results

In this section, we first compare the proposed CSS-MOEA with several MOEAs, namely CMOPSO [32], MPMMEA [28], OSPNSDE [9], NMPSO [18]. At last, the contributions of proposed operator, namely multi-objective surrogate model are tested.

Remark, we provided several recent surrogate-assisted MOEAs in the introduction. The reason we did not compare many surrogate-assisted MOEAs is that *1*) our CSS-MOEA does not need large sampling data and model training process, the complexity is low and computation speed is fast, however KTA2, EDN-ARMOEA, and KRVEA, etc. are slow in running time without GPU support. We compared the OSPNSDE instead as a reference. OSPNSDE is also a surrogate model-based method that has an acceptable running time cost. *2*) Meanwhile, we consider CSS as a general operator rather than a complex surrogate model due to its properties of calculation, hence we compared CSS-MOEA with other MOEAs. The speed of running CSS-MOEA for one problem is as fast as CMOPSO, as the k-d tree size is small, neighbor searching time is linear function to the tree size.

Table 1. Statistical results (Mean and Standard Deviation) of IGD values on WFG problems calculated by CMOPSO [32], NMPSO [18], MPMMEA [28], OSPNSDE [9] and CSS-MOEA.

Test Problem suite. WFG/DTLZ

Year			2018	2018	2020	2019	
Problem	M	D	CMOPSO	NMPSO	MPMMEA	OSPNSDE	CSS-MOEA
WFG1	2	11	8.3023e-1 (6.63e-2) ≈	3.6018e-1 (6.70e-2) −	2.5166e-1 (9.31e-2) +	1.4376e+0 (5.84e-2) −	8.2276e-1 (8.74e-2) −
WFG2	2	11	1.3207e-2 (6.43e-4) +	1.6599e-1 (5.57e-2) −	3.7512e-2 (8.93e-3) −	6.5667e-2 (7.92e-2) ≈	3.6707e-2 (4.49e-2) −
WFG3	2	11	1.4749e-2 (5.25e-4) +	1.4431e-2 (5.20e-4) +	4.4836e-2 (6.61e-3) −	4.1318e-1 (1.97e-1) −	1.9841e-2 (7.66e-3) −
WFG4	2	11	5.2932e-2 (1.13e-2) −	2.8760e-2 (3.18e-3) ≈	4.0016e-2 (4.93e-3) −	1.6547e-1 (7.88e-2) −	2.7307e-2 (5.20e-3) −
WFG5	2	11	6.7704e-2 (1.95e-3) ≈	7.7996e-2 (6.02e-3) −	7.3923e-2 (2.74e-3) −	1.9880e-1 (5.62e-1) ≈	6.8272e-2 (2.37e-3) −
WFG6	2	11	1.9007e-2 (6.63e-3) +	2.1309e-1 (5.03e-2) −	2.9980e-1 (1.45e-3) −	2.9290e-1 (2.73e-1) −	5.5401e-2 (1.43e-2) −
WFG7	2	11	1.4831e-2 (4.48e-4) −	3.1938e-2 (4.51e-3) −	3.6706e-2 (3.44e-3) −	6.7374e-1 (4.12e-1) −	1.3357e-2 (3.13e-4) −
WFG8	2	11	1.2031e-1 (3.99e-3) ≈	1.2622e-1 (5.14e-3) −	1.3267e-1 (4.23e-3) −	6.2317e-1 (2.59e-1) −	1.2234e-1 (5.37e-3) −
WFG9	2	11	2.8494e-2 (2.52e-3) −	1.2574e-1 (1.00e-1) −	2.3663e-1 (3.22e-3) −	1.3718e-1 (1.53e-1) −	2.4500e-2 (3.97e-3) −
DTLZ1	2	6	1.3912e+0 (1.18e+0) ≈	4.9026e-1 (6.11e-1) +	9.1528e-2 (1.74e-1) +	1.5780e+0 (1.38e+0) ≈	1.8563e+0 (1.50e+0) −
DTLZ2	2	11	4.4047e-3 (6.09e-5) −	9.5319e-3 (9.20e-4) −	1.1794e-2 (1.43e-3) −	5.1298e-3 (1.96e-4) −	4.1898e-3 (4.66e-5) −
DTLZ3	2	11	5.3637e+1 (2.95e+1) −	4.6959e+1 (1.55e+1) −	7.5847e+0 (3.85e+0) +	2.4915e+1 (2.25e+1) −	2.5844e+1 (1.63e+1) −
DTLZ4	2	11	1.8889e-1 (3.28e-1) ≈	2.2922e-1 (3.44e-1) +	1.2100e-2 (1.60e-3) ≈	5.4334e-3 (2.09e-4) ≈	2.9938e-1 (3.71e-1) −
DTLZ5	2	11	4.4003e-3 (5.86e-5) −	9.8074e-3 (1.06e-3) −	1.1452e-2 (9.60e-4) −	5.1103e-3 (2.41e-4) −	4.1802e-3 (3.02e-5) −
DTLZ6	2	11	4.1265e-3 (2.82e-5) +	9.4242e-3 (7.79e-4) ≈	1.1539e-2 (2.77e-3) ≈	4.1133e-3 (2.54e-5) +	4.1058e-1 (5.51e-1) −
DTLZ7	2	21	4.8294e-2 (1.35e-1) −	7.1626e-3 (3.60e-4) −	1.1992e-2 (9.92e-4) +	1.4345e-1 (2.04e-1) ≈	3.0060e-2 (9.78e-2) −
+/=/−			4/5/7	4/3/9	4/2/10	1/6/9	

Real world application Problem

Problem	M	D	CMOPSO	NMPSO	MPMMEA	OSPNSDE	CSS-MOEA
RE22	2	3	9.82e+6 (4.37e+7) ≈	1.79e+2 (5.24e-2) −	1.80e+2 (4.95e+0) −	NaN	1.68e+2 (2.65e-2)
RE23	2	4	5.11e+2 (1.24e+1) +	4.93e+2 (2.99e+1) +	6.16e+2 (1.95e+2) ≈	6.85e+3 (4.58e+2) −	5.15e+2 (1.22e+0)
RE24	2	2	6.08e+1 (1.79e-3) +	7.21e+1 (5.79e+0) −	6.08e+1 (3.72e-3) +	6.08e+1 (5.61e-3) +	6.20e+1 (1.81e-3)
RE25	2	3	5.27e+2 (5.57e+6) −	3.55e+1 (3.64e+6) −	7.55e+6 (6.98e+5) −	8.58e-1 (1.77e-1) ≈	8.67e-1 (2.28e-16)
+/=/−			2/0/2	1/0/3	1/1/2	1/1/2	

+/=/−(win/tie/lose) indicates the results is significantly better, similar to, significantly worse than that obtained by CSS-MOEA

We implemented our method in jMetal 5.10 [6] by Java. An algorithm may obtain different results (e.g., fitness values) with the Matlab, Python, and Java codes. This is due to the precision threshold of the double-precision floating-point format. Nevertheless, with the evaluation criteria tested on both jMetal and PlatEMO, such difference does not influence the indicator comparison much among the algorithms. The solution results of CMOPSO, NMPSO, MPMMEA, and OSPNSDE are obtained by PlatEMO [27] source from website[2]. At the same time, all compared MOEAs use the default setting in the PlatEMO. Then we calculate the IGD values on all algorithms by jMetal. In this experiment, we utilize widely used test suite WFG [13] and DTLZ [5], WFG1~WFG9, DTLZ1~DTLZ7 in their default settings. The population size is 100 and the number of fitness evaluation is set to 20000 (on WFG, DTLZ problems). We employ Inverted Generational Distance (IGD) [33] performance metrics for comparison as papers in CMOPSO [32], OSPNSDE [9] also used IGD.

We also test the proposed method on real-world application problems from source [25], in which 4 bi-objective problems are solved by CSS-MOEA. The detail of the application problem is as follows: RE22 is the Reinforced concrete beam design problem; RE23 is the Pressure vessel design problem; RE24 is the Hatch cover design problem, and RE25 is the Coil compression spring design problem. The population size is 100 and the number of fitness evaluation is set to 60000 (for RE problems) for all algorithms.

Wilcoxon's signed-rank test provides the comparison results at a significance level of 0.05. All cells of the tables with the symbol '+/ = /−' indicate that the compared algorithm provides better/equal/worse results than CSS-MOEA with confidence. Each performance indicator value is calculated in mean & standard deviation mode. To easily analyze metric value results, a grey-colored background is used to identify best-performed solutions.

We list the figure results on bi-objective and three-objective problems by CSS-MOEA. Fig. 3 gives a visualization of the non-dominated solutions obtained by the compared algorithms and CSS-MOEA on two objective problems. From Fig. 3, it can be observed that MPMMEA performs worse on WFG7,8 and WFG9 problems, for the blue color nodes represented solutions are far away from the Pareto front solutions in red. The figure's illustration is consistent with table results (Table 1) where the MPMMEA loses to others on WFG7,8 and WFG9 problems. CMOPSO performs worse on the DTLZ3 problem, in which the figure (DTLZ3) shows the green nodes represented solutions locating positions away from PF from Fig. 3 (a).

Table 1 gives the results of real-world problems at the bottom of this table. The results are obtained by CSS-MOEA and compared algorithms. 'NaN' indicates the algorithm can not solve the relevant problem repeatedly successively. CMOPSO shows the competitive performance with CSS-MOEA on real-world application problems. And other algorithms perform worse than the proposed one, where NMPSO, MPMMEA, and OSPNSDE lose 3, 2, and 2 out of four problems to CSS-MOEA, respectively.

---

[2] https://github.com/BIMK/PlatEMO.

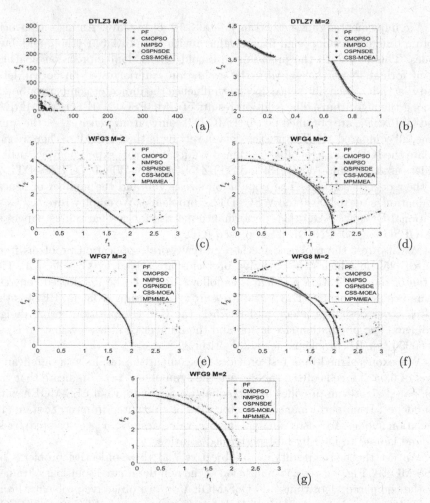

Fig. 3. Visualization of the non-dominated solutions obtained by the compared algorithms and CSS-MOEA on DTLZ3, DTLZ7, WFG3, WFG4, WFG7, WFG8 and WFG9 with two objectives. PF represents the Pareto front solutions.

Contributions of Proposed Operator. This section studies the contributions of the proposed components. As an example, we implement rand search as a variant of CSS-MOEA without the surrogate model selection, which picks randomly from multiple offspring. We consider 20000 maximum allowed fitness evaluation for each WFG on six to three objective problems. Based on the obtained results in terms of IGD and HV values over 20 independent runs, their Wilcoxon signed-rank test is summarized in Table 2 for all functions WFG1∼WFG9.

Table 2. Wilcoxon test results on problems WFG1 WFG9 with 3, 4, 5 and 6 objectives of IGD, and HV [36] performance.

M	CSS-MOEA, IGD Vs.									CSS-MOEA, HV Vs.									
6	Random search	−	▽	▲	−	−	▲	−	−	−	Random search	▽	▽	▽	▽	−	▽	▽	▽	▽
5	Random search	−	▽	▽	▽	−	▽	▽	▽	▽	Random search	▽	▽	▽	▽	−	▽	▽	▽	▽
4	Random search	−	▽	−	−	▽	▽	−	−	▽	Random search	▽	▽	▽	▽	▽	▽	▽	▽	▽
3	Random search	▽	▽	▽	▽	▽	▽	▽	▽	▽	Random search	▽	▽	▽	▽	▽	▽	▽	▽	▽

▲,−, ▽ indicate the result is better, similar, worse than
the obtained by CSS-MOEA.
[win/tie/lose] Random search vs CSS-MOEA: 2/15/55

Therefore, we can conclude that the proposed CSS operator plays the main role in density searching with better IGD performance. Additionally, the appropriate number of neighbors chosen in the surrogate model can slightly improve the performance due to more selection pressure on better solutions.

5   Conclusion

In this paper, we propose a SS-based MOEA, namely CSS-MOEA, based on the proposed cheap surrogate selection. In this algorithm, we can exploit multiple potential optimal spaces, directed by projection matrices of different ranks (different projected sub-space). In this algorithm, CSS is employed in the preselection operation to assist in reducing the number of function evaluations required for several candidate solutions which are generated in different projected sub-spaces. For this purpose, CSS employs a density probability algorithm to select one candidate solution from a pool of candidates, which relies on current neighbors' information rather than extensive sampling data, in order to select one candidate solution. Hence, the model does not require a training process and maintenance. As a result, its computation cost is very low compared to conventional surrogate models. Moreover, with the goal of reducing the complexity necessary for finding neighbors, we construct a k-d tree with a sorted index tuple as a node in the tree, which thus allows building the tree at a cost of at most $\mathcal{O}(MN \log N)$ and finding k neighbors at a cost of at most $\mathcal{O}(k \log N)$.

The empirical results demonstrate that the proposed method produces good results on popular problems, such as WFG and DTLZ in the comprehensive evaluation test. Nevertheless, the projection matrix dependence on its parameters still needs to be explored further. This CSS estimates the density of solutions and plays a role in candidate filtering, which does not offer accurate fitness evaluation to a candidate. The application of SS to other operators in different EAs is still a worthwhile study that needs to be pursued further.

Acknowledgement. This work was supported by the Ministry of Education, Youth and Sports of the Czech Republic in project META MO-COP, reg. No. LTAIN19176.

References

1. Chugh, T., Jin, Y., Miettinen, K., Hakanen, J., Sindhya, K.: A surrogate-assisted reference vector guided evolutionary algorithm for computationally expensive many-objective optimization. IEEE Trans. Evol. Comput. **22**(1), 129–142 (2016)
2. Corne, D.W., Jerram, N.R., Knowles, J.D., Oates, M.J.: Pesa-ii: region-based selection in evolutionary multiobjective optimization. In: Proceedings of the 3rd Annual Conference on Genetic and Evolutionary Computation, pp. 283–290. Morgan Kaufmann Publishers Inc. (2001)
3. Deb, K., Jain, H.: An evolutionary many-objective optimization algorithm using reference-point-based nondominated sorting approach, part i: solving problems with box constraints. IEEE Trans. Evol. Comput. **18**(4), 577–601 (2013)
4. Deb, K., Pratap, A., Agarwal, S., Meyarivan, T.: A fast and elitist multiobjective genetic algorithm: NSGA-II. IEEE Trans. Evol. Comput. **6**(2), 182–197 (2002)
5. Deb, K., Thiele, L., Laumanns, M., Zitzler, E.: Scalable test problems for evolutionary multiobjective optimization. In: Abraham, A., Jain, L., Goldberg, R. (eds.) Evolutionary multiobjective optimization, pp. 105–145. Springer, London (2005). https://doi.org/10.1007/1-84628-137-7_6
6. Durillo, J.J., Nebro, A.J., Alba, E.: The jmetal framework for multi-objective optimization: Design and architecture. In: IEEE Congress on Evolutionary Computation, pp. 1–8. IEEE (2010)
7. Emmerich, M., Beume, N., Naujoks, B.: An EMO algorithm using the hypervolume measure as selection criterion. In: Coello Coello, C.A., Hernández Aguirre, A., Zitzler, E. (eds.) EMO 2005. LNCS, vol. 3410, pp. 62–76. Springer, Heidelberg (2005). https://doi.org/10.1007/978-3-540-31880-4_5
8. Gong, W., Zhou, A., Cai, Z.: A multioperator search strategy based on cheap surrogate models for evolutionary optimization. IEEE Trans. Evol. Comput. **19**(5), 746–758 (2015)
9. Guerrero-Pena, E., Araújo, A.F.R.: Multi-objective evolutionary algorithm with prediction in the objective space. Inf. Sci. **501**, 293–316 (2019)
10. He, C., Huang, S., Cheng, R., Tan, K.C., Jin, Y.: IEEE transactions on cybernetics **51**(6), 3129–3142 (2020)
11. Hong, W., Tang, K., Zhou, A., Ishibuchi, H., Yao, X.: A scalable indicator-based evolutionary algorithm for large-scale multiobjective optimization. IEEE Trans. Evol. Comput. **23**(3), 525–537 (2018)
12. Hua, Y., Jin, Y., Hao, K.: A clustering-based adaptive evolutionary algorithm for multiobjective optimization with irregular pareto fronts. IEEE Trans. Cybern. **49**(7), 2758–2770 (2018)
13. Huband, S., Barone, L., While, L., Hingston, P.: A scalable multi-objective test problem toolkit. In: Coello Coello, C.A., Hernández Aguirre, A., Zitzler, E. (eds.) EMO 2005. LNCS, vol. 3410, pp. 280–295. Springer, Heidelberg (2005). https://doi.org/10.1007/978-3-540-31880-4_20
14. Kang, F., Xu, Q., Li, J.: Slope reliability analysis using surrogate models via new support vector machines with swarm intelligence. Appl. Math. Model. **40**(11–12), 6105–6120 (2016)
15. Knowles, J.: ParEGO: a hybrid algorithm with on-line landscape approximation for expensive multiobjective optimization problems. IEEE Trans. Evol. Comput. **10**(1), 50–66 (2006)
16. Kumar, A., Das, S., Zelinka, I.: A self-adaptive spherical search algorithm for real-world constrained optimization problems. In: Proceedings of the 2020 Genetic and Evolutionary Computation Conference Companion, pp. 13–14 (2020)

17. Kumar, A., Misra, R.K., Singh, D., Mishra, S., Das, S.: The spherical search algorithm for bound-constrained global optimization problems. Appl. Soft Comput. **85**, 105734 (2019)

18. Lin, Q., et al.: Particle swarm optimization with a balanceable fitness estimation for many-objective optimization problems. IEEE Trans. Evol. Comput. **22**(1), 32–46 (2016)

19. Marron, J., Nolan, D.: Canonical kernels for density estimation. Stat. Probab. Lett. **7**(3), 195–199 (1988)

20. Pan, L., He, C., Tian, Y., Wang, H., Zhang, X., Jin, Y.: A classification-based surrogate-assisted evolutionary algorithm for expensive many-objective optimization. IEEE Trans. Evol. Comput. **23**(1), 74–88 (2018)

21. Panichella, A.: An adaptive evolutionary algorithm based on non-euclidean geometry for many-objective optimization. In: Proceedings of the Genetic and Evolutionary Computation Conference, pp. 595–603 (2019)

22. Phan, D.H., Suzuki, J.: R2-ibea: R2 indicator based evolutionary algorithm for multiobjective optimization. In: 2013 IEEE Congress on Evolutionary Computation, pp. 1836–1845. IEEE (2013)

23. Su, G., Peng, L., Hu, L.: A gaussian process-based dynamic surrogate model for complex engineering structural reliability analysis. Struct. Saf. **68**, 97–109 (2017)

24. Sun, Y., Yen, G.G., Yi, Z.: IG indicator-based evolutionary algorithm for many-objective optimization problems. IEEE Trans. Evol. Comput. **23**(2), 173–187 (2018)

25. Tanabe, R., Ishibuchi, H.: An easy-to-use real-world multi-objective optimization problem suite. Appl. Soft Comput. **89**, 106078 (2020)

26. Tian, Y., Cheng, R., Zhang, X., Cheng, F., Jin, Y.: An indicator-based multiobjective evolutionary algorithm with reference point adaptation for better versatility. IEEE Trans. Evol. Comput. **22**(4), 609–622 (2017)

27. Tian, Y., Cheng, R., Zhang, X., Jin, Y.: PlatEMO: A MATLAB platform for evolutionary multi-objective optimization [educational forum]. IEEE Comput. Intell. Mag. **12**(4), 73–87 (2017)

28. Tian, Y., Liu, R., Zhang, X., Ma, H., Tan, K.C., Jin, Y.: A multi-population evolutionary algorithm for solving large-scale multi-modal multi-objective optimization problems. University of Surrey, Technical Report (2020)

29. Trivedi, A., Srinivasan, D., Sanyal, K., Ghosh, A.: A survey of multiobjective evolutionary algorithms based on decomposition. IEEE Trans. Evol. Comput. **21**(3), 440–462 (2016)

30. Wang, H., Zhu, X., Du, Z.: Aerodynamic optimization for low pressure turbine exhaust hood using kriging surrogate model. Int. Commun. Heat Mass Trans. **37**(8), 998–1003 (2010)

31. Zhang, Q., Li, H.: Moea/d: a multiobjective evolutionary algorithm based on decomposition. IEEE Trans. Evol. Comput. **11**(6), 712–731 (2007)

32. Zhang, X., Zheng, X., Cheng, R., Qiu, J., Jin, Y.: A competitive mechanism based multi-objective particle swarm optimizer with fast convergence. Inf. Sci. **427**, 63–76 (2018)

33. Zhou, A., Jin, Y., Zhang, Q., Sendhoff, B., Tsang, E.: Combining model-based and genetics-based offspring generation for multi-objective optimization using a convergence criterion. In: 2006 IEEE International Conference on Evolutionary Computation, pp. 892–899. IEEE (2006)

34. Zhou, L., Zhou, A., Zhang, G., Shi, C.: An estimation of distribution algorithm based on nonparametric density estimation. In: 2011 IEEE Congress of Evolutionary Computation (CEC), pp. 1597–1604. IEEE (2011)
35. Zitzler, E., Laumanns, M., Thiele, L.: SPEA2: improving the strength pareto evolutionary algorithm. TIK-report **103** (2001)
36. Zitzler, E., Thiele, L.: Multiobjective evolutionary algorithms: a comparative case study and the strength pareto approach. IEEE Trans. Evol. Comput. **3**(4), 257–271 (1999)

Empirical Similarity Measure
for Metaheuristics

Jair Pereira Junior[1]([⊠])[iD] and Claus Aranha[2][iD]

[1] Graduate School of Systems and Information Engineering, University of Tsukuba,
Tsukuba, Japan
pereira-junior.ua.ws@alumni.tsukuba.ac.jp
[2] Faculty of Engineering, Information and Systems, University of Tsukuba,
Tsukuba, Japan
caranha@cs.tsukuba.ac.jp

Abstract. Metaheuristic Search is a successful strategy for solving optimization problems, leading to over two hundred published metaheuristic algorithms. Consequently, there is an interest in understanding the similarities between metaheuristics. Previous studies have done theoretical analyses based on components and search strategies, providing insights into the relationship between different algorithms. In this paper, we argue that it is also important to consider the classes of optimization problems that the algorithms are capable of solving. To this end, we propose a method to measure the similarity between metaheuristics based on their performance on a set of optimization functions. We then use the proposed method to analyze the similarity between different algorithms as well as the similarity between the same algorithm but with different parameter settings. Our method can show if parameter settings of the same algorithm are more similar between themselves than to other algorithms and suggest a clustering based on the performance profile.

Keywords: Metaheuristic search algorithms · Optimization
problems · Algorithm similarity

1 Introduction

Metaheuristic Search is a strategy for solving optimization problems. Generally, it iterates through three steps: 1) Generate a set of candidate solutions to the target optimization problem; 2) Evaluate the quality of those solutions; and 3) Modify the solution set based on the evaluation. Representative Metaheuristics include Genetic Algorithms, Particle Swarm Optimization, and Differential Evolution.

In recent years, Metaheuristics have been useful for several hard optimization problems, especially in cases where the problems are multimodal and noncontinuous. Because of this, many Metaheuritisc algorithms have been developed, aiming at specific optimization problem classes. Today, over two hundred

M. Mernik et al. (Eds.): BIOMA 2022, LNCS 13627, pp. 69–83, 2022.
https://doi.org/10.1007/978-3-031-21094-5_6

Metaheuristic algorithms have been published in the literature [2], and there is a concern in the community that a majority of those might be minor variations of each other, providing little to no contribution to the field [10,15].

Consequently, there is an interest in understanding metaheuristic algorithms in terms of how similar or different they are and what classes of optimization problems they can solve. Some studies have proposed breaking down these algorithms in terms of their components [1,9,10]. The analyses in these works have provided insight into the relationship between different algorithms and indicate the possibility of automatically generating variants for specific problems. On the other hand, the manual analysis in these works is not scalable to a larger amount of algorithms nor easily connected to automatic algorithm generation.

In this context, we propose a method to measure the similarity between metaheuristics based on their performance on a set of optimization functions. By choosing the optimization functions carefully so that they have different fitness landscapes, we hope to find differences in the performance profiles of the algorithms under analysis. This approach not only enables comparing the algorithms on their performance in these specific classes of problems, but it may also give insight into how close the algorithms are to each other by clustering the performance results. This can aid in creating algorithm ensembles, creating algorithm portfolios, and choosing a set of operators for automatic algorithm design.

We validate our method by examining seven metaheuristics on 24 benchmark functions. We register the performances of each algorithm under five different parameter settings and use two different metrics to evaluate the performance profiles of the algorithms.

The first metric is the silhouette score [14] used in clustering, which measures how well a data point matches its own cluster compared to other clusters. Since this metric depends on a clustering process, we propose a second metric called *performance similarity* to compare two algorithm instances directly.

We use both metrics to answer the following questions: (1) how similar are instances of the same algorithm with different parameter settings? (2) how similar are instances of different algorithms that were fine-tuned on the same benchmark function? (3) What can we learn from the cluster organizations of algorithms, parameter instances, and problems?

We found that for some algorithms, their instances tend to cluster well, while for others, most instances have different performance profiles. We also observed that tuning the parameter of different algorithms on the same problem does not result in similar performance profiles between the instances. Moreover, instances of the same algorithm tend to cluster together while also having relatively good similarity scores. Lastly, we observe low scores when comparing instances of different algorithms.

The remainder of this paper is structured as follows. Section 2 presents related works on metaheuristics comparison; Sect. 3 the necessary background to understand our method to compare metaheuristic algorithms; Sect. 4 describes the proposed method; Sect. 5 presents the results and discussion; Sect. 6 concludes this work. Finally, all code and data used in our experiments are freely available online at our Github repository[1].

2   Related Works

The continuous creation of metaheuristic algorithms has been harshly criticized. The reason is that many recent algorithms share multiple similarities with older ones and are often described in terms of metaphor, making it hard to understand how they work.

After these issues were raised, research has been done to analyze metaheuristics beyond the metaphor. As such, Lones [9] identifies the metaheuristic's underlying strategies of well-known algorithms, which is an essential step in bringing cohesion and unified language to the field. For example, the author categorizes crossover as Intermediate Search, which explores the region between two or more candidate solutions, and mutation as Neighbourhood Search, which explores the nearby region of a candidate solution. Note that there are many ways to implement crossover and mutation, potentially changing these categories.

Those concepts are then used in the following work [10] to discuss common characteristics between some recent metaheuristics to the Particle Swarm Optimization (PSO). The author discusses that most recent algorithms have similarities with PSO (mainly the directional search). However, they also have some differences due to not having the same components and implementing other components of their own. This paper also goes on that the PSO community explored some ideas of the recent metaheuristics at an earlier or later date.

Furthermore, de Armas et al. [1] expand the analysis of search strategies, focusing on swarm-based metaheuristics. They propose a Pool Template to measure the similarity between two metaheuristics. In their analysis of strategies, the Directional Search, which identifies productive directions in the search space [9] is now divided into how the direction is obtained (global best, its own best, elite, and neighbor). Moreover, their Pool Template allows us to decompose and compare metaheuristics based on their components/strategies and measure their similarity. They concluded by acknowledging that novelty is indeed an issue, and some studied algorithms are special cases of others.

Together, these works paved the way for metaheuristics description, decomposition, and theoretical comparison. Expanding further, we argue that it is also important to compare algorithms empirically. This is important because two algorithms that implement similar components may have different performance profiles across multiple problems. This difference can happen due to many factors, for example, the order of the components, numerical parameters, and interaction between components.

---

[1] Code available at the following Github repository: https://github.com/jair-pereira/mhcmp/tree/bioma2022.

Algorithm 1. Metaheuristics General Template

1: $X \leftarrow$ initialize a set of random candidate solutions
2: evaluate all candidate solutions in X
3: **while** termination criteria is false **do**
4:     $X \leftarrow$ update X
5:     evaluate all candidate solutions in X
6: **end while**

3   Preliminaries

In this section, we introduce topics essential to understanding our proposed method to empirically measure similarity between the metaheuristics.

3.1   Metaheuristic Algorithms

We test our method on seven metaheuristic algorithms. We chose 3 well-known algorithms: Particle Swarm Optimization [8], Differential Evolution [5], and population-based Simulated Annealing [16]. And four recent algorithms: Artificial Tribe Algorithm [3], Firefly Algorithm [17], Gravity Search Algorithm [13], Roach Infestation Algorithm [7]. We chose these recent metaheuristics because they have some common components with PSO. All metaheuristics were implemented based on their base paper.

All metaheuristics follow the general structure shown in Algorithm 1, diverging on how they perform the update step. The search strategies of the metaheuristics in this work are summarized in Table 1 which is an adaptation of [1]. These strategies can result in different search behavior due to using or not random numbers, numerical parameters, and interaction with other components.

Note that we are proposing a framework to compare metaheuristic algorithms empirically. Thus, we use some algorithms to test this framework. This is not an extensive analysis of the chosen metaheuristics.

3.2   Benchmark Functions

Our approach requires a benchmark containing several functions with different characteristics. For this reason, we chose the BBOB testbed (numbbo/COCO [6]). BBOB is a continuous numerical black-box optimization benchmark that implements 24 noiseless functions divided into five groups. Functions within a group have similar landscape characteristics. These groups are: (1) separable functions, (2) functions with low or moderate conditioning, (3) functions with high conditioning and unimodal, (4) multi-modal functions with adequate global structure, and (5) multi-modal functions with weak global structure. All functions are dimensionally scalable (2, 3, 5, 10, 20, and 40D) and have 15 instances (artificial shift on the function space). In this benchmark, an algorithm solves the problem when the error to the optimum is 1e-8 or lower. The BBOB testbed

Table 1. Metaheuristic Algorithms: Search strategies of the chosen algorithms

Search strategy	PSO	ATA	GSA	RIO	FFA	DE	SA
Follow the best	x	x					
Follow its own best	x	x		x			
Follow elite			x	x	x		
Follow a random solution		x					
Random move					x		x
Guided move (previous solution)	x	x	x	x	x		
Guided move(previous direction)	x		x	x			
Replace fora random new solution				x			
Replace all	x	x	x	x	x		
Replace if better	x	x		x		x	x
Changing behavior		x					
Intermediate search						x	

is a useful benchmark because the characteristics of every function are documented in [4], providing insights into how the algorithm behaves in different search spaces. Table 2 shows all the functions and their main characteristics summarized from [4].

3.3 Parameter Tuning

Parameter tuning is an essential aspect of metaheuristic algorithms because parameters' values highly impact the algorithms' behavior and performance. A standard approach to configuring algorithms is to explore the algorithm's parameter space looking for values that perform best on the given problem instances.

iRace [12] is a tool that implements iterated racing to configure algorithms. To briefly explain, iRace repeats the following procedures: samples new configurations based on a particular distribution, discards the statistically worse ones, and updates the sampling distribution [11].

4 Proposed Comparison Method

Our goal is to measure similarity between metaheuristic algorithms based on how well they solved a set of problems with diverse landscape characteristics. To achieve this goal, we propose a three-steps method that includes (1) generating algorithm instances through parameter tuning, (2) performance profiling all generated instances on benchmark problems with different landscape characteristics, and (3) comparing the algorithm's instances by using silhouette score and performance similarity. Figure 1 presents an overview of our method, where the details are in the subsections below.

Table 2. Main characteristics of all functions in the BBOB testbed.

ID	Name	Modality	Separable	Conditioning	Global structure
1	Sphere	Uni	Yes	Low	Highly symmetric
2	Ellipsoidal	Uni	Yes	High	Smooth, local irregularities
3	Rastrigin	High	Yes	Low	Regular
4	Buche-Rastrigin	High	Yes	Low	Asymmetric
5	Linear slope	Uni	Yes		Linear
6	Attractive sector	Uni	No	Low/Moderate	Asymmetric
7	Step ellipsoidal	Uni	No	Low	Many plateaus
8	Rosenbrock, original	Low	No	Low/Moderate	
9	Rosenbrock, rotated	Low	No	Low/Moderate	
10	Ellipsoidal	Uni	No	High	Smooth local irregularities
11	Discus	Uni	No	High	Local irregularities
12	Bent	Uni	No	High	Smooth and narrow ridge
13	Sharp	Uni	No	High	Non-smooth ridge
14	Different powers	Uni	No	High	
15	Rastrigin	Multi	No	Low	Adequate
16	Weierstrass	Multi	No		Adequate
17	Schaffers	Multi	No	Low	Adequate
18	Schaffers ill-conditioned	Multi	No	Moderate	Adequate
19	Composite Griewank-Rosenbrock	Multi	No		Adequate
20	Schwefel	Multi	No		Weak
21	Gallagher's 101-me	Multi	No	Low	Weak
22	Gallagher's 21-hi	Multi	No	Moderate	Weak
23	Katsuura	Multi	No		Weak
24	Lunacek bi-Rastrigin	Multi	No		Weak

4.1 Algorithm Instances

We first generate five algorithms instances through parameter tuning. The reason is that the performance of metaheuristic algorithms depends on the choice of numerical parameters. Suitable parameters for one problem may not work well for another. For this reason, we want to generate instances of each chosen metaheuristics by tuning its parameters on different problems.

Since BBOB divides its 24 functions into five categories based on the functions' characteristics, we picked exactly five functions for parameter tuning, one from each group (functions 01, 06, 10, 16, and 23).

Then, we generate five instances of each metaheuristics using iRace. This process is done one function at a time, resulting in five parameter settings per algorithm. We run iRace with a budget of 2000 comparisons, and each comparison evaluates the target function a total of 1e+5 times. At the end of this step, we have five instances in regards to parameter settings for all seven chosen metaheuristics.

The parameter setting for each algorithm instance is omitted due to the page limit; however, it is available in the Github repository with the experimental code. Note that we may tune an algorithm on two different problems and obtain similar parameter values. This may happen because the values may work well enough for both problems or are local optima in the parameter space.

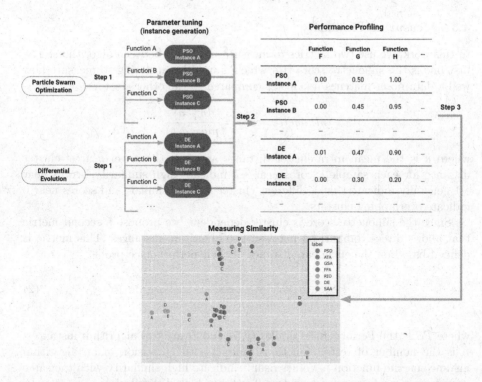

Fig. 1. Proposed method to compare metaheuristics empirically. In the first step, algorithm instances are generated through parameter tuning on functions with different landscape characteristics. In the second step, the performance of all generated instances on many benchmark functions is recorded (performance profile). In the last step, we use the performance profile to measure similarity between metaheuristics.

4.2   Algorithm Profiling

We define the *performance profile* of an algorithm instance as the array of its performances across all testing functions. To record the performance profile, we run all algorithm instances on the remaining 19 BBOB functions with 1e+6 number of functions evaluated, recording the average performance of five repetitions per problem per algorithm instance. In other words, a performance profile is the 19 features that an algorithm instance has, where each feature is its performance on a specific benchmark function. We use the performance to measure similarity between algorithms instances because this can tell how similar two instances are on a practical level. By comparing the performance profile, we can tell if two instances solve problems with similar performances despite having different components or parameter settings.

4.3   Measuring Similarity

In this work, we use two metrics to measure similarity between algorithms. The first one is the silhouette score [14] which is used in clustering to measure how well a data point matches its cluster compared to the other clusters.

$$silhouette\_score = \frac{(b-a)}{max(a,b)} \tag{1}$$

where a is the mean intra-cluster distance and b is the mean nearest-cluster distance for each sample. Scores near $+1$ indicate high similarity, scores near -1 generally indicate that a different cluster is more similar, and scores near 0 indicate overlapping clusters.

Since the silhouette score is cluster-dependent, we propose a second metric that allows direct comparison between two algorithm instances. This metric is defined based on the euclidean distance of their performance profile:

$$PS(A,B) = \frac{1}{1 + \sqrt{\sum_{i=1}^{n}(e_i^A - e_i^B)^2}} \tag{2}$$

where PS is the Performance Similarity, A and B are two algorithm instances, n is the number of tested functions, and e is the precision of an algorithm instance on the function i. Values near 1 indicate high similarity with the same performance profile, and values near 0 indicate high dissimilarity.

The precision value is the log10 of the average error to the optimum:

$$e_i^A = log_{10}(fvalue_i - fopt_i) \tag{3}$$

where $fvalue$ is the average of the best function value obtained across repetitions by the algorithm A on problem i and $fopt$ is the optimum value of function i.

5   Results

Our goal is to measure how similar algorithms are empirically. Before we start, let us recall that the performance profile is the set of performances of an algorithm instance across multiple problems. Silhouette score measures how well one algorithm instance matches its cluster compared to the other clusters. Performance similarity measures the similarity of two instances by comparing their performance profile, where values near 1 indicate that the performance profiles are similar, not necessarily that the algorithms use the same search strategies.

In this section, we present three analyses. First, we compare instances of the same algorithm. Second, we compare instances of different algorithms tuned on the same problem. Third, we use K-Means to cluster the algorithm instances and the similarity metrics to evaluate the quality of the clusters.

Table 3. Silhouette score: how similar an algorithm instance is to all other instances of the same algorithm. Each row contain different instances of the same algorithm. Ranges from +1 (high similarity) to -1 (high dissimilarity).

| | INSTANCES | | | | | |
	1	2	3	4	5	Average
PSO	−0.30	−0.18	−0.44	0.02	−0.38	−0.25
ATA	−0.07	0.29	0.17	−0.11	0.26	0.11
GSA	−0.33	−0.31	0.05	0.35	0.35	0.02
FFA	**0.74**	**0.77**	**0.74**	0.41	**0.75**	**0.68**
RIO	0.48	−0.33	0.17	0.53	0.53	0.28
DE	−0.41	0.41	0.29	0.4	−0.06	0.12
SAA	0.36	0.35	0.44	0.44	0.36	0.39

5.1 Comparing Instances of the Same Algorithm

In this first analysis, we examine the similarity between instances of the same algorithm. To do so, we use the silhouette score to measure how similar an instance is to all other instances of the same algorithm. Scores near +1 indicate high similarity, scores near −1 generally indicate that a different algorithm is more similar, and scores near 0 indicate overlapping algorithms. Then, we do the same analysis but use the performance similarity, which ranges from 0 (low similarity) to 1 (identical performance profile).

Starting with the silhouette scores in Table 3, most instances have positive scores, showing that instances of the same algorithm are more similar to themselves than to other algorithms. This may indicate that the algorithm's operators affect more their performance profile than the parameter values found during tuning. Although most instances have positive scores, those are not near +1. A not-perfect match is expected due to differences caused by the parameter values and stochastic behavior.

Stochasticity seems to have a big effect since all SAA instances have extremely similar parameter settings but an average silhouette score of 0.39. On the other hand, FFA instances have very different settings with an average silhouette score of 0.68, meaning that FFA instances are more strongly related to each other than to other algorithms.

Interestingly, PSO is the only algorithm with an average negative score of −0.25, meaning that different PSO instances could belong to clusters of other algorithms. This shows that PSO's parameter settings affect its performance profile and may reinforce the argument that many recent metaheuristics are special cases of PSO.

Moving to the performance similarity in Fig. 2, we observe that most instances are not similar ($\approx$0.1), despite being the same algorithm. This is true, especially for the PSO instances with near 0.1 scores.

A few instances score 1.0, but all have identical parameter settings (GSA 4 and 5; SAA 3 and 4; and SAA 1 and 5). On the opposite way, instances with similar parameter settings have scores in the lower range. One case is ATA 2 and

Table 4. Silhouette score: similarity of instances of different algorithms tuned on the same problem. Each row contain instances of different algorithms tuned on the same function. Ranges from +1 (high similarity) to -1 (high dissimilarity).

Cluster	PSO	ATA	GSA	FFA	RIO	DE	SAA	Average
1	−0.04	+0.04	+0.13	+0.02	+0.06	−0.13	−0.01	+0.01
2	−0.05	−0.18	−0.24	−0.28	−0.18	−0.11	−0.25	−0.19
3	−0.15	−0.15	−0.21	−0.24	−0.33	−0.09	−0.16	−0.19
4	−0.08	−0.33	−0.18	−0.34	−0.33	−0.16	−0.19	−0.23
5	−0.00	−0.07	−0.10	−0.21	−0.27	−0.01	−0.26	−0.13

5 with a score of 0.3 while having the same population size (400), near values of c (1.96 and 1.99), and tolerance (0.73 and 0.77). Another case is SAA 1 and 2, scoring 0.4 with the same population size (25) and near alpha values (0.01 and 0.02).

In summary, these results indicate that instances of the same algorithm do not always cluster well. This is observed in the low silhouette score of PSO (−0.25), GSA (0.02), ATA (0.10), and DE (0.12). It is also observed low-performance similarity between instances of these algorithms, being the exception ATA 1 and 4 (0.3), ATA 2 and 5 (0.3), and DE 2 and 4 (0.4). On the other hand, FFA (silhouette score = 0.68) clustered well, where most of its instances have performance similarity scores between 0.3 and 0.5. Lastly, SAA and RIO are in the middle, with silhouette scores of 0.39 and 0.28, respectively. These two algorithms have few instances with high-performance similarity.

5.2   Comparing Instances of the Same Tuning Function

In the second analysis, we compare instances of different algorithms tuned on the same benchmark function. Here, we assess if using the same function to parameter tune different algorithms would result in a similar performance profile.

The silhouette scores are shown in Table 4. Surprisingly, all instances have negative or near 0 scores. Likewise, the similarity scores (data not shown) between all instances are low (<0.2).

These results together indicate that using the same function to tune the parameter of multiple algorithms does not result in a similar performance profile.

5.3   Clustering the Algorithms' Instances Based on Similarity

In this third analysis, we check if the algorithm instances form other types of clusters. To do so, we use the silhouette score to find the optimal number of cluster k for K-Means (k = 13 with score = 0.49 or k = 15 with score = 0.50). Then, we use K-Means with k = 15 to cluster all the algorithm instances and compute the silhouette score for each sample.

All clusters have positive scores, as shown in Table 5, excluding the ones with a single algorithm instance each (clusters 2, 5, 11, 12, and 14). Clusters 4 and 10

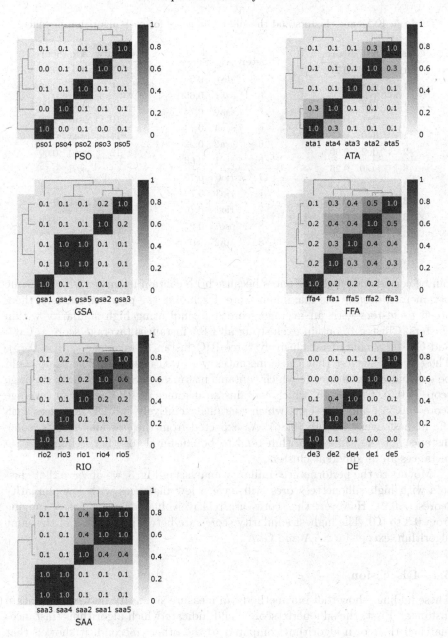

Fig. 2. Performance Similarity: heatmap showing the similarity between instances of the same algorithm. Values near 1 indicate high-performance profile similarity.

Table 5. K-Means clusters and the silhouette score for each algorithm instance.

Cluster	Alg	Score
1	ffa2	0.71
1	ffa4	0.71
1	ffa3	0.20
1	ffa1	0.73
1	ffa0	0.65
2	de2	0.00
3	ata0	0.26
3	rio2	0.13
3	ata3	0.25
3	gsa0	0.12
4	saa2	1.00
4	saa3	1.00

Cluster	Alg	Score
5	de4	0.00
6	saa1	0.65
6	saa0	0.74
6	saa4	0.74
6	gsa2	0.22
6	gsa1	0.66
7	rio0	0.70
7	rio3	0.79
7	rio4	0.76
8	pso1	0.27
8	ata2	0.05

Cluster	Alg	Score
9	de1	0.86
9	de3	0.86
10	gsa3	1.00
10	gsa4	1.00
11	pso3	0.00
12	pso0	0.00
13	pso4	0.14
13	ata4	0.51
13	ata1	0.50
14	de0	0.00
15	rio1	0.58
15	pso2	0.58

can be considered single instances because both contain instances with the same parameters, despite the maximum score. Excluding the previous clusters, there are five clusters with an average score>0.5, indicating high similarity within clusters. Cluster 1 contain exclusively all FFA instances (average score = 0.60) and Cluster 7 contains exclusively three RIO instances (average score = 0.75). These scores suggest that those instances were well clustered. The same could be concluded for Cluster 6, which contains instances of SAA and GSA (average score = 0.6), Cluster 15, which contains an instance of RIO and PSO (average score = 0.58), and Cluster 9, which contains exclusively two DE instances with the highest average score (0.86). As expected from the previous results, some instances of the same algorithm tend to be clustered together, while all PSO instances are in different clusters.

Moving to the performance similarity analysis in Fig. 3, we observe that clusters with high silhouette scores still have a few instances with low similarity scores ($\approx$0.2). However, the scores are relatively high in most cases, ranging from 0.3 to 0.6. The highest similarity scores are between instances of the same algorithm, except for SAA and GSA.

5.4 Discussion

These findings show that our method can measure similarity between algorithm instances. First, the silhouette score could indicate which algorithms' instances matched their own algorithm compared to the others. Second, it showed that algorithms instances tuned on the same problem did not cluster well. Third, the silhouette score together with K-Means showed which instances of all algorithms clustered well. Moreover, the performance similarity score allows direct comparison between two algorithms' instances. It could show the algorithms with the most diverse performance profiles and which instances are the most similar.

In the same way that we use the performance profile to compare algorithms, we can use this same data to compare the similarity of benchmark functions. This

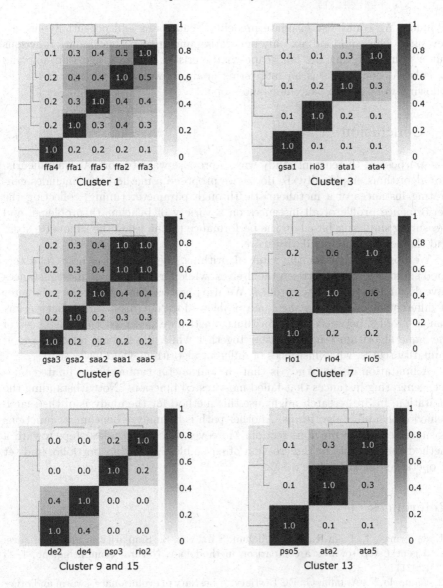

Fig. 3. Performance Similarity: heatmap showing the similarity between instances for each cluster created by K-Means. Values near 1 indicate high-performance profile similarity.

paper considers that an algorithm instance has performance values on different benchmark functions. Instead, we can consider that a benchmark function has the performance of many algorithms. This analysis could give many insights into the benchmark. To cite a few, we could evaluate if landscape characteristics negatively impact the algorithm's performances or the ones that do not impact so much. We could also evaluate predefined categories and potentially suggest new ones. This could aid not only in curating and creating benchmark functions but also in analyzing the landscape characteristics that an algorithm exploits or fails. Although this is an interesting analysis, we chose to leave it for the following work due to page limit and scope.

6   Conclusion

The purpose of the current study was to propose ways to compare metaheuristics algorithms empirically. To do so, we proposed a method that includes generating instances of a metaheuristic through parameter tuning, collecting the performance profile of all instances on a variety of benchmark problems, and measuring similarity based on the performance profile using the silhouette score and the performance similarity score.

We observed that for the same algorithms, their instances have relatively good similarity scores between themselves, while for some others most instances have dissimilar performance profiles. We also observed that tuning the parameter of different algorithms on the same problem does not result in similar performance profiles between the algorithms. Lastly, we observed that instances of the same algorithm tend to cluster together while also having relatively good similarity scores, while instances of different algorithms yield lower scores.

A limitation of this study is that the parameter tuning had a limited budget, generating instances that failed most tested functions. Notwithstanding the limitation, future research might use this method for the analysis of the search behavior of similar algorithms. Another path is to analyze benchmark functions as suggested in the previous section. Moreover, this study takes steps toward a method that can aid in the creation of ensembles, algorithm portfolio, and set of operators for automatic design.

References

1. de Armas, J., Lalla-Ruiz, E., Tilahun, S.L., Voß, S.: Similarity in metaheuristics: a gentle step towards a comparison methodology. Natural Computing, pp. 1–23 (2021)
2. Campelo, F., Aranha, C.: EC Bestiary: A bestiary of evolutionary, swarm and other metaphor-based algorithms, June 2018. https://doi.org/10.5281/zenodo.1293352. https://doi.org/10.5281/zenodo.1293352
3. Chen, T., Wang, Y., Li, J.: Artificial tribe algorithm and its performance analysis. J. Softw. **7**(3), 651–656 (2012)
4. Finck, S., Hansen, N., Ros, R., Auger, A.: Real-parameter black-box optimization benchmarking 2010: Presentation of the noisy functions. Tech. rep, Citeseer (2010)
5. Fleetwood, K.: An introduction to differential evolution. In: Proceedings of Mathematics and Statistics of Complex Systems (MASCOS) One Day Symposium, 26th November, Brisbane, Australia, pp. 785–791 (2004)
6. Hansen, N., et al.: COmparing Continuous Optimizers: numbbo/COCO on Github (2019)

7. Havens, T.C., Spain, C.J., Salmon, N.G., Keller, J.M.: Roach infestation optimization. In: 2008 IEEE Swarm Intelligence Symposium, pp. 1–7. IEEE (2008)
8. Kennedy, J., Eberhart, R.: Particle swarm optimization. In: Proceedings of ICNN'95-international conference on neural networks. vol. 4, pp. 1942–1948. IEEE (1995)
9. Lones, M.A.: Metaheuristics in nature-inspired algorithms. In: Proceedings of the Companion Publication of the 2014 Annual Conference on Genetic and Evolutionary Computation, pp. 1419–1422 (2014)
10. Lones, M.A.: Mitigating metaphors: a comprehensible guide to recent nature-inspired algorithms. SN Comput. Sci. **1**(1), 1–12 (2020)
11. López-Ibáñez, M., Cáceres, L.P., Dubois-Lacoste, J., Stützle, T.G., Birattari, M.: The irace package: User guide. Institut de Recherches Interdisciplinaires et de Développements en ... IRIDIA (2016)
12. López-Ibáñez, M., Dubois-Lacoste, J., Cáceres, L.P., Birattari, M., Stützle, T.: The irace package: iterated racing for automatic algorithm configuration. Operations Res. Perspectives **3**, 43–58 (2016)
13. Rashedi, E., Nezamabadi-Pour, H., Saryazdi, S.: Gsa: a gravitational search algorithm. Inf. Sci. **179**(13), 2232–2248 (2009)
14. Rousseeuw, P.J.: Silhouettes: a graphical aid to the interpretation and validation of cluster analysis. J. Comput. Appl. Math. **20**, 53–65 (1987)
15. Sörensen, K.: Metaheuristics-the metaphor exposed. Int. Trans. Oper. Res. **22**(1), 3–18 (2015)
16. Van Laarhoven, P.J., Aarts, E.H.: Simulated annealing. In: Simulated annealing: Theory and applications, pp. 7–15. Springer (1987)
17. Yang, X.-S.: Firefly algorithms for multimodal optimization. In: Watanabe, O., Zeugmann, T. (eds.) SAGA 2009. LNCS, vol. 5792, pp. 169–178. Springer, Heidelberg (2009). https://doi.org/10.1007/978-3-642-04944-6_14

Evaluation of Parallel Hierarchical Differential Evolution for Min-Max Optimization Problems Using SciPy

Margarita Antoniou[1,2]([✉]) [iD] and Gregor Papa[1,2] [iD]

[1] Computer Systems Department, Jožef Stefan Institute,
Jamova c. 39, Ljubljana, Slovenia
{margarita.antoniou,gregor.papa}@ijs.si
[2] Jožef Stefan International Postgraduate School, Jamova c. 39, Ljubljana, Slovenia

Abstract. When optimization is applied in real-world applications, optimal solutions that do not take into account uncertainty are of limited value, since changes or disturbances in the input data may reduce the quality of the solution. One way to find a robust solution and consider uncertainty is to formulate the problem as a min-max optimization problem. Min-max optimization aims to identify solutions which remain feasible and of good quality under even the worst possible scenarios, i.e., realizations of the uncertain data, formulating a nested problem. Employing hierarchical evolutionary algorithms to solve the problem requires numerous function evaluations. Nevertheless, Evolutionary Algorithms can be easily parallelized. This work investigates a parallel model for differential evolution using SciPy, to solve general unconstrained min-max problems. A differential evolution is applied for both the design and scenario space optimization. To reduce the computational cost, the design level optimization is parallelized. The performance of the algorithm is evaluated for a different number of cores and different dimensionality of four benchmark test functions. The results show that, when the right parameters of the algorithm are selected, the parallelization can be of high benefit to a nested differential evolution.

Keywords: Min-max optimization · Parallelization · Differential evolution

1 Introduction

Any kind of real-world optimization problem contains to a degree uncertainty in its data, be it by inherent stochasticity or due to errors. One way to consider this uncertainty and find a robust solution to the problem is to formulate it as a min-max optimization problem [8]. This formulation aims to find the best worst-case solution and hence the most robust.

M. Mernik et al. (Eds.): BIOMA 2022, LNCS 13627, pp. 84–98, 2022.
https://doi.org/10.1007/978-3-031-21094-5_7

Numerous approaches have been developed to solve the min-max optimization problem. Among the traditional ones -which require a specific mathematical formulation- one can find branch-and-bound algorithms [16] or approximation methods [2]. When such a formulation is not applicable, Evolutionary Algorithms (EA) can be used. One of the first such EAs can be found in [6,7], where a genetic algorithm is used in a coevolutionary fashion, successfully solving min-max problems that hold specific conditions. Another EA approach is using the inherent hierarchical formulation, leading to nested algorithms and such a nested Particle Swarm Optimization algorithm can be found in [13]. What is common knowledge about the EAs though, is that they require numerous iterations and function evaluations, making the computational cost prohibitively high. A popular way to mitigate this problem is the use of surrogates, as suggested in [24], with some cost in the accuracy though. A surrogate-assisted min-max multifactorial EA is proposed in [23] employing evolutionary multitasking optimization and surrogate techniques. A min-max Differential Evolution (DE) is proposed in [18] with many improvements to the original DE, designed specially for min-max problems. Another nested approach was presented in [5], where a DE with an estimation of distribution algorithm is proposed and a priori knowledge of the previous generations is utilized to reduce the computational expense.

When dealing with populations, parallelism is inherent since each of the individuals who make up the population is an independent component [14]. There are numerous studies that deal with different parallelization of EAs [1,3,20]. More specifically for DE -a popular EA that we use in this study- parallel models can be found in [12,17,21]. Most of them refer to EAs that deal with single level or multiobjective problems. Two parallel models of bilevel DE are suggested in [15], that reduced drastically the computational time in a number of test functions. There is a study of a co-evolutionary DE algorithm in C-CUDA for solving min-max problems in [10]. To the best of our knowledge, there is no study that researches a parallel model for a purely hierarchical (nested) DE for min-max problems.

One popular and user-friendly implementation of DE is in the SciPy python package [22]. It gives the users the option to use different workers and evaluate the population in parallel. Given its popularity and simplicity, it is a suitable choice to use this framework for this work. We apply a DE algorithm for both the design and scenario space, using the parallelization option for the design space. In this way, the design space population is evaluated in parallel, meaning that the second-level DE of the scenario space is run in parallel.

We then proceed to test the method in four benchmark test-functions with different properties, known from the literature. To test the scalability of the results, we scaled the test-functions up to 10 dimensions.

Our research questions are the following:

RQ 1: What kind of speedup do we achieve for different test-functions and different dimensionality of the problem?

RQ 2: Does the population size of the design space affect the speedup and how?

RQ 3: Does the DE mutation strategy of the design space affect the runtime?

In this study, we approach the first question and scratch the surface for the other two.

The organization of this paper is as follows. Section 2 defines the min-max optimization problem and Sect. 3 gives an overview of the Differential Evolution and its nested form for solving min-max problems and explains how it is parallelized in this work. The experimental setup, the test functions used, along with the performance of the method are described in Sect. 4. Finally, Sect. 5 summarizes the paper and gives some ideas for future work.

2   Definition of the Problem

The general unconstrained min-max problem can be described as:

$$\min_{x \in X} \max_{s \in S} f(x, s) \tag{1}$$

where x is a solution selected from search space X and s a scenario chosen from the scenario space S. The objective is to locate a solution $x* \in X$ that minimizes the worst-case objective $\max_{s \in S} f(x, s)$. The problem is considered symmetrical when the following condition is true

$$\min_{x \in X} \max_{s \in S} f(x, s) = \max_{s \in S} \min_{x \in X} f(x, s)$$

Symmetrical problems have independent feasible regions of the search and scenario space, making their solution more tractable.

3   Differential Evolution for MinMax Problems

3.1   Overview of Differential Evolution

The traditional definition of Differential Evolution can be found in [19]. Following the standard evolutionary algorithm schema, a population of candidate solutions undergoes the operations of mutation, crossover, and selection. There exist numerous strategies, that DE can apply to mutate the individuals through a difference mechanism. The most regularly used variant is to select two random individuals from the current population and add their scaled vector difference to the individual vector to be mutated (rand/1/bin). Another popular variant is best/1/bin, where as base vector the best vector in the population is used. The pseudocode of the DE algorithm can be seen in Algorithm 1.

Algorithm 1: Pseudocode of DE.

```
Input  : Population Size (NP), Max Generations (MaxGen), CR, F, Dimension D
Output: Optimal Solution
//Initialization
Generate NP individuals randomly
while budget condition do
    for k = 1 to NP do
    |   calculate fit(xₖ)
    end
    for k = 1 to NP do
    |   //mutation
    |   Generate three random indexes r₁, r₂ and r₃, where r₁ ≠ r₂ ≠ r₃ ≠ k
    |   Vₖᴳ = Xᵣ₁ᴳ + F * (Xᵣ₂ᴳ − Xᵣ₃ᴳ) /* rand/1/bin                          */
    |   /* best/1/bin: Vₖᴳ = X_bestᴳ + F * (Xᵣ₁ᴳ − Xᵣ₂ᴳ)                      */
    |   //crossover
    |   for i = 1 to n do
    |   |   if rand(0,1)< CR then
    |   |   |   U[i] = Vₖ[i]
    |   |   else
    |   |   |   U[i] = Xₖ[i]
    |   |   end
    |   end
    end
    //selection
    if fit(Uₖᴳ) ≤ fit(Xₖᴳ) then
    |   Xₖᴳ⁺¹ = Uₖᴳ
    end
end
```

3.2   Hierarchical (Nested) Differential Evolution and Parallel Model

The main steps of the nested algorithm can be seen in Fig. 1. Each time a Design Space individual has to be evaluated, the algorithm runs a nested DE in order to find the optimal maximization solution in the scenario space. This can be seen with red dotted lines in the figure, where in the sequential manner, the algorithm has to wait till the nested DE is done to proceed and evaluate the next individual.

The model implemented in this work makes use of the workers argument in scipy.optimize.differential_evolution of the optimization package. The population is subdivided into workers sections and evaluated in parallel, by using Python multiprocessing.Pool[1]. It should be noted, that only the objective function evaluations are carried out in parallel, after the new population has evolved.

The model focuses on parallelization of the design space population, and it is similar to the parallel upper-level model for bilevel problems presented in [15][2]. The population is subdivided into workers sections and evaluated in parallel. Denoting $Npop$ the design space population size and $nproc$ the number of processes, each section handles $Npop/nproc$ individuals. Each process handles

---

[1] https://docs.scipy.org/doc/scipy/reference/generated/scipy.optimize.differential_ev olution.html.

[2] The specific model uses the traditional synchronous parallelization of the DE, where also the operators are applied in parallel to produce the population.

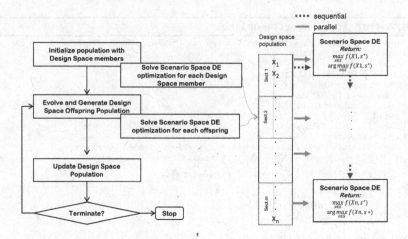

Fig. 1. Flowchart of nested Differential Evolution Algorithm (DE).

a section of the evolved population and solves the scenario space problem with DE in order to evaluate the new individuals. In Fig. 1, the procedure is shown with green lines, where the population is divided into sections and evaluated in parallel.

4   Experimental Setup and Results

All cases have been independently run 20 times for each test instance on an Intel(R) Xeon(R) with 2 CPU E5-2680 v3 @ 2.50 GHz that have 12 cores each and the Ubuntu 21.04 operating system. The algorithms are implemented in Python 3.7 (Python SciPy library [22]), using the parallelization of differential evolution for SciPy. The relevant code can be found in [4].

4.1   Benchmark Test Functions

The performance of the proposed algorithm is tested on 4 benchmark problems of min-max optimization, as found in [11]. The functions are the following:

Test function f_1 :

$$\min_{x \in X} \max_{s \in S} f_1(x, s) = (x_1 - 5)^2 - (s_1 - 5)^2 \tag{2}$$

with $x \in [0, 10], s \in [0, 10]$. The known solution is $x^* = 5$ and $s^* = 5$ with an optimal value of $f_1(x^*, s^*) = 0$. This test function is a saddle point function. The function along with the known optimum is plotted in Fig. 2a and it serves as an example of a symmetric function.

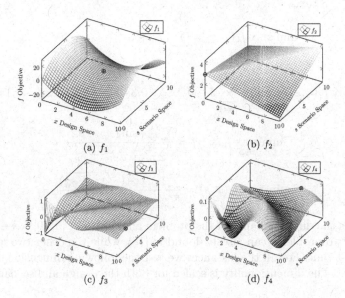

(a) f_1 (b) f_2

(c) f_3 (d) f_4

Fig. 2. 3D mesh plots of the test functions. The black circle corresponds to the known optimum.

Test function f_2:

$$\min_{x \in X} \max_{s \in S} f_2(x, s) = \min \left\{ 3 - 0.2x_1 + 0.3s_1, 3 + 0.2x_1 - 0.1s_1 \right\} \quad (3)$$

with $x \in [0, 10], s \in [0, 10]$. The optimal points are $x^* = 0$ and $s^* = 0$ and the optimal value is approximated at $f_2(x^*, s^*) = 3$. It is a two-plane asymmetrical function. The 3-D plot of this function, along with the known optima, are shown in Fig. 2b.

Test function f_3:

$$\min_{x \in X} \max_{s \in S} f_3(x, s) = \frac{\sin(x_1 - s_1)}{\sqrt{x_1^2 + s_1^2}} \quad (4)$$

with $x \in [0, 10], s \in [0, 10]$. The known solution is $x^* = 10$ and $s^* = 2.1257$ with an optimal value of $f_3(x^*, s^*) = 0.097794$. It is a damped sinus asymmetrical function, as shown in Fig. 2c.

Test function f_4:

$$\min_{x \in X} \max_{s \in S} f_4(x, s) = \frac{\cos(\sqrt{x_1^2 + s_1^2})}{\sqrt{x_1^2 + s_1^2 + 10}} \quad (5)$$

with $x \in [0, 10], s \in [0, 10]$. The known optimal solutions are $x^* = 7.0441$ and $s^* = 10$ or $s^* = 0$ and the optimal value is $f_4(x^*, s^*) = 0.042488$. It is a damped cosine wave asymmetrical function, as shown in Fig. 2d.

To further evaluate the performance, the 4 test functions are modified to be scalable as follows

$$\min_{x \in X} \max_{s \in S} f_1(x, s) = \sum_{n=1}^{D} (x_n - 5)^2 - (s_n - 5)^2 \tag{6}$$

$$\min_{x \in X} \max_{s \in S} f_2(x, s) = \min \left\{ \sum_{n=1}^{D} 3 - 0.2x_n + 0.3s_n, \sum_{n=1}^{D} 3 + 0.2x_n - 0.1s_n \right\} \tag{7}$$

$$\min_{x \in X} \max_{s \in S} f_3(x, s) = \frac{\sum_{n=1}^{D} \sin(x_n - s_n)}{\sum_{n=1}^{D} \sqrt{x_n^2 + s_n^2}} \tag{8}$$

$$\min_{x \in X} \max_{s \in S} f_4(x, s) = \frac{\sum_{n=1}^{D} \cos(\sqrt{x_n^2 + s_n^2})}{\sum_{n=1}^{D} \sqrt{x_n^2 + s_n^2} + 10} \tag{9}$$

where D is the dimensionality of the problem and with $x \in [0, 10]^D, s \in [0, 10]^D$. The first two problems can also be found in [18], while the other two are scaled for the first time. For all the instances we test the following dimensionality: D = [1, 2, 5, 10]. The dimensionality is scaled for both the design and scenario space.

4.2   Parameter Settings

The control parameter values used are reported in Table 1, unless stated otherwise. We kept the default SciPy crossover, mutation, and strategy values. The values of population and generation size are selected to be close to the ones in similar experiments done in [9]. Note that to reach better accuracy for the higher dimensionality, a larger number of population and/or generation sizes are needed. Nevertheless, we kept the sizes the same for reasons of computational budget and clear comparison in terms of running times. It is -in any case- not in the scope of this paper to examine the accuracy.

Table 1. Selected control parameters.

	Design space	Scenario space
Crossover rate	0.7	0.7
Mutation rate	U(0.5, 1)	U(0.5, 1)
Population size	20	10
Number of generations	5	10
Strategy	Best1bin	Best1bin

4.3   Results and Discussion

4.3.1   Case Study 1: Evaluation of Different Test Functions and Dimensionality

In Tables 2, 3, 4 and 5 the statistical results for the test-functions are reported. More specifically, we report the median and standard deviation of the runtime in seconds and the speedup. The speedup measures the ratio between the sequential and the parallel execution times (runtime).

In Fig. 3 the runtime and the speedup curve for each test function and instance with respect to the different number of cores used are depicted. More specifically, the Figs. 3b, 3d, 3f and 3h show the speedup curve for the test functions under analysis and the different dimensionality. It is noticed that for f_1, f_3 and f_4 the speedup considering up to 16 processors is increasing for each dimensionality. Same for f_2, but only for 2/2D,5/5D, and 10/10D. A different behavior is spotted for f_2 and 1/1D, as already for 2 cores, the speedup is decreasing, only to start improving again till we reach the 8 cores. For all test functions, for the 1 dimension case, the speedup decreases after the 16 processors. In general, the speedup increases with the increase of the dimension as it is expected, since calculating the objective function is more expensive. This can also be seen in Figs. 3a, 3c, 3e and 3g showing the running times in seconds for each function and each dimensionality with respect to the number of cores used. It is clear the required running times are also "scaled" to the dimension for all the test functions.

For the higher dimension of the problems, meaning 5/5D and 10/10D, the speedup is always increasing, except for f_4 and 10/10D, which decreases slightly for 24 cores. It is worth noting that for 5/5D and 10/10D dimensionality, there is almost a 10 times speedup when using 24 processors for f_1. For example, in the sequential case for f_1 and 10/10D, the algorithm needs 111.7 s or almost 2 min to run, while after the parallelization needs only 10 s. For the rest of the test functions the speedup for 5/5D and 10/10D ranges around 4–5. As an example, for f_4 and 10/10D, the sequential algorithm needs around 260 s or almost 4 min to run, while after the parallelization needs almost 1 min. This indicates the positive effects of parallelization on the min-max problem especially when the dimension is higher.

For f_3 and f_4, 5/5D shows greater improvement with respect to the number of cores used than 10/10D. These results are very close though, and more experiments and sample runs are needed to correct the accuracy. For lower dimensions, especially for 1/1D, the performance drops and the speedup is worsening with the number of cores.

To show that the parallelization has little effect on the accuracy of the results found, in Tables 2, 3, 4 and 5 we report the median accuracy in % for all the test functions and instances. We calculate the error rate as the absolute differences between the best objective function values provided by the algorithm and the known global optimal objective values of each test function. The accuracy formula provides accuracy as a difference of error rate from 100%. This is expressed as

$$Acc\% = 100 - |f' - f^*| * 100/f^* \tag{10}$$

where f' and f^* are the best found and the true optimal values, respectively. As is expected, the accuracy achieved in all cases does not significantly fluctuate among different cores. The accuracy for f_4 and for 10/10D is in general lower (almost half). Asymmetrical test functions constitute a more complicated problem and a different parametrization of the DE might be needed (e.g. larger population size, more generations, etc.) Nevertheless, the algorithm yields to the known global optima in the other cases, meaning the nested approach can solve both symmetrical and asymmetrical problems.

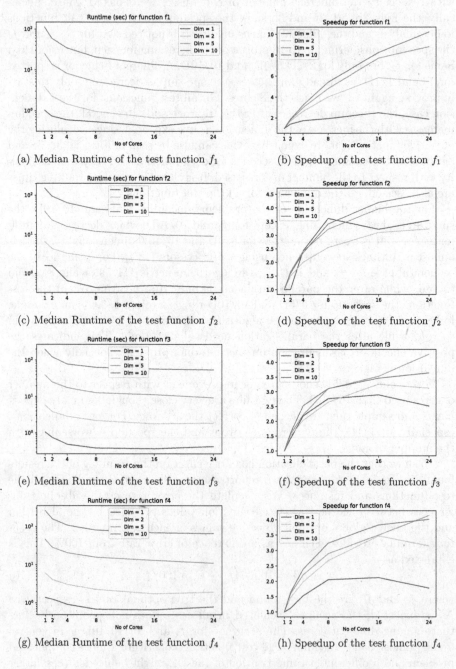

(a) Median Runtime of the test function f_1 (b) Speedup of the test function f_1

(c) Median Runtime of the test function f_2 (d) Speedup of the test function f_2

(e) Median Runtime of the test function f_3 (f) Speedup of the test function f_3

(g) Median Runtime of the test function f_4 (h) Speedup of the test function f_4

Fig. 3. Runtime and speedup plots of the test-functions.

Table 2. Running time results in seconds and median accuracy for the f_1.

Dimension	Cores	Median	Std.dev.	Speedup	Median Acc %
1	1	1.030820	0.104606	–	99.99
	2	0.730346	0.124426	1.411412	99.99
	4	0.592372	0.097924	1.740156	99.99
	8	0.531529	0.116901	1.939347	99.99
	16	0.412243	0.082602	2.500514	99.99
	24	0.549554	0.104442	1.875738	99.99
2	1	3.876020	0.038887	–	100.0
	2	2.277507	0.044546	1.701870	100.0
	4	1.464805	0.048612	2.646099	100.0
	8	0.962618	0.189352	4.026541	100.0
	16	0.664449	0.039447	5.833436	100.0
	24	0.706010	0.086775	5.490039	100.0
5	1	24.739887	0.136860	–	99.99
	2	13.712247	0.146743	1.804218	99.99
	4	8.226063	0.114659	3.007500	99.99
	8	5.129729	0.109766	4.822844	100.0
	16	3.207961	0.146968	7.712028	99.99
	24	2.811062	0.128049	8.800903	99.99
10	1	111.715032	1.033029	–	99.99
	2	59.729019	0.520408	1.870364	99.99
	4	34.562721	0.520408	3.232241	99.99
	8	20.576803	0.492896	5.429174	99.99
	16	13.552745	0.520146	8.242982	99.99
	24	10.929520	0.637464	10.221403	99.99

Table 3. Running time results in seconds and median accuracy for the f_2.

Dimension	Cores	Median	Std.dev.	Speedup	Median Acc %
1	1	1.596429	0.204669	–	100.0
	2	1.584016	0.300120	1.007836	99.96
	4	0.658811	0.278364	2.423197	100.0
	8	0.441545	0.269987	3.615554	100.0
	16	0.496514	0.329453	3.215277	99.99
	24	0.450400	0.268212	3.544468	100.0
2	1	6.620913	1.047487	–	98.88
	2	4.412415	0.717296	1.500519	98.90
	4	3.308857	0.868970	2.000967	98.76
	8	2.593604	0.842654	2.552785	98.76
	16	2.137756	0.640007	3.097133	99.17
	24	2.051248	0.961823	3.227749	99.21
5	1	38.835937	2.556689	–	85.28
	2	23.445822	3.648188	1.656412	83.84
	4	16.113869	1.323346	2.410094	85.42
	8	12.053100	1.986224	3.222070	85.08
	16	9.801519	1.462326	3.962236	85.07
	24	9.129921	2.965911	4.253699	83.98
10	1	152.548321	3.390158	–	39.78
	2	95.240152	9.687155	1.601723	42.61
	4	64.895123	10.153455	2.350690	41.13
	8	44.688494	5.504860	3.413593	44.40
	16	36.344648	7.485144	4.197271	41.55
	24	33.554359	5.100844	4.546304	45.15

Table 4. Running time results in seconds and median accuracy for the f_3.

Dimension	Cores	Median	Std.dev.	Speedup	Median Acc %
1	1	1.066014	0.111083	–	99.99
	2	0.751404	0.059305	1.418697	99.99
	4	0.488195	0.061816	2.183584	99.99
	8	0.385446	0.087810	2.765667	99.99
	16	0.368403	0.063252	2.893611	99.99
	24	0.423447	0.14840	2.517468	99.97
2	1	5.818274	0.672975	–	96.86
	2	3.851147	0.495885	1.510790	98.08
	4	2.861794	0.544807	2.033086	97.97
	8	2.196644	0.623258	2.648710	97.12
	16	1.784786	0.588991	3.259928	97.14
	24	1.860299	0.584155	3.127601	97.00
5	1	39.398736	1.696303	–	98.20
	2	25.141406	1.523187	1.567086	97.97
	4	17.753222	1.242934	2.219244	97.68
	8	12.665715	1.918927	3.110660	98.52
	16	11.165424	1.870048	3.528638	97.44
	24	9.216848	1.711269	4.274643	97.08
10	1	197.311041	12.023453	–	97.91
	2	122.471174	8.766168	1.611081	97.60
	4	78.167577	10.112983	2.524206	97.87
	8	63.370734	5.421352	3.113599	97.41
	16	56.940333	7.925534	3.465225	98.21
	24	54.158177	7.496917	3.643236	97.64

Table 5. Running time results in seconds and median accuracy for the f_4.

Dimension	Cores	Median	Std.dev.	Speedup	Median Acc %
1	1	1.398317	0.171784	–	94.14
	2	1.246892	0.255854	1.121442	93.94
	4	0.913910	0.189214	1.530038	97.63
	8	0.680622	0.194613	2.054470	97.19
	16	0.659713	0.161889	2.119585	97.53
	24	0.785280	0.204446	1.780661	97.86
2	1	6.071163	0.607564	–	96.89
	2	4.111252	0.752849	1.476719	97.67
	4	3.060104	0.416516	1.983972	97.57
	8	2.269734	0.754214	2.674834	98.52
	16	1.748137	0.437059	3.472932	97.68
	24	1.926225	0.412750	3.151845	97.31
5	1	45.444606	2.641320	–	96.83
	2	28.333389	3.517377	1.603924	98.03
	4	20.544451	2.007804	2.212014	97.92
	8	15.674804	2.192337	2.899214	97.79
	16	12.987283	2.854605	3.499162	96.18
	24	10.972779	2.310476	4.141577	97.86
10	1	260.356936	14.329717	–	94.35
	2	157.981917	11.599310	1.648017	93.44
	4	108.636768	14.879940	2.396582	95.48
	8	80.369076	10.209701	3.239516	95.98
	16	67.907552	7.578134	3.833991	94.72
	24	64.801424	8.651000	4.017766	96.68

4.3.2 Case Study 2: Influence of the Population Size

To evaluate the effect of the design space population size on the speedup of the algorithm, test function f_1 with 1/1D is run for different population sizes for 20 independent runs. The population size = 4, 8, 16, 24, 48 were selected to match the number of cores, as the design space population is the one that is divided into a number of processes and then calculated in parallel and might give some insight. Also, the population sizes of 5, 10, 20, and 30 are tested, as they are more commonly selected values.

In Fig. 4 the speedup is reported. Also, in Fig. 5 the stacked bar charts of speedup for the different population sizes and the number of cores are shown. In this graph, each value of the different population size speedup is placed after the previous one and the total value of each bar is all the segment values added together. As is expected, the higher the population size, the higher the speedup achieved by adding cores. Moreover, in most cases, as the number of cores exceeds the population size, the speedup is worsened. This can be specifically noted for $npop = 8$ and the number of $cores > 8$, as well as for $npop = 20$ and the number of $cores > 16$. The combination of the number of cores and the population size of the design space is affecting the speedup of the parallel model. There is a significant decrease in the speedup for $npop = 48$ and $cores = 24$, which might be due to the communication costs. More experiments along with profiling tools are needed to showcase the exact influence of selecting population size in analogy to the available cores, especially to higher dimensionality and larger population sizes, which are expected to show larger differences in the runtimes and are the cases that will most benefit from parallelization.

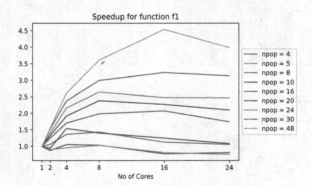

Fig. 4. Speedup plots of the function f_1 for dim = 1 and different design space population size.

4.3.3 Case Study 3: Influence of the Strategy

Since the implementation of the parallelization is synchronous, it would be interesting to see if there is any influence of the design space strategy. Therefore, we tested the algorithm by changing the strategy to rand1bin - the most commonly used. The runs refer to function f_1 and 1/1D. In Fig. 6 the speedup bar chart is shown for the different strategies and dimensionality with respect to the number of cores. In the cases that there is a difference, rand1bin shows a larger speedup. There is an inherent sequential operation of finding the best individual of the current generation, and this result agrees with what was noted also in [10]. More experiments are needed to reach safe conclusions and are in our future steps.

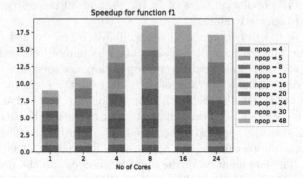

Fig. 5. Speedup stacked bar charts of the function f_1 for dim $= 1$ and different design space population size (*npop*).

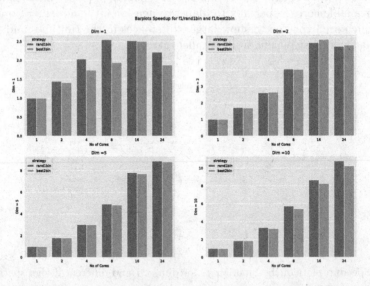

Fig. 6. Speedup bar charts of the function f_1 for dim $= 1$ and different strategy.

5   Conclusion and Future Work

In this work, the parallel model for solving min-max optimization problems via the user-friendly Python SciPy library was evaluated. The approach was tested for a nested DE and on four test functions from the literature. The four test functions were symmetrical and asymmetrical to test the behavior of the model on different min-max problems. Moreover, the problems were scaled to test the effect of the model also on different dimensions of the problems. In addition, a first insight was given about the effect of the population size and the strategy used on the speedup. The results show that the model is drastically reducing the computational time when the correct combination of the number of cores and population size of the design space is selected. The results especially indicated the large decrease of computational time on problems of higher dimensionality and when a larger population size is needed. This can motivate research of large-scale min-max problems via metaheuristics.

As noted above, the effect of the population size and the strategy on the speedup is our ongoing research. Though SciPy is used in this work, other more sophisticated parallel frameworks, such as CUDA with multi-CPU and multi-GPU can be used to further take advantage of the natural parallelization of DE of both levels. Finally, it would be interesting to test the current implementation on engineering applications and real-world problems.

Acknowledgements. This work was supported by the European Commission's H2020 program under the Marie Skłodowska-Curie grant agreement No. 722734 (UTOPIAE) and by the Slovenian Research Agency (research core funding No. P2-0098). The authors would like to thank Peter Korošec and Gašper Petelin for interesting discussions regarding this work.

References

1. Adamidis, P.: Parallel evolutionary algorithms: a review. In: Proceedings of the 4th Hellenic-European Conference on Computer Mathematics and its Applications (HERCMA 1998), Athens, Greece. Citeseer (1998)
2. Aissi, H., Bazgan, C., Vanderpooten, D.: Min-max and min-max regret versions of combinatorial optimization problems: a survey. Eur. J. Oper. Res. **197**(2), 427–438 (2009)
3. Alba, E.: Parallel evolutionary algorithms can achieve super-linear performance. Inf. Process. Lett. **82**(1), 7–13 (2002)
4. Antoniou, M. (2022). https://github.com/MargAnt0/ParallelMinMax_Scipy
5. Antoniou, M., Papa, G.: Differential evolution with estimation of distribution for worst-case scenario optimization. Mathematics **9**(17), 2137 (2021)
6. Barbosa, H.J.: A genetic algorithm for min-max problems. In: Goodman, E.D. (ed.) Proceedings of the First International Conference on Evolutionary Computation and Its Applications, pp. 99–109 (1996)
7. Barbosa, H.J.: A coevolutionary genetic algorithm for constrained optimization. In: Proceedings of the 1999 Congress on Evolutionary Computation-CEC99 (Cat. No. 99TH8406), vol. 3, pp. 1605–1611. IEEE (1999)
8. Beyer, H.G., Sendhoff, B.: Robust optimization-a comprehensive survey. Comput. Methods Appl. Mech. Eng. **196**(33–34), 3190–3218 (2007)
9. Cramer, A.M., Sudhoff, S.D., Zivi, E.L.: Evolutionary algorithms for minimax problems in robust design. IEEE Trans. Evol. Comput. **13**(2), 444–453 (2008)

10. Fabris, F., Krohling, R.A.: A co-evolutionary differential evolution algorithm for solving min-max optimization problems implemented on GPU using C-CUDA. Expert Syst. Appl. **39**(12), 10324–10333 (2012)
11. Jensen, M.T.: A new look at solving minimax problems with coevolutionary genetic algorithms. In: Metaheuristics: Computer Decision-Making, pp. 369–384. Springer, Boston (2003). https://doi.org/10.1007/978-1-4757-4137-7_17
12. Kozlov, K., Samsonov, A.: New migration scheme for parallel differential evolution. In: Proceedings of the International Conference on Bioinformatics of Genome Regulation and Structure, pp. 141–144 (2006)
13. Laskari, E.C., Parsopoulos, K.E., Vrahatis, M.N.: Particle swarm optimization for minimax problems. In: Proceedings of the 2002 Congress on Evolutionary Computation, CEC 2002 (Cat. No. 02TH8600), vol. 2, pp. 1576–1581. IEEE (2002)
14. Luong, T.V., Melab, N., Talbi, E.G.: GPU-based island model for evolutionary algorithms. In: Proceedings of the 12th Annual Conference on Genetic and Evolutionary Computation, GECCO 2010, pp. 1089–1096. Association for Computing Machinery, New York (2010). https://doi.org/10.1145/1830483.1830685
15. Magalhães, T., Krempser, E., Barbosa, H.: Parallel models of differential evolution for bilevel programming. In: Proceedings of the 2018 International Conference on Artificial Intelligence (2018)
16. Montemanni, R., Gambardella, L.M., Donati, A.V.: A branch and bound algorithm for the robust shortest path problem with interval data. Oper. Res. Lett. **32**(3), 225–232 (2004)
17. Pedroso, D.M., Bonyadi, M.R., Gallagher, M.: Parallel evolutionary algorithm for single and multi-objective optimisation: differential evolution and constraints handling. Appl. Soft Comput. **61**, 995–1012 (2017)
18. Qiu, X., Xu, J.X., Xu, Y., Tan, K.C.: A new differential evolution algorithm for minimax optimization in robust design. IEEE Trans. Cybern. **48**(5), 1355–1368 (2017)
19. Storn, R., Price, K.: Differential evolution-a simple and efficient heuristic for global optimization over continuous spaces. J. Global Optim. **11**(4), 341–359 (1997)
20. Sudholt, D.: Parallel evolutionary algorithms. In: Kacprzyk, J., Pedrycz, W. (eds.) Springer Handbook of Computational Intelligence, pp. 929–959. Springer, Heidelberg (2015). https://doi.org/10.1007/978-3-662-43505-2_46
21. Tasoulis, D.K., Pavlidis, N.G., Plagianakos, V.P., Vrahatis, M.N.: Parallel differential evolution. In: Proceedings of the 2004 Congress on Evolutionary Computation (IEEE Cat. No. 04TH8753), vol. 2, pp. 2023–2029. IEEE (2004)
22. Virtanen, P., et al.: SciPy 1.0 contributors: SciPy 1.0: fundamental algorithms for scientific computing in Python. Nat. Methods **17**, 261–272 (2020). https://doi.org/10.1038/s41592-019-0686-2
23. Wang, H., Feng, L., Jin, Y., Doherty, J.: Surrogate-assisted evolutionary multitasking for expensive minimax optimization in multiple scenarios. IEEE Comput. Intell. Mag. **16**(1), 34–48 (2021)
24. Zhou, A., Zhang, Q.: A surrogate-assisted evolutionary algorithm for minimax optimization. In: IEEE Congress on Evolutionary Computation, pp. 1–7. IEEE (2010)

Explaining Differential Evolution Performance Through Problem Landscape Characteristics

Ana Nikolikj[1,2]([✉]) [iD], Ryan Lang[3], Peter Korošec[1] [iD], and Tome Eftimov[1] [iD]

[1] Computer Systems Department, Jožef Stefan Institute, 1000 Ljubljana, Slovenia
{ana.nikolikj,peter.korosec,tome.eftimov}@ijs.si
[2] Jožef Stefan International Postgraduate School, 1000 Ljubljana, Slovenia
[3] Computer Science Division Stellenbosch University, Stellenbosch, South Africa
langr@sun.ac.za

Abstract. Providing comprehensive details on how and why a stochastic optimization algorithm behaves in a particular way, on a single problem instance or a set of problem instances is a challenging task. For this purpose, we propose a methodology based on problem landscape features and explainable machine learning models, for automated algorithm performance prediction. Performing this for ten different configurations of the Differential evolution (DE) algorithm on the 24 COCO benchmark problems, it can be estimated which set of landscape features contributes the most to the correct performance prediction of each DE configuration. From the results we concluded that different landscape features are important when predicting the performance of the different DE configurations. However, for DE configurations with similar performance on the benchmark problems, the regression models identify similar feature portfolios as important for the performance prediction task. We provided explanations for the behaviour of the algorithms in this scenario, by identifying the set of most important features and further using this information to compare the different algorithms and find algorithms with similar exploration and exploitation capabilities.

Keywords: Automated algorithm performance prediction ·
Exploratory landscape analysis · Algorithm behaviour

1   Introduction

Many real-world problems require optimisation, which is finding the most profitable, least expensive solution or just improving the current practise. Very often these problems are too complex to be mathematically modeled, where the analytic form or any properties of the problem are unknown. Under these circumstances one typically resorts to stochastic optimization, where iterative sampling-based algorithms dominate the field. They guide the search towards the optimal solution of a certain problem by sampling candidate solutions, evaluating them

M. Mernik et al. (Eds.): BIOMA 2022, LNCS 13627, pp. 99–113, 2022.
https://doi.org/10.1007/978-3-031-21094-5_8

(assessing their quality) and using the gained information to select the new solutions for the next iteration. The process proceeds in iterations until converging to an estimated optimum. In the last decades, many iterative optimization algorithms have been designed and improved upon [1–3].

The algorithms show very different performance on different problems, and the main weakness that remains open is that they are treated as black-box algorithms, without understanding their behaviour on a specific optimization problem instance or a group of problem instances. Such behaviour is completely expected when taking into account the diversity of real world problems in terms of their properties. If one possesses a good overview over the properties that make a problem instance easy or hard for a specific algorithm, one may choose the "right" algorithm from a portfolio to solve it efficiently. This task is known as Algorithm Selection (AS) [4]. Even more, sometimes different hyper-parameters values of the same optimization algorithm can lead to different performance. Therefore, it is also worth to investigate the behaviour of different hyper-parameter settings of the algorithms to find the promising one. This is another task known as Algorithm Configuration (AC). However to perform both, linking the properties of the problem instances to the performance of the algorithm achieved on them is needed.

In this direction, first the problem instance features need to be extracted so that a problem instance can be characterised/represented by them. In continuous single-objective optimization the properties/features of the problem instances are derived based on analysis of its fitness landscape [5,6]. With the goal of understanding more about the characteristics of the fitness landscape of an unknown continuous optimization problem, the Exploratory Landscape Analysis (ELA) features were first introduced in [6]. These are numerical features that quantify the properties of the optimization problem instances. The next step is to link them to the algorithm's performance. State-of-the-art studies preform the linking by training a supervised machine learning model (i.e. classifier/regressor) on a set of benchmark problem instances [7–10].

One algorithm that is well established for solving continuous single-objective optimization problem instances is Differential Evolution (DE). DE is a population-based metaheuristic that belongs to the family of Evolutionary Algorithms. It was introduced by Storn et al. in [1] and has been so far extensively studied in the field of continuous optimization. It uses the iterative improvement technique where the solution candidate is changed based on an evolutionary process. The algorithm is based on the idea of generating new offspring by applying recombination operators on the scaled difference vectors of the existing solutions in the population. There are many basic strategies which define how the offspring is computed (based on scaling factor, F, and crossover rate, Cr). In the selection phase the current solution is replaced if it is outperformed by the offspring.

Our Contribution and Results: To explore algorithm behaviour, we propose a methodology based on explainable automated algorithm performance prediction. The methodology is based on ML regression models for predicting the performance and ML techniques that explain models' predictions [11]. To show the utility of the methodology, a suite of benchmark problems and ten randomly generated DE configurations is used. To link the benchmark problem instance landscape features to the performance of the algorithm configuration achieved on them, a Random Forest (RF) is utilized, as the underlying ML algorithm for learning an regression model for performance prediction of each configuration separately. The models were further analyzed with the SHAP method [12], to provide explanations how the landscape features contribute the to the end predictions. Performing the analysis across all ten DE configurations we are able to identify the most important landscape feature sets for each configuration. Further, we represent the configurations with the feature importance and project the representations in a new common embedded space, where the different configurations can be compared and their behaviour analyzed.

Outline: The reminder of the paper is organized as follows: Section 2 presents the related work. Section 3 presents the ML pipeline for identifying the most important ELA features, the landscape and performance data involved in this study, together with the hyper-parameters used for training of the regression models. The results and discussion are presented in Sect. 4. Finally, the conclusions are presented in Sect. 5.

2   Related Work

The importance of the ELA features for classifying continuous single-objective optimization problems using ML models was investigated in [13]. In [14], automated algorithm performance prediction of several algorithms was analyzed by random forest, extreme gradient boosting, recursive partitioning, regression trees, and kernel-based support vector machine as ML regression models. This was done in combination with four classical feature selection techniques: greedy forward-backward selection and backward-forward selection, and variants of (10+5) genetic algorithm [15].

The ELA features are calculated from the trajectory data of the algorithm (trajectory-based ELA) and prediction of the performance of the CMA-ES algorithm is done in a fix-budget scenario, in [16]. It was shown that classical feature selection techniques in combination with Random Forest as ML model, with no hyper-parameter tuning does not lead to improvement in the performance prediction task.

Recently, a study [17] tried to select the most important features that contribute to an optimization algorithm performance prediction by using the SHAP method to estimate feature importance. Performance regression model was trained for each of 16 modular CMA-ES variants (i.e., one regression model per each CMA-ES variant). The direction of impact of each feature to the end prediction, for each configuration, was also presented. Further, these explanations were used to find similar/dissimilar CMA-ES configurations. Inspired by this study, here we focus on performing a study to investigate the behaviour of DE configurations and see if/how the results generalize for a different algorithm portfolio.

3   Experimental Setup

Here, the experimental design is explained in more detail starting with the benchmark problem portfolio, landscape data, the algorithm portfolio and the performance data. Next, the ML models are explained together with the evaluation procedure used for the ML experiments. At the end, the method for feature importance is presented.

3.1   Benchmark Problem Portfolio

The COCO benchmark suite [18] is used as the problem portfolio. It is available on the COCO platform [19]. The COCO benchmark suite consists of 24 continuous single-objective noiseless optimization problems. Different problem instances can be defined by transforming the base problems with scaling and translation in the objective space. The first five instances from each problem are used, resulting in a total of 120 benchmark problem instances in this work. Lastly, the problem dimension D was set to 10.

3.2   Landscape Data

ELA are numerical computer-generated features, based on sampling the solution space rather than relying on expert designed features, to characterize a problem. For calculating the features, the improved Latin Hypercube sampling technique was used as sampling technique, with sample size $800 \times D$ (8000). The sampling technique and the sample size was selected based on conclusions from previous studies, so that they lead to robust calculation of the ELA features [20]. The calculation was repeated 30 times, as it is a stochastic process and the median value was taken as the final feature value. The R package "flacco" [21] was utilized for calculation of the features.

In its original, the features were grouped into six so-called low-level properties (Convexity, Curvature, y-Distribution, Levelset, Local Search and Meta Model). These (numerical features) were used to characterize (usually categorical and expert-designed) high-level properties, such as the Global Structure, Multi-modality or Variable Scaling. A detailed description of the features can be found in [6]. Since then, a number of new features have been developed with the aim of improving the strategy's effectiveness. We selected all the ELA features which are cheap to calculate with regard to sample size, and do not require additional sampling as suggested in [14]. A total of 64 features were calculated. The selected features are coming from the following groups: classical ELA (y-distribution measures, level-set, meta-model) [6], Dispersion [22] Information Content [23], Nearest Better Clustering [24] and Principal Component Analysis [25].

3.3 Algorithm Portfolio

As an algorithm portfolio ten different, randomly selected DE configurations were tested, as presented in the Table 1. We indexed these configurations starting from DE_1 till DE_{10} for easier notation during the results.

Table 1. Selected Differential Evolution configurations.

Index	strategy	F	Cr
DE_1	Best/2/Bin	0.024	0.850
DE_2	Best/3/Bin	0.533	0.810
DE_3	Best/2/Bin	0.863	0.993
DE_4	RandToBest/1/Exp	0.139	0.517
DE_5	Best/2/Exp	0.626	0.996
DE_6	Best/1/Bin	0.617	0.515
DE_7	Rand/Rand/Bin	0.516	0.687
DE_8	Rand/2/Exp	0.330	0.028
DE_9	Rand/2/Bin	0.134	0.365
DE_{10}	Best/1/Bin	0.451	0.173

3.4 Performance Data

We focus on the fixed-budget performance scenario, where the performance met--ric for measuring the DE performance is the target precision of an algorithm (i.e., the distance between the best solution reached after a certain budget of function evaluations and the estimated optimal solution). Further, we calculate the logarithm of the precision, as suggested in [8]. The intuition behind doing so lies in the fact that we want to make use of the information captured in the exponent of the actual precision value, which can be interpreted as a distance level to the optimum (for two algorithms with actual target precision of 10^{-2} and 10^8, we can say that the latter is 6 distance levels closer to the optimum than the former).

DE is an iterative population based meta-heuristic. The *population size* of DE is set to equal D (10). In this way, a single iteration represents 10 function evaluations and a single run contains multiple iterations. The stopping criteria of a run was set to *population size* $* 10,000$ (100,000) evaluations. For each run the trace of the optimization was recorded by storing the population solutions from all algorithm iterations. All the DE configurations were run 30 times, because the algorithms are stochastic in nature, and by taking the median of the reached precision over the multiple runs we can get better approximation of the algorithms performance. The precision after $10D$, $30D$, $50D$, $100D$, $300D$, and $500D$ function evaluations was used as target variable to the ML regression models.

3.5   Regression Models

The main idea behind using a regression model for automated performance prediction is linking the landscape features of a set of problem instances, to the algorithms' performance achieved on them. Regression as supervised learning technique have been studied in the context of landscape-aware algorithm, as it predicts continuous numerical values, which is what the performance of the optimization algorithms essentially is. For the learning process we have utilized Random Forest (RF) [26]. The RF models have been trained with the following hyper-parameters: $number\ of\ estimators = 25$, $maximum\ depth = 25$, and $criterion =$ "MAE" has been selected to measure the quality of the splits performed by the ML model. The definition of MAE is provided in the next section. Finding the best hyper-parameters for the RF models was not subject to this study. The algorithms were implemented using the $scikit - learn$ package in Python.

The models are trained in a single-target regression (STR) model i.e., the target data comes from a single optimization algorithm (i.e., in our case from a single DE configuration). When performances of multiple algorithms are predicted, then a separate STR model is trained for predicting the performance on each optimization algorithm. We train a model for each algorithm and budget combination. As opposed to Multi-Target Regression (MTR), where the objective is to make predictions about two or more continuous variables at once (i.e., from multiple DE configuration), using the same feature set.

3.6   Leave-One Instance Out Validation

The RF regression models are evaluated in a stratified k-fold cross-validation scenario [27]. Cross-validation is a statistical approach used to estimate the performance of a ML model on unseen data and when the training data is limited. The process involves randomly dividing the set of observations into k groups, or folds, of approximately equal size. One of the folds is treated as a test set, and the model is trained on the remaining $k - 1$ folds. The process is repeated until every fold has been used as a test set. In this way, the model is trained on slightly different data every time. Therefore, it can be evaluated how the performance of the model changes with regard to this change. The final performance of the model is reported as the average of the performances on the k test folds. The approach generally results in a less biased or less optimistic estimate of the model performance than other methods.

In this case, the evaluation of the algorithm performance prediction models was done using leave-one-instance out cross-validation. This means that one of the five instances available per problem was left out for testing, and the remaining four were used for training of the models. The approach results in five different folds. Leave-one-instance out cross-validation was implemented, since recent studies showed that leave-one-problem out (i.e., including all its instances) does not provide transferable results [28].

In order to assess and compare the performance of the regression models, the Mean Absolute Error (MAE) was used as model performance metric. The Mean Absolute Error (MAE) is calculated as the sum of model prediction errors divided by the total number of data instances, where the prediction errors are the absolute difference between the predicted algorithm precision on a problem instance and the actual precision that the algorithm achieved on a problem instance. We reported the performance of the models as the average Mean Absolute Error(MAE) across all five test folds.

3.7   SHAP Explanations

To provide explanations how a regression model comes to the end prediction, a common technique is to use Shapley values as feature importances, calculated with the SHAP (SHapley Addictive exPlanations) method. This method comes from the coalition game theory, where the feature values (i.e., ELA values) of a data instance (i.e., problem instance) are considered as players in a coalition. More details about SHAP can be found in [12]. By performing the SHAP analysis, we can find the contribution of each ELA feature to the end prediction of each DE configuration. With regard to the level on which explanations are provided, we can distinguish between global and local explanations. Global explanations provide the contribution of the ELA feature to the performance prediction across all benchmark problem instances used as learning data, while the local explanations provide the contribution of each ELA feature on a specific problem instance. In our analysis we use the global explanations as feature importance, representing the behaviour of different DE configurations across the set of benchmark problem instances.

4   Results and Discussion

Here, first we compare the performance of the algorithm portfolio on the BBOB benchmark problems. Next, the results from the predictive performance of the ML models are shown. At the end the conclusions from feature importance analysis of each DE configuration are presented.

4.1   Optimization Algorithms Performance

We measure the performance as the logarithm of the precision of the reached target, after a fixed budget of function evaluations. Figure 1 shows the performance of the algorithms averaged across all 24 problems, for the different budgets. From these results, we can understand how the configurations behave overall and further compare them in terms of performance. In general, the precision improves as the number of function evaluations increases, for most of the algorithms. For small budgets up to $50D$, the performances are very similar. And then we can observe groups of configurations with similar performances e.g., DE_4, DE_7, DE_9 and DE_1. This indicates that for small budgets all the configurations are still performing exploration (i.e., searching for regions with good solutions). While for larger budgets, there is an indication that some configurations have better exploitation power and manage to converge towards an estimated optimum.

Figures 2 and 3 present the precision obtained by each DE configuration, on the 24 COCO benchmark problems separately. The log precision after $30D$ and $500D$ function evaluations is displayed. From these results, we can understand how the configurations behave on the different problems. Algorithm performances are significantly worse for some problems such as 2, 10 and 12, for all algorithms. In $30D$ going across problems the performance is not so diverse, meaning that the DE configurations show similar performance. In 500D the best performing configurations start to differentiate, such as DE_7 and DE_4.

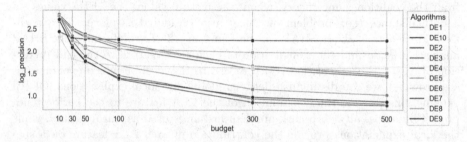

Fig. 1. Log precision (across all 24 problems) obtained by the DE configurations for budget of 30D, 50D, 100D, 300D, 500D function evaluations.

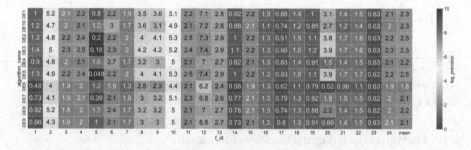

Fig. 2. Log precision obtained by the DE configurations on each of the COCO benchmark problems, for budget $30D$

4.2   Performance Prediction

In Fig. 4, the Mean Absolute Error (MAE) obtained by the Random Forest models when predicting the performance, is presented. MAE is averaged across all 24 problems for all different budgets. From the figure, we can conclude that

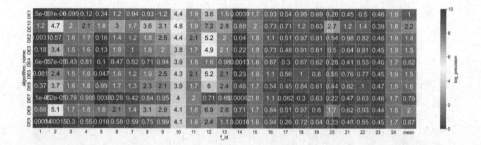

Fig. 3. Log precision obtained by the DE configurations on each of the COCO benchmark problems, for budget $500D$

the RF models achieve the lower errors for smaller budgets. However, the errors tend to increase noticeably with the budget for configurations DE_4, DE_7, DE_9 and DE_1.

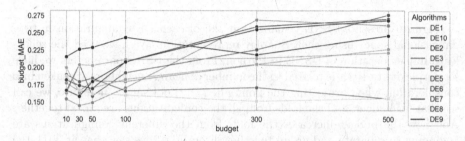

Fig. 4. Mean Absolute Error (MAE) (across all 24 problems) obtained by the Random Forest models in predicting the performance of each DE configuration, for the different budgets

In Fig. 5, the boxplots show the distribution of the mean absolute error (MAE) per benchmark problem, for each DE configuration. From the results, we can see that the RF models are consistent when predicting across problems, most of the MAEs are around 0.25–0.50. Another thing to point out here are the outliers, which means that there are high errors when predicting the performance only on some benchmark problems. RF achieves the lowest errors and there is low deviation when predicting the performance for DE_6 and DE_8, for all budgets.

4.3 Linking ELA Features to DE Performance

To represent each DE configuration through the contribution of the ELA features to its performance, we aggregated the Shapley values for each ELA feature across all problem instances from each training fold. This means that each DE configuration is represented with five different vectors, each one coming from the five training folds. The analysis was repeated as described for different budgets.

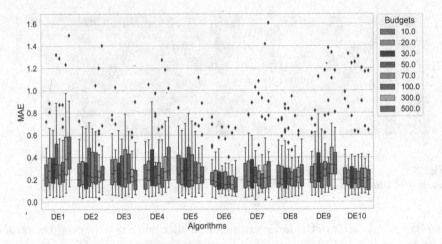

Fig. 5. The Mean Absolute Error (MAE) for each COCO benchmark problem obtained by the Random Forest models for all algorithms and budgets.

Figure 6 presents t-SNE visualization [29] of each DE configuration in two-dimensional space using their Shapley representation. We have used the scikit-learn implementation of t-SNE in Python. From the t-SNE parameters we have set perplexity to 10 (it is related to the number of nearest neighbors are taken into consideration when projecting the points in $2D$). Looking at the figure, we can see that for budget $30D$ the folds from all the configurations are mixed together. And how the budget increases the folds from the different configurations are beginning to separate and group together. From Fig. 2 we saw that at $30D$, the algorithms are still exploring the problem instances and achieve similar performance, so the ML models basically use the same ELA features and additionally similar targets, so the resulting SHAP values (i.e., the most contributing ELA features) are also similar for all configurations. At $500D$ the DE configuration representations are consistent across different folds within the same configuration, since their Shapley representations place them close together. This means that to predict the performance of an DE configuration, the importance of the ELA features utilized by the RF model is the same across the folds. This indicates the most important ELA features for each DE configuration are consistent. Further, we can spot two groups of configurations on the plot. We concluded that algorithms with similar performance are put in the same group (see Fig. 1, for budget $500D$) and the ML models identify similar feature portfolios as useful for them. The difference between the groups of configurations points that different ELA features are utilized (i.e., their importance changes), which points to the fact that they have different exploration and exploitation capabilities. Such kind of representation can help further to analyze algorithms' behaviour and find more similar configurations.

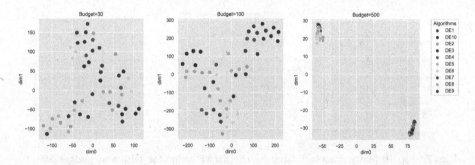

Fig. 6. t-SNE visualization of the RF models, trained per each fold, for budget $30D$, $100D$ and $500D$. The models are represented as vectors of 64 Shapley values (i.e., one per each ELA feature).

To see how the most important features differentiate in the two groups we have shown the most important features for two configurations, DE_1 from the upper group, and DE_6 from the lower group. Figure 7 presents the impact each feature has on the models prediction. A single dot in the plot represents the ELA feature value for a corresponding data instance from the training data. Red dots represent high values and blue dots represent low values of the features. The location on the on the x-axis represents the magnitude of impact that the value of the feature has on the prediction of the target variable. If we average the impacts for all data instances we will get the final importance of a feature. The features are ordered top-down by their importance averaged over all data instances for a single fold. Because of the way we evaluated the models, we can see the features importance for each fold, and also the last subplot presents feature importance if we average the importance across all folds. We can see that the results differ across folds. The results indicate that involving different instances from the problems in the training data results in selecting some different features. This means that some features are affected and are not invariant on the transformations (i.e., shifting and scaling applied to generate the instances from the same problem. The results confirm what has already been shown about the invariance of the ELA features with regard to different transformations of the problem instances [30]. However, we estimate the most important features, on algorithm level, as averaged feature importance across all five folds. The final portfolio of most important features is displayed in the last subplots in Figures 7 and 8). If then we compare them, we can see that they use different features.

In Fig. 9 we plot the percentage of overlap between the top 5 features of the final feature portfolio for each pair of DE configurations for $30D$ and $500D$ function evaluations. Each differently colored line in the plot refers to a percentage of overlapping of a specific DE configuration with all others. Using it, each configuration has 100% overlapping with it self, since we have an intersection of the same feature portfolio. It is also visible that the overlapping in $20D$ is high for all configurations, while in $500D$ there is higher overlapping with configurations from the same group, as indicated from the scatter plot earlier. This indicate that

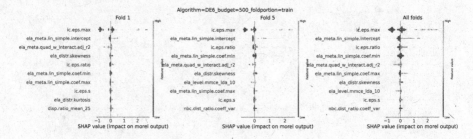

Fig. 7. Top 10 features as indicated by the SHAP value for DE$_6$ and budget 500D

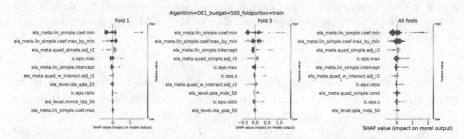

Fig. 8. Top 10 features as indicated by the SHAP value for DE$_1$ and budget 500D

in 30D the configurations are still doing exploration and similar problem properties contribute to the prediction, while in 500D the problem characteristics that contribute to the exploitation of the DE configuration make the difference between better the two groups of algorithms behaviour.

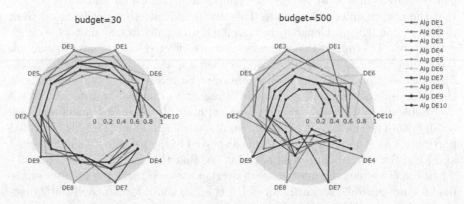

Fig. 9. Intersection (%) of top 10 features as indicated from the Shapley values, between the different algorithms, for budget 30D and 500D

5   Conclusions

Understanding the behaviour of an stochastic optimization algorithm is still an open question and in most cases the algorithm is treated as a black-box system. Even more, it is also worth to explore the impact of different hyper-parameters within the same algorithm to its end performance. To provide more details about algorithm behaviour, we proposed an approach based on benchmark problem landscape features and supervised machine learning models for automated algorithm performance prediction, which link the landscape properties of the problem instances to the algorithm performance. By additionally explaining the predictions of the learned models using the SHAP method and we were able to link algorithm performance to the most important landscape properties of the problem instances for this fixed experimental setup. We evaluated the approach on the five instances from the 24 COCO benchmark problems in combination with ten DE configurations by learning a Random Forest regression model.

From the analysis we can conclude that different landscape features are important for predicting the performance of the different algorithm configurations. However, for configurations with achieve similar performance on the benchmark problems, the regression models identify similar feature portfolios as important for the performance prediction task, (i.e. for algorithms with similar performance similar representation is discovered).

This kind of representation can lead to improved algorithm selection and configuration. In the future, a more diverse algorithm portfolio can also be considered. The methodology can be applied to any set of configurations of the same algorithm, or even further to a portfolio of different algorithms, to incorporate no idea of "global" performance. We also now analysed different budgets for single problem dimension, but extension of the analysis can be done to different problem dimensions, to analyze patterns that show how features importance behave when the problem dimension increases.

References

1. Storn, R., Price, K.: Differential evolution-a simple and efficient heuristic for global optimization over continuous spaces. J. Global Optim. **11**(4), 341–359 (1997)
2. Hansen, N., Ostermeier, A.: Completely derandomized self-adaptation in evolution strategies. Evol. Comput. **9**(2), 159–195 (2001)
3. Kennedy, J., Eberhart, R.: Particle swarm optimization. In: Proceedings of ICNN 1995-International Conference on Neural Networks, vol. 4, pp. 1942–1948. IEEE (1995)
4. Rice, J.R.: The algorithm selection problem. In: Advances in Computers, vol. 15, pp. 65–118. Elsevier (1976)
5. Malan, K.M., Engelbrecht, A.P.: A survey of techniques for characterising fitness landscapes and some possible ways forward. Inf. Sci. **241**, 148–163 (2013)
6. Mersmann, O., Bischl, B., Trautmann, H., Preuss, M., Weihs, C., Rudolph, G.: Exploratory landscape analysis. In: Proceedings of the 13th Annual Conference on Genetic and Evolutionary Computation, pp. 829–836 (2011)

7. Derbel, B., Liefooghe, A., Vérel, S., Aguirre, H., Tanaka, K.: New features for continuous exploratory landscape analysis based on the soo tree. In: Proceedings of the 15th ACM/SIGEVO Conference on Foundations of Genetic Algorithms, pp. 72–86 (2019)

8. Jankovic, A., Doerr, C.: Landscape-aware fixed-budget performance regression and algorithm selection for modular CMA-ES variants. In: Proceedings of the 2020 Genetic and Evolutionary Computation Conference, pp. 841–849 (2020)

9. Kerschke, P., Hoos, H.H., Neumann, F., Trautmann, H.: Automated algorithm selection: survey and perspectives. Evolut. Comput. **27**(1), 3–45 (2019)

10. Eftimov, T., Jankovic, A., Popovski, G., Doerr, C., Korošec, P.: Personalizing performance regression models to black-box optimization problems. In: Proceedings of the Genetic and Evolutionary Computation Conference, pp. 669–677 (2021)

11. Roscher, R., Bohn, B., Duarte, M.F., Garcke, J.: Explainable machine learning for scientific insights and discoveries. IEEE Access **8**, 42200–42216 (2020)

12. Marcílio, W.E., Eler, D.M.: From explanations to feature selection: assessing SHAP values as feature selection mechanism. In: 2020 33rd SIBGRAPI Conference on Graphics, Patterns and Images (SIBGRAPI), pp. 340–347. IEEE (2020)

13. Renau, Q., Dreo, J., Doerr, C., Doerr, B.: Towards explainable exploratory landscape analysis: extreme feature selection for classifying BBOB functions. In: Castillo, P.A., Jiménez Laredo, J.L. (eds.) EvoApplications 2021. LNCS, vol. 12694, pp. 17–33. Springer, Cham (2021). https://doi.org/10.1007/978-3-030-72699-7_2

14. Kerschke, P., Trautmann, H.: Automated algorithm selection on continuous black-box problems by combining exploratory landscape analysis and machine learning. Evol. Comput. **27**(1), 99–127 (2019)

15. Babatunde, O.H., Armstrong, L., Leng, J., Diepeveen, D.: A genetic algorithm-based feature selection (2014)

16. Jankovic, A., Eftimov, T., Doerr, C.: Towards feature-based performance regression using trajectory data. In: Castillo, P.A., Jiménez Laredo, J.L. (eds.) EvoApplications 2021. LNCS, vol. 12694, pp. 601–617. Springer, Cham (2021). https://doi.org/10.1007/978-3-030-72699-7_38

17. Trajanov, R., Dimeski, S., Popovski, M., Korošec, P., Eftimov, T.: Explainable landscape-aware optimization performance prediction. In: 2021 IEEE Symposium Series on Computational Intelligence (SSCI), pp. 01–08. IEEE (2021)

18. Hansen, N., Auger, A., Finck, S., Ros, R.: Real-parameter black-box optimization benchmarking 2010: experimental setup. Ph.D. thesis, INRIA (2010)

19. Hansen, N., Auger, A., Ros, R., Mersmann, O., Tušar, T., Brockhoff, D.: COCO: a platform for comparing continuous optimizers in a black-box setting. Optim. Methods Softw. **36**(1), 114–144 (2021)

20. Škvorc, U., Eftimov, T., Korošec, P.: The effect of sampling methods on the invariance to function transformations when using exploratory landscape analysis. In: 2021 IEEE Congress on Evolutionary Computation (CEC), pp. 1139–1146. IEEE (2021)

21. Kerschke, P., Hanster, C., Dagefoerde, J.: Maintainer Pascal Kerschke. Package 'flacco' (2020)

22. Lunacek, M., Whitley, D.: The dispersion metric and the CMA evolution strategy. In: Proceedings of the 8th Annual Conference on Genetic and Evolutionary Computation, pp. 477–484 (2006)

23. Muñoz, M.A., Kirley, M., Halgamuge, S.K.: Exploratory landscape analysis of continuous space optimization problems using information content. IEEE Trans. Evol. Comput. **19**(1), 74–87 (2014)

24. Kerschke, P., Preuss, M., Wessing, S., Trautmann, H.: Detecting funnel structures by means of exploratory landscape analysis. In: Proceedings of the 2015 Annual Conference on Genetic and Evolutionary Computation, pp. 265–272 (2015)
25. Kerschke, P., Trautmann, H.: Comprehensive feature-based landscape analysis of continuous and constrained optimization problems using the R-Package Flacco. In: Bauer, N., Ickstadt, K., Lübke, K., Szepannek, G., Trautmann, H., Vichi, M. (eds.) Applications in Statistical Computing. SCDAKO, pp. 93–123. Springer, Cham (2019). https://doi.org/10.1007/978-3-030-25147-5_7
26. Biau, G., Scornet, E.: A random forest guided tour. TEST **25**(2), 197–227 (2016). https://doi.org/10.1007/s11749-016-0481-7
27. Refaeilzadeh, P., Tang, L., Liu, H.: Cross-validation. Encyclopedia Database Syst. **5**, 532–538 (2009)
28. Škvorc, U., Eftimov, T., Korošec, P.: Transfer learning analysis of multi-class classification for landscape-aware algorithm selection. Mathematics **10**(3) 2022
29. Van der Maaten, L., Hinton, G.: Visualizing data using T-SNE. J. Mach. Learn. Res. **9**(11) (2008)
30. Škvorc, U., Eftimov, T., Korošec, P.: Understanding the problem space in single-objective numerical optimization using exploratory landscape analysis. Appl. Soft Comput. **90**, 106138 (2020)

Genetic Improvement of TCP Congestion Avoidance

Alberto Carbognin, Leonardo Lucio Custode, and Giovanni Iacca[✉]

Department of Information Engineering and Computer Science, University of Trento,
Trento, Italy
alberto.carbognin@studenti.unitn.it,
{leonardo.custode,giovanni.iacca}@unitn.it

Abstract. The Transmission Control Protocol (TCP) protocol, i.e., one of the most used protocols over networks, has a crucial role on the functioning of the Internet. Its performance heavily relies on the management of the congestion window, which regulates the amount of packets that can be transmitted on the network. In this paper, we employ Genetic Programming (GP) for evolving novel congestion policies, encoded as C++ programs. We optimize the function that manages the size of the congestion window in a point-to-point WiFi scenario, by using the NS3 simulator. The results show that, in the protocols discovered by GP, the Additive-Increase-Multiplicative-Decrease principle is exploited differently than in traditional protocols, by using a more aggressive window increasing policy. More importantly, the evolved protocols show an improvement of the throughput of the network of about 5%.

Keywords: Genetic programming · NS3 · TCP · Network protocols

1 Introduction

In the era of the Internet of Things (IoT), networked systems have become a crucial part of our everyday lives. Network protocols, which describe the interactions that can occur in a networked system, are traditionally modeled by means of automata, which require: a) complete knowledge about the environment, and b) strict assumptions on the interactions that can occur. In this scenario, several works proposed formal methods that, given a set of service specifications, perform automatic synthesis of the network protocols [1–4].

As an alternative to this approach, bio-inspired techniques can be used to evolve network protocols by simulating their behavior, i.e., without any need for formalizing all the protocol requirements. So, even though the computational budget required these approaches is higher than the one needed for formal methods, they have the advantage that there is no need for a complete knowledge of the environment. Thus, bio-inspired techniques allow to evolve protocols for scenarios that are hard to model analytically. Moreover, protocols discovered by

ⓒ The Author(s), under exclusive license to Springer Nature Switzerland AG 2022
M. Mernik et al. (Eds.): BIOMA 2022, LNCS 13627, pp. 114–126, 2022.
https://doi.org/10.1007/978-3-031-21094-5_9

means of bio-inspired approaches allow to perform continual learning and adaptation, which allows for: a) an improvement of the performance of the protocol over time; and b) adaptation to changing domains.

Among the various protocols at the bases of modern Internet, of the most important ones is the Transmission Control Protocol (TCP). The key element of TCP is the so-called congestion avoidance mechanism, which makes use of a congestion window to avoid overloading the link between the sender and the receiver. The size of the congestion window is traditionally managed by means of an Additive-Increase-Multiplicative-Decrease approach. However, it may be possible to adopt alternative, automatically generated congestion avoidance mechanisms.

In this paper, we apply Genetic Programming (GP) [5] for the automatic synthesis of a congestion window management protocol. We employ the NS3 simulator [6] to evaluate the effectiveness of the protocols evolved in a point-to-point WiFi scenario. In our numerical experiments, we observe that the evolved protocols are able to obtain approximately a 5% improvement in performance with respect to the corresponding baseline protocols.

The rest of the paper is structured as follows. In the next section, we present the background concepts on TCP. Then we make a brief overview of the related works in Sect. 3. In Sect. 4, we describe our methods. In Sect. 5, we present our experimental setup and numerical results. Finally, in Sect. 6 we conclude this work.

2   Background

The TCP is one of the most used communication protocols together, with the User Datagram Protocol (UDP) in the Transport Layer of the Internet Protocol Suite. TCP is well-known for its reliability rather than speed performance, indeed it is able to detect the loss of data packets, request missing segments, and guarantee that all the information is transmitted and delivered to the receiver. This behavior, however, reduces the available bandwidth; in fact, a potential issue that may arise by applying these reliability features is the congestion of the network. Besides the protocol implementation, network congestion can be caused by many factors, the most common being the low amount of bandwidth available from the channel and a not properly designed network infrastructure. TCP has the duty of preventing and mitigating network congestion by using ad-hoc strategies.

In order to achieve a stable and reliable connection between two hosts, it is required that the transmission is somewhat controlled at both ends. For instance, propagation delays due to the network infrastructure could affect negatively the overall throughput. Congestion control algorithms have been developed to avoid and recover from this kind of network degradation.

A congested network can quickly result in very low performance. Traditional congestion control algorithms can be divided into two categories: end-to-end and network assisted. While in the former only information about the sender and the receiver is needed, in the latter metrics regarding the network infrastructure are used to take decisions [7].

The challenge, for end-to-end algorithms, is to use implicit signals to infer the congestion status of the network. For instance, for packet loss-based approaches, the objective is to increase throughput by exploiting the bandwidth. In general, if the sender does not receive back the acknowledgment from the receiver after a certain amount of time, the sender may "infer" that the packet is lost. On the other hand, delay-based approaches are better suited for networks that need high speed and flexibility [7], but also in this case calculating the exact transmission delay is tricky; other paths have been researched and some hybrid algorithms have been proposed such as [8].

3   Related Works

Networks have now evolved into very complex systems, where one specific solution may be suitable for one network but ineffective in another one. For this reason, research has focused in solutions that make use of various Artificial Intelligence algorithms, including Evolutionary Computation and Machine Learning, to improve flexibility and performance of protocols. We briefly discuss some of these works below.

Two rather comprehensive surveys on the application of bio-inspired techniques to the evolution of network protocols can be found in [9,10]. Most of the existing approaches focus on offline optimization. For instance, in [11], the authors employ the Particle Swarm Optimization algorithm for the routing path selection. In [12], the authors propose the ant routing method for optimizing routing models. In [13], the authors propose for the first time an EA to evolve protocols. After this work, several papers have tried to use an EA to evolve a variety of network protocols: MAC access protocols [14,15], e.g. through Finite State Machines (FSMs) [16–18]; wireless protocols [19]; aggregation protocols [20–22]; and protocol adaptors [23].

Another line of work consists in using distributed EAs (including GP) to evolve some elements of the network, e.g. through distributed GP [24,25] to evolve the nodes' parameters and functioning logics of WSNs, or through distributed optimization algorithms, including EAs and single-solution algorithms, such as simulated annealing, as in [26], and other optimization paradigms [27–29].

Finally, online learning approaches have been proposed, which allow the network elements to reconfigure at runtime. Su and Van Der Schaar, in [30], propose a learning approach in which each node, by observing the others' behavior, tries to predict the other nodes' reaction to its actions. STEM-Net [31] is a method that equips each node with an EA, that allows to reconfigure each layer of the node, depending on the current state. In [32], the authors propose an approach where protocols are formed as a combination of "fraglets". This concept is similar to those presented in [33,34]. Another recent work on online optimization over networks has been presented in [35].

Other works have applied Machine Learning to predict congestion signals by available data. For instance, the Biaz algorithm is able to distinguish a wireless loss from a congestion loss [36]. Another algorithm is ZigZag [37], which is able to work with different networks infrastructures. The key advantage over common congestion control algorithms for wireless networks is the ability to take into consideration multiple parameters. In [38] a Bayesian detector has been developed and implemented by modifying the TCP New Reno algorithm; the experimental results reported that the model was able to infer the distribution of the round-trip time degradation caused by packet reordering and congestion. A critical point of these models is the difficulty in finding a suitable trade-off between network performance improvements and network resources consumption.

4   Method

In this work, we employ Genetic Programming (GP) [5] to evolve congestion control policies in the form of C++ programs. The function set is shown in Table 1, while the terminal set is shown in Table 2. The parameters used for the GP algorithm are shown in Table 3.

Note that, besides the selection, crossover, and mutation operators, another evolutionary operator is introduced: the Stagnation-Driven Extinction Protocol (SDEP) [39], which controls the extinction of the individuals in the evolutionary process. It makes use of the following hyperparameters:

- p_{sdep}: the extinction probability
- t_{sdep}: threshold used to control the individuals affected by extinction
- k_{sdep}: the number of stagnating generations that, once reached, triggers the operator

In this work, we employ a modified version of the Targeted extinction approach proposed in [39], where p_{sdep} is modified over time:

$$p^k_{sdep} = p^0_{sdep} + f_{sdep}(k) \tag{1}$$

where $f_{sdep}(k)$ is defined as:

$$f_{sdep}(k) = \frac{10\sqrt{k}}{1 + e^{-k}} \tag{2}$$

Moreover, instead of sorting the individuals by fitness, as done in [39], we employ a threshold-based approach to control the individuals affected by extinction. This allows us to reduce the computational complexity of the extinction protocol to a linear complexity. For this purpose, we make use of a threshold computed as follows:

$$\tau = F_{elite}(1 - t_{sdep}) \tag{3}$$

where F_{elite} is the fitness of the elite individual, and all the individuals that have fitness below τ are affected by extinction.

Table 1. Non-terminals used, their corresponding C++ code, argument types and return types.

Non-terminal	C++ code	Argument types	Return type
assignment	arg1 = arg2;	*variable, exp*	body
IfThenElse	if (arg1){ arg2 };	*condition, body*	*body*
lt	(arg1 < arg2)	*condition, condition*	*condition*
lte	(arg1 <= arg2)	*condition, condition*	*condition*
gt	(arg1 > arg2)	*condition, condition*	*condition*
gte	(arg1 > = arg2)	*condition, condition*	*condition*
eq	(arg1 == arg2)	*condition, condition*	*condition*
neq	(arg1 != arg2)	*condition, condition*	*condition*
expression	arg1	*body*	*body*
mul	arg_1, ..., arg_n	*body*	*body*
sum	arg_1, ..., arg_n	*body*	*body*
sub	arg_1, ..., arg_n	*body*	*body*
div	arg_1, ..., arg_n	*body*	*body*
ReduceCwnd	ReduceCwnd(arg_1)	*body*	*body*
CongestionAvoidance	TcpLinuxCongestionAvoidance(arg_1, arg_2)	*body*	*body*

Table 2. Terminals used, their corresponding C++ code and type.

Terminal	C++ code	Type
cnt	arg1	*body*
segmentsAcked	arg1	*body*
tcb->m_cWnd	arg1	*body*
tcb->m_segmentSize	arg1	*body*
tcb->m_ssThresh	arg1	*body*

Table 3. Parameter setting (Koza-style tableau) of the Genetic Programming algorithm.

Parameter	Value
Objective	Throughput
Function set	See Table 1
Terminal set	See Table 2
Population size	30
Number of generations	50
Max lines of code	100
Mutation	Operator flip, prob: 0.6
	Switch branches, prob: 0.3
	Switch expression, prob: 0.7
	Truncate node, prob: 0.25
	Max mutations: 10 mutations

4.1 Code Simplification Procedure

To simplify the evolved trees, we created a procedure that parses the rendered code and generates a more compact version of it. The procedure performs multiple tasks that can be summarized as follows:

- remove empty lines: loops through the code lines and removes the empty one after applying the function .strip;
- gathering variable names: loop through code lines and detects the declaration of variable int and float[1];
- remove if unused: creates an empty list of used variables, loops through the code lines to check if they are used in expression or `IfThenElse` condition blocks, delete the variables that are not inside the list of the used ones;
- clean empty "IfThenElse": loop through code lines and removes branches that are empty;
- simplify expression: loops through the code lines and detects the expression, if they only contains constant values they are simplified;
- compressing to "for" loop: loops through the code lines and detects the equal code lines, it then compress them inside a for loop.

5 Experimental Results

To evaluate our method, we employ the NS3 simulator [40,41]. The Network topology used in our experiments consists of two hosts connected through WiFi with an application data rate of 100 Mbps, a payload size of 1500 bytes and simulation time of 5 s. The position of the hosts is assumed to be fixed during the network simulation. The code we used for our experiments is available at https://carbogninalberto.github.io/pyGENP/.

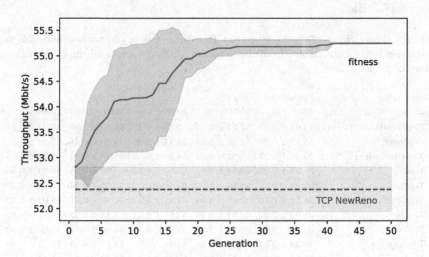

Fig. 1. Fitness trend (mean ± std. dev. across 10 runs) of the protocols evolved from TCP New Reno (blue) vs. the baseline throughput of TCP New Reno (red). (Color figure online)

---

[1] The variables inside the expression are not detected.

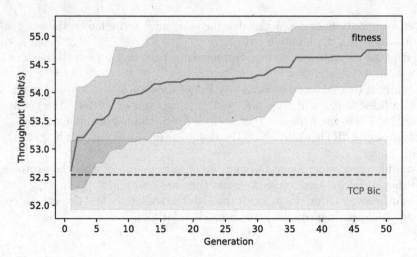

Fig. 2. Fitness trend (mean ± std. dev. across 10 runs) of the protocols evolved from TCP Bic (blue) vs. the baseline throughput of TCP Bic (red). (Color figure online)

First of all, from Figs. 1 and 2 we can see that, for both TCP New Reno and Bic, the evolutionary process quickly outperforms the corresponding baseline values of throughput. On average, we can see that the evolved protocols achieve a 5% improvement on the baseline value of throughput in 50 generations. Moreover, we observe that the evolutionary process seems to stabilize faster (and more robustly across runs) in the case of TCP New Reno with respect to the case of TCP Bic.

Table 4. Throughput of the best evolved TCP protocols in comparison with different congestion control algorithms from the literature (average across 10 runs), for different values of payload size.

Algorithm	Payload size (bytes)				
	250	1500	3000	7500	15000
TcpNewReno (ours)	14.05 ± 0.02	55.24 ± 0.04	53.73 ± 0.49	54.45 ± 0.50	55.30 ± 0.51
TcpNewReno	35.17 ± 0.26	52.38 ± 0.44	53.44 ± 0.46	54.70 ± 0.41	55.20 ± 0.41
TcpBic (ours)	35.44 ± 0.27	54.57 ± 0.02	55.39 ± 0.02	56.58 ± 0.02	57.74 ± 0.03
TcpBic	35.36 ± 0.25	52.54 ± 0.62	53.39 ± 0.54	54.86 ± 0.54	55.25 ± 0.56
TcpHybla	35.28 ± 0.21	52.58 ± 0.97	53.65 ± 0.53	54.65 ± 0.24	55.08 ± 0.30
TcpHighSpeed	35.30 ± 0.16	52.71 ± 0.60	53.45 ± 0.35	54.99 ± 0.67	55.02 ± 0.32
TcpHtcp	35.06 ± 0.26	52.53 ± 0.46	53.67 ± 0.61	55.13 ± 0.52	55.43 ± 0.43
TcpVegas	30.49 ± 3.06	53.89 ± 1.50	55.68 ± 0.17	55.96 ± 0.68	55.36 ± 0.62
TcpScalable	35.22 ± 0.34	52.27 ± 0.46	53.71 ± 0.66	54.69 ± 0.54	55.21 ± 0.45
TcpVeno	35.37 ± 0.24	52.53 ± 0.54	53.68 ± 0.38	54.67 ± 0.66	55.14 ± 0.46
TcpYeah	35.47 ± 0.23	52.58 ± 0.31	53.39 ± 0.82	54.72 ± 0.32	54.88 ± 0.39
TcpIllinois	35.16 ± 0.13	52.50 ± 0.48	53.79 ± 0.77	54.61 ± 0.47	55.08 ± 0.22
TcpWestwood	35.12 ± 0.27	52.77 ± 0.33	53.70 ± 0.50	54.74 ± 0.37	55.18 ± 0.52
TcpWestwoodPlus	35.17 ± 0.18	52.59 ± 0.38	53.78 ± 0.68	54.82 ± 0.37	55.30 ± 0.37
TcpLedbat	35.13 ± 0.16	52.45 ± 0.46	53.81 ± 0.48	54.78 ± 0.50	55.15 ± 0.46

In Table 4, we report the performance metrics for each algorithm available in the NS3 simulator. While our protocols have been evolved on a payload of 1500 bytes, we test them with different payload sizes, to understand whether the resulting protocols are biased towards the payload size used for the evolutionary process. We set as maximum payload size 15000 bytes, which is approximately 1/4 of the maximum theoretical payload size allowed by TCP (65535 bytes). By analyzing the results in the table, it seems reasonable to say that the evolved TCP New Reno protocol appears biased on the payload size used during the evolutionary process (1500 bytes); indeed, it performs comparably or worse than the original TCP New Reno for all the other payload sizes. On the other hand, while the evolved TCP Bic has less performance gain with respect to the original TCP Bic protocol, on average it performs better for all the payload sizes above 1500 bytes. Of note, the throughput reached by the evolved TCP Bic with a payload size of 15000 bytes is the highest among all the other compared congestion control protocols.

Listing 1.1 reports the code of one of the best evolved individuals obtained in the case of TCP New Reno; the solution sets the segmentsAcked variable to a fixed value of 175. It then calls the ReduceCwnd function that is updating the CWND as $CWND = max(\frac{CWND}{2}, segmentSize)$ and then calls the "TcpLinuxCongestionAvoidance" function. The interesting part of this protocol is the fact that this solution always sets the *segmentAcked* variable to a fixed value, thus removing the loss feedback that should be used by the TCP New Reno to signal possible congestion of the network. Moreover, it always reduces the congestion window before executing the congestion avoidance. The logic of this last code is to increase the congestion window by the segment size if the congestion window counter is equal or bigger than the number of segment sizes contained in the CWND, the variable w. Moreover, it always updates the counter by the segmentsAcked which is a static value. Then, it further updates the CWND if the congestion window counter is bigger than the variable w. Further investigations must be done to understand if in this case the static segmentsAcked is behaving if the network is congested; in the evolution environment, the simple network packets are lost according to the Friis propagation loss model [42,43]. The algorithm may have exploited some specific properties of the simulated scenario; for this reason, future work should also include an analysis of the packet loss rates.

The solutions obtained were also very different from each other across runs. For instance, the one reported in Listing 1.2 shows a more complex logic even though, in terms of throughput, it achieves the same result as the one showed in Listing 1.1. This might indicate that the metric used to optimized the protocol may not be able to correctly discriminate solutions of different complexities.

In Listing 1.3, we report one of the best solutions obtained in the case of the TCP Bic protocol; the evolved logic in this case is a bit more complicated than the ones found in the case of TCP New Reno.

```
1  segmentsAcked = (int)175.271;
2  ReduceCwnd(tcb);
3  TcpLinuxCongestionAvoidance(tcb, segmentsAcked);
```

Listing 1.1. Best evolved individual for **TCP New Reno** with payload size 1500 bytes.

```
1  if (tcb->m_segmentSize > tcb->m_ssThresh) {
2      tcb->m_ssThresh = (int)1.0;
3      ReduceCwnd(tcb);
4  } else {
5      tcb->m_ssThresh = (int)35.63;
6      tcb->m_ssThresh = (int)2706.002;
7  }
8  tcb->m_segmentSize = (int)(85.733 - (56.436) - (tcb->
       m_ssThresh) - (92.142) - (70.956) - (6.654));
9  float hnPsxuBCtPVMMYBm = (float)(36.2 - (8.073) -
       (16.935) - (78.417) - (21.996) - (tcb->
       m_segmentSize));
10 if (tcb->m_segmentSize >= segmentsAcked) {
11     segmentsAcked = (int)(hnPsxuBCtPVMMYBm * (68.195) *
       (tcb->m_cWnd) * (tcb->m_ssThresh) * (9.219) *
       (96.226) * (85.611) * (31.971) * (18.886));
12     ReduceCwnd(tcb);
13 } else {
14     segmentsAcked = (int)(72.07 - (tcb->m_ssThresh) - (
       segmentsAcked));
15 }
16 TcpLinuxCongestionAvoidance(tcb, segmentsAcked);
17 if (tcb->m_cWnd <= hnPsxuBCtPVMMYBm) {
18     hnPsxuBCtPVMMYBm = (float)(92.554 - (74.251) -
       (81.969) - (27.667) - (segmentsAcked) - (54.344) -
       (64.616) - (12.799));
19     segmentsAcked = SlowStart(tcb, segmentsAcked);
20     tcb->m_cWnd = (int)(41.965 + (69.396) + (26.746) +
       (30.182) + (tcb->m_ssThresh) + (12.114) + (tcb->
       m_ssThresh));
21 } else {
22     hnPsxuBCtPVMMYBm = (float)(tcb->m_segmentSize *
       3643.109);
23     tcb->m_cWnd = (int)-34.453;
24 }
25
26 for (int i = 0; i < 2; i++) {
27     TcpLinuxCongestionAvoidance(tcb, segmentsAcked);
28 }
```

Listing 1.2. Another best evolved individual for **TCP New Reno** with payload size 1500 bytes.

```
 1 if (cnt != segmentsAcked) {
 2     tcb->m_segmentSize = (int)(13.704 * (53.117) *
       (74.527) * (segmentsAcked) * (61.69) * (17.898));
 3     if (m_cWndCnt > cnt) {
 4         tcb->m_cWnd += tcb->m_segmentSize;
 5         m_cWndCnt = 0;
 6     }
 7 } else {
 8     tcb->m_segmentSize = (int)-352.836;
 9     tcb->m_ssThresh = (int)1.133;
10 }
11 int MkycqeZLOKenojJc = (int)1.549;
12 cnt = (int)(94.326 * (89.844) * (7.283) * (47.081) * (
       tcb->m_ssThresh) * (94.293) * (segmentsAcked) *
       (72.366) * (25.407));
13 ReduceCwnd(tcb);
14 if (tcb->m_ssThresh > cnt) {
15     tcb->m_cWnd = (int)(segmentsAcked + (68.779) +
       (83.102) + (85.846) + (cnt) + (9.069));
16     if (m_cWndCnt > cnt) {
17         tcb->m_cWnd += tcb->m_segmentSize;
18         m_cWndCnt = 0;
19     }
20     MkycqeZLOKenojJc = (int)(23.82 - (79.771) -
       (14.523) - (27.086) - (65.009) - (0.513) - (49.232)
       - (tcb->m_ssThresh));
21 } else {
22     tcb->m_cWnd = (int)(52.023 - (70.074) - (19.636) -
       (tcb->m_ssThresh) - (47.417) - (55.579) - (
       MkycqeZLOKenojJc));
23 }
24 ReduceCwnd(tcb);
```

Listing 1.3. Best evolved individual for **TCP Bic** with payload size 1500 bytes.

6   Conclusions and Future Work

Networks have become ubiquitous in our everyday lives. To increase the efficiency of such networks, it is crucial to efficiently manage the size of the TCP congestion window depending on the scenario at hand. In this paper, we propose a bio-inspired approach to the optimization of congestion control algorithms for a point-to-point WiFi scenario. As shown in Sect. 5, we were able to evolve protocols that increase the performance up to about 5% with respect to the baseline protocols from the literature. This result indicates that the proposed approach is a promising alternative for optimal protocol design.

Future work may focus, among the other things, on: the modification of the fitness evaluation process, to take into account different payload sizes; the evolution of protocols that are able to work well with different packet loss models; the extension of the function set used in GP, in order to include loops and operators with arity greater than 2; the study of the GP parameter effect on the resulting protocols, e.g. through the irace [44] or the ParamILS [45] packages.

References

1. Saleh, K., Probert, R.: Automatic synthesis of protocol specifications from service specifications. In: International Phoenix Conference on Computers and Communications, pp. 615–621. IEEE, New York (1991)
2. Probert, R.L., Saleh, K.: Synthesis of communication protocols: survey and assessment. Trans. Comput. **40**(4), 468–476 (1991)
3. Carchiolo, V., Faro, A., Giordano, D.: Formal description techniques and automated protocol synthesis. Inf. Softw. Technol. **34**(8), 513–521 (1992)
4. Saleh, K.: Synthesis of communications protocols: an annotated bibliography. SIGCOMM Comput. Commun. Rev. **26**(5), 40–59 (1996)
5. Koza, J.R.: Genetic programming: on the programming of computers by means of natural selection. In: Complex Adaptive Systems. MIT Press, Cambridge (1992)
6. Riley, G.F., Henderson, T.R.: The NS-3 network simulator. In: Wehrle, K., Güneş, M., Gross, J. (eds.) Modeling and Tools for Network Simulation, pp. 15–34. Springer (2010). https://doi.org/10.1007/978-3-642-12331-3_2
7. Jiang, H., et al.: When machine learning meets congestion control: a survey and comparison. arXiv:2010.11397 [cs], October 2020
8. Tan, K., Song, J., Zhang, Q., Sridharan, M.: A compound TCP approach for high-speed and long distance networks. In: Proceedings IEEE INFOCOM 2006, 25TH IEEE International Conference on Computer Communications, pp. 1–12, April 2006. ISSN: 0743-166X
9. Nakano, T.: Biologically inspired network systems: a review and future prospects. Trans. Syst. Man Cybern. Part C (Appl. Rev.) **41**(5), 630–643 (2010)
10. Dressler, F., Akan, O.B.: A survey on bio-inspired networking. Comput. Netw. **54**(6), 881–900 (2010)
11. Guo, K., Lv, Y.: Optimizing routing path selection method particle swarm optimization. Int. J. Pattern Recogn. Artif. Intell. **34**(12), 2059042 (2020)
12. Zhang, X., Li, J., Qiu, R., Mean, T.-S., Jin, F.: Optimized routing model of sensor nodes in internet of things network. Sens. Mater. **32**(8), 2801–2811 (2020)
13. El-Fakih, K., Yamaguchi, H., Bochmann, G.: A method and a genetic algorithm for deriving protocols for distributed applications with minimum communication cost. In: International Conference on Parallel and Distributed Computing and Systems, Calgary, AB, Canada, IASTED, pp. 1–6 (1999)
14. Lewis, T., Fanning, N., Clemo, G.: Enhancing IEEE802.11 DCF using genetic programming. In: Vehicular Technology Conference, vol. 3, pp. 1261–1265. IEEE, New York (2006)
15. Roohitavaf, M., Zhu, L., Kulkarni, S., Biswas, S.: Synthesizing customized network protocols using genetic programming. In: Genetic and Evolutionary Computation Conference Companion, pp. 1616–1623. ACM, New York (2018)
16. Sharples, N., Wakeman, I.: Protocol construction using genetic search techniques. In: Cagnoni, S. (ed.) EvoWorkshops 2000. LNCS, vol. 1803, pp. 235–246. Springer, Heidelberg (2000). https://doi.org/10.1007/3-540-45561-2_23

17. Hajiaghajani, F., Biswas, S.: Feasibility of evolutionary design for multi-access MAC protocols. In: Global Communications Conference, pp. 1–7. IEEE, New York (2015)

18. Hajiaghajani, F., Biswas, S.: MAC protocol design using evolvable state-machines. In: International Conference on Computer Communication and Networks, pp. 1–6. IEEE, New York (2015)

19. Tekken-Valapil, V., Kulkarni, S.S.: Derivation of network reprogramming protocol with Z3 (2017)

20. Weise, T., Geihs, K., Baer, P.A.: Genetic programming for proactive aggregation protocols. In: Beliczynski, B., Dzielinski, A., Iwanowski, M., Ribeiro, B. (eds.) ICANNGA 2007. LNCS, vol. 4431, pp. 167–173. Springer, Heidelberg (2007). https://doi.org/10.1007/978-3-540-71618-1_19

21. Weise, T., Zapf, M., Geihs, K.: Evolving proactive aggregation protocols. In: O'Neill, M., et al. (eds.) EuroGP 2008. LNCS, vol. 4971, pp. 254–265. Springer, Heidelberg (2008). https://doi.org/10.1007/978-3-540-78671-9_22

22. Weise, T., Tang, K.: Evolving distributed algorithms with genetic programming. Trans. Evolut. Comput. **16**(2), 242–265 (2011)

23. Van Belle, W., Mens, T., D'Hondt, T.: Using genetic programming to generate protocol adaptors for interprocess communication. In: Tyrrell, A.A.M., Haddow, P.C., Torresen, J. (eds.) ICES 2003. LNCS, vol. 2606, pp. 422–433. Springer, Heidelberg (2003). https://doi.org/10.1007/3-540-36553-2_38

24. Johnson, D.M., Teredesai, A.M., Saltarelli, R.T.: Genetic programming in wireless sensor networks. In: Keijzer, M., Tettamanzi, A., Collet, P., van Hemert, J., Tomassini, M. (eds.) EuroGP 2005. LNCS, vol. 3447, pp. 96–107. Springer, Heidelberg (2005). https://doi.org/10.1007/978-3-540-31989-4_9

25. Valencia, P., Lindsay, P., Jurdak, R.: Distributed genetic evolution in WSN. In: International Conference on Information Processing in Sensor Networks, pp. 13–23. ACM/IEEE, New York (2010)

26. Iacca, G.: Distributed optimization in wireless sensor networks: an island-model framework. Soft. Comput. **17**(12), 2257–2277 (2013). https://doi.org/10.1007/s00500-013-1091-x

27. Wang, S., Li, C.: Distributed robust optimization in networked system. IEEE Trans. Cybern. **47**(8), 2321–2333 (2017)

28. Ning, B., Han, Q., Zuo, Z.: Distributed optimization of multiagent systems with preserved network connectivity. IEEE Trans. Cybern. **49**(11), 3980–3990 (2019)

29. Wang, D., Yin, J., Wang, W.: Distributed randomized gradient-free optimization protocol of multiagent systems over weight-unbalanced digraphs. IEEE Trans. Cybern. **51**(1), 473–482 (2021)

30. Su, Y., Van Der Schaar, M.: Dynamic conjectures in random access networks using bio-inspired learning. J. Sel. Areas Commun. **28**(4), 587–601 (2010)

31. Aloi, G., et al.: STEM-Net: an evolutionary network architecture for smart and sustainable cities. Trans. Emerging Telecommun. Technol. **25**(1), 21–40 (2014)

32. Yamamoto, L., Schreckling, D., Meyer, T.: Self-replicating and self-modifying programs in Fraglets. In: Workshop on Bio-Inspired Models of Network, Information and Computing Systems, pp. 159–167. IEEE, New York (2007)

33. Tschudin, C., Yamamoto, L.: Self-evolving network software. Praxis der Informationsverarbeitung und Kommunikation **28**(4), 206–210 (2005)

34. Miorandi, D., Yamamoto, L.: Evolutionary and embryogenic approaches to autonomic systems. In: International Conference on Performance Evaluation Methodologies and Tools, pp. 1–12. ACM, New York (2008)

35. Yaman, A., Iacca, G.: Distributed embodied evolution over networks. Appl. Soft Comput. **101**, 106993 (2021)
36. Biaz, S., Vaidya, N.: Discriminating congestion losses from wireless losses using inter-arrival times at the receiver. In: Proceedings 1999 IEEE Symposium on Application-Specific Systems and Software Engineering and Technology, ASSET 1999 (Cat. No.PR00122), pp. 10–17, March 1999
37. Cen, S., Cosman, P.C., Voelker, G.M.: End-to-end differentiation of congestion and wireless losses. IEEE/ACM Trans. Netw. **11**(5), 703–717 (2003)
38. I'onseca, N., Crovella, M.: Bayesian packet loss detection for TCP. In: Proceedings IEEE 24th Annual Joint Conference of the IEEE Computer and Communications Societies, vol. 3, pp. 1826–1837. IEEE, Miami (2005)
39. Ye, G.Z., Kang, D.K.: Extended evolutionary algorithms with stagnation-based extinction protocol. Appl. Sci. **11**(8), 3461 (2021)
40. Kurkowski, S., Camp, T., Colagrosso, M.: Manet simulation studies: the Incredibles. SIGMOBILE Mob. Comput. Commun. Rev. **9**(4), 50–61 (2005)
41. Stojmenovic, I.: Simulations in wireless sensor and ad hoc networks: matching and advancing models, metrics, and solutions. IEEE Commun. Mag. **46**(12), 102–107 (2008)
42. Friis, H.T.: A note on a simple transmission formula. Proc. IRE **34**(5), 254–256 (1946)
43. Stoffers, M., Riley, G.: Comparing the NS-3 propagation models. In: IEEE 20th International Symposium on Modeling, Analysis and Simulation of Computer and Telecommunication Systems, IEEE 2012, pp. 61–67 (2012)
44. López-Ibáñez, M., Dubois-Lacoste, J., Cáceres, L.P., Birattari, M., Stützle, T.: The irace package: iterated racing for automatic algorithm configuration. Oper. Res. Perspect. **3**, 43–58 (2016)
45. Hutter, F., Hoos, H.H., Leyton-Brown, K., Stützle, T.: ParamILS: an automatic algorithm configuration framework. J. Artif. Intell. Res. **36**, 267–306 (2009)

Hybrid Acquisition Processes in Surrogate-Based Optimization. Application to Covid-19 Contact Reduction

Guillaume Briffoteaux[1,2]([⊠]), Nouredine Melab[2], Mohand Mezmaz[1], and Daniel Tuyttens[1]

[1] Mathematics and Operational Research Department, University of Mons, Mons, Belgium
{guillaume.briffoteaux,mohand.mezmaz,daniel.tuyttens}@umons.ac.be
[2] University of Lille, Inria, UMR 9189 - CRIStAL, Lille, France
nouredine.melab@univ-lille.fr

Abstract. Parallel Surrogate-Assisted Evolutionary Algorithms (P-SAEAs) are based on surrogate-informed reproduction operators to propose new candidates to solve computationally expensive optimization problems. Differently, Parallel Surrogate-Driven Algorithms (P-SDAs) rely on the optimization of a surrogate-informed metric of promisingness to acquire new solutions. The former are promoted to deal with moderately computationally expensive problems while the latter are put forward on very costly problems. This paper investigates the design of hybrid strategies combining the acquisition processes of both P-SAEAs and P-SDAs to retain the best of both categories of methods. The objective is to reach robustness with respect to the computational budgets and parallel scalability.

1 Introduction

To solve black-box expensive optimization problems where the objective function is computationally costly to evaluate, Parallel Surrogate-Based Optimization Algorithms (P-SBOAs) are built by leveraging parallel computing and machine learning. Two categories of P-SBOAs arise: Parallel Surrogate-Assisted Evolutionary Algorithms (P-SAEAs) and Parallel Surrogate-Driven Algorithms (P-SDAs). Both families of algorithms differ by their Acquisition Process (AP), the mechanism in charge of suggesting new promising candidate solutions. On the one hand, the AP from P-SDAs aims at quickly reaching good solutions, consequently providing a strong improvement that faints after few cycles. On the other hand, the AP from P-SAEAs is more exploratory, thus making the improvement slighter but more durable. In a previous study, we observed that P-SAEAs are generally recommended in the context of moderately expensive problems and that P-SDAs are usually preferred to deal with very expensive problems [1]. Moderately expensive problems are characterized by a budget greater than 1000

M. Mernik et al. (Eds.): BIOMA 2022, LNCS 13627, pp. 127–141, 2022.
https://doi.org/10.1007/978-3-031-21094-5_10

simulations or a simulation lasting less than 5 min. In this article, we present a hybrid method retaining the best of both P-SAEAs and P-SDAs. A strategy previously proposed in [2] already relies on hybridization of APs but shows a serious limitation regarding parallel scalability. The challenge we address in this study is to come up with a strategy that is robust with respect to the computational budget allocated to the search and that scales with multiple computing cores.

For a moderately expensive objective function, the computational budget may allow a meaningful number of expensive evaluations. Consequently, the database of exactly evaluated solutions may grow significantly enough for the surrogate training to become non-negligible in terms of computations. In this situation, it is not convenient to express the computational budget only as a limited number of objective function evaluations as it is commonly done in the field of surrogate-based optimization [3–7]. Indeed, the cost related to surrogate training would be hidden. Instead, we chose to define the budget as a limited duration on a limited number of computing cores.

Benefiting from numerous computing cores raises concerns to the performance of P-SBOAs. The AP emphasized in [8] outputs $q = 4$ new candidates per iteration, consequently triggering $q = 4$ parallel evaluations. The low value attributed to q points the difficulty of conserving a relevant degree of diversity when numerous new candidates are sampled at once, thus preventing to efficiently leverage more computing cores. In [2], the proposed Surrogate Model Based Optimization + Evolutionary Algorithm (SMBO+EA) demonstrates a superiority compared to state-of-the-art P-SDAs for a number of computing cores $n_{cores} < 10$. However, this hybrid method performs similarly to a surrogate-free parallel evolutionary algorithm for $n_{cores} \geqslant 10$.

The main contribution of this paper is the HSAP strategy (Hybrid Successive Acquisition Processes) that employs successively two APs during the search, thus providing robustness with respect to the computational budgets and efficient use of multiple computing cores. The numerical experiments consider an objective function based on a black-box simulator of Covid-19 transmission. The related moderately expensive real-world optimization problem consists in finding the best contact reduction strategy to minimize the number of deaths while attaining herd immunity.

The paper is organized as follows. In Sect. 2 a background on surrogate-based optimization is proposed and the Covid-19-related problem is presented in Sect. 3. The new algorithms based on hybrid APs are dissected in Sect. 4 and are compared with state-of-the-art methods through numerical experiments whose outcomes are reported in Sect. 5. Finally, conclusions and future research directions are pointed out in Sect. 6.

2   Background on Surrogate-Based Optimization

The surrogate-model is based on a machine learning algorithm for interpolation or regression in order to imitate the expensive objective function. The two models considered in this study are a Gaussian Process with a Radial Basis Function kernel (GP_RBF) and a Bayesian Neural Network approximated by Monte-Carlo Dropout (BNN_MCD).

The general idea of the GP_RBF is to model the influence of one point x on the prediction at another point x' by the kernel function defined by:

$$k(x, x') = \sigma \exp\left(-\frac{\|x - x'\|^2}{2s^2}\right) \tag{1}$$

where σ and s are hyper-parameters called the scale and the length scale respectively. By considering the observations as random variables and by applying the Bayes theorem, the GP_RBF provides a prediction $\hat{f}(x')$ and a predictive standard deviation $\hat{s}(x')$ at an unknown point x'. The operation of training a GP is cubic to the number of training samples. More thorough details about GPs are given in [9].

The main principle behind BNN_MCD is to sample n_{sub} sub-networks $\hat{f}_i$ from a global artificial neural network and to use the n_{sub} predictions to compute an average prediction and a standard deviation:

$$\hat{f}(x') = \frac{1}{n_{sub}} \sum_{i=1}^{n_{sub}} \hat{f}_i(x') \qquad \hat{s}(x') = \sqrt{\frac{1}{n_{sub}} \sum_{i=1}^{n_{sub}} (\hat{f}_i(x') - \hat{f}(x'))^2} \tag{2}$$

The sub-networks are sampled by randomly deactivating neurons in the global network. It has been proven in [10] that this technique amounts to perform an approximated Bayesian training. The operation of training a BNN_MCD is linear to the number of training samples.

The two categories of P-SBOAs, namely P-SAEAs and P-SDAs, differ by the coupling between the surrogate-model and the optimizer [1]. In P-SAEAs, the surrogate is attached to the Evolutionary Algorithm (EA) by means of an Evolution Control (EC) that defines the promisingness of new candidate solutions. The EA carries out the search by evolving a population of candidates through the stages of selection, reproduction and replacement. The surrogate is introduced at any stage to replace the expensive objective function [11]. In P-SDAs, a metric of promisingness called the Infill Criterion (IC) is optimized to locate new potential candidates [12]. The IC and the EC are based on the predictive objective value ($\hat{f}$) and/or predictive standard deviation ($\hat{s}$) delivered by the surrogate. The difference between the two concepts of IC and EC is thin. The EC is defined as a comparison operator while the IC is a real-valued metric. It is straightforward to convert an IC into an EC and a dedicated EA can be set up to optimize an IC corresponding to a given EC (by basing the selection and replacement on the EC) [13].

To build the new hybrid acquisition processes, we rely on two actual surrogate-optimizer couplings. The first one is denoted SaaF (Surrogate as a Filter) and comes from a P-SAEA [11]. The corresponding AP consists in generating multiple new solutions by reproduction and to filter them through the EC to retain the q most promising candidates. The second coupling is derived from a P-SDA and is denoted *cl-mean* (Constant Liar with Mean) [14]. In *cl-mean*, to generate q new candidates, the IC is optimized q times and the surrogate model is updated between each optimization. The surrogate update is based on the new proposed candidates associated to the mean of the objective values observed in the database of already simulated points. As soon as the q new candidates are available, they are simulated in parallel.

The question of what is a promising solution is answered by defining the promisingness. In this work, we focus on two ensembles of ECs: the voting committee of ECs and the dynamic inclusive ensemble of ECs. The voting committee already presented in [15], consists in making the ECs vote for the candidates. A solution receives one vote if one EC considers it as promising and the candidates gathering the more votes are the most promising ones. The actual committee highlighted here is denoted *com-spf* and embeds three ECs, the minimization of $\hat{f}$ (favoring exploitation), the maximization of $\hat{s}$ (promoting exploration) and a third EC based on the Pareto dominance between exploitation and exploration similar to the one exhibited in [16]. The dynamic inclusive ensemble of EC denoted *dyn-df-incl* comprises two ECs, the maximization of the distance d to the database of known solutions (favoring exploration) and the minimization of $\hat{f}$. The contribution of each EC to constitute the batch of the q new candidates varies during the search. Indeed, the search is decomposed into 5 equal periods and the proportions of the contribution for (max d, min $\hat{f}$) are consecutively (100%, 0%), (75%, 25%), (50%, 50%), (25%, 75%) and (0%, 100%). In other terms, exploration is favored at the beginning of the search and exploitation is reinforced at latter stages.

3   COVID-19 Contact Reduction Problem

At the beginning of the Covid-19 crisis, when no vaccines were available, governments of the affected countries adopted different strategies to contain the spread of the virus. While some countries imposed lockdown and physical distancing, others, bet on reaching herd immunity by natural transmission. This approach has not proven to be effective during the first two years of the epidemic [17]. However, at the time, studying the possible consequences of this strategy was of importance. Recently, the new Omicron variant of the coronavirus and the deployment of vaccines revive the debate about herd immunity [18].

The problem consists in optimizing the contact reduction strategy to minimize the number of Covid-19-related deaths in Spain while reaching herd immunity. The Spanish population is divided into 16 age-categories and the decision variables represent the contact mitigation factors to apply to each category. For a decision vector $x \in [0,1]^{16}$, $f_1(x)$ represents the simulated number of deaths after the considered period and $f_2(x) \in \{0,1\}$ is a simulated boolean variable indicating whether herd immunity has been reached. The optimization problem consists in finding x^* such that:

$$x^* = \underset{x \in [0,1]^{16} \text{ s.t. } f_2(x)=1}{\arg \min} f_1(x) \tag{3}$$

According to [19], handling constrained problems with EAs can be realized by different means. For our problem, it is not known how to generate feasible candidates so designing repairing operators or specific reproduction operators is impossible. Rejecting infeasible individuals would prevent to keep knowledge about the infeasible region location, besides, this technique works only if the search space is convex, that is probably not the case. Adding the amount of infeasibility as an additional objective would increase the complexity of the problem because the new objective would be boolean. Finally, we opt for the penalization of the infeasible candidates to handle the constraint of the Covid-19 contact reduction problem. The penalty value is set to the approximate Spanish population size (46,000,000) as it is the only *a priori* known upper bound for f_1. A higher value would more likely prevent the search to visit the boundary region between the feasible and infeasible search spaces.

Therefore, the problem is re-formulated as an unconstrained optimization problem by applying a penalty to the objective f_1 when herd immunity is not reached. The re-formulated problem consists thus in finding x^* such that:

$$x^* = \underset{x \in [0,1]^{16}}{\arg \min} \tilde{f}(x) \tag{4}$$

where:

$$\tilde{f}(x) = \begin{cases} f_1(x) \text{ if } f_2(x) = 1 \\ f_1(x) + 46,000,000 \text{ if } f_2(x) = 0 \end{cases} \tag{5}$$

The impact of the contact reduction strategy is simulated thanks to the AuTuMN simulator available at https://github.com/monash-emu/AuTuMN/ [20]. This simulator is developed by the Department of Public Health and Preventive Medicine at Monash University in Melbourne, Australia, to study epidemic transmission. Both quantities f_1 and f_2 are obtained *via* resolution of differential equations governing the flow of individuals in a compartmental model where the population is divided according to the disease state (Susceptible, Exposed, Infectious, Recovered) [21]. The graph of f_1 is expected to be multi-modal with flat regions according to the prior knowledge issued by the developers of AuTuMN. The simulation takes place in three phases. First, the past dynamic of the epidemic is analysed by calibrating uncertain parameters according to past information. Second, the contact reduction strategy is applied during a period of 12 months. After the 12-month period, mobility restrictions are lifted and herd immunity is recognized if incidence still decreases after two weeks while assuming persistent immunity for recovered individuals [22]. The degrees of contact between individuals are integrated into the model through the contact matrix C provided by [23] where the populations are divided into 16 age-categories. $C_{i,j}$ is the average number of contacts per day that an individual of age-group j makes with individuals of age-group i. The decision variables representing the mitigation factors are applied to matrix C such that $C_{i,j}$ is replaced by $x_i.x_j.C_{i,j}$. A decision variable $x_i = 0$ impedes any contact to individuals from age-category i while setting $x_i = 1$ lets the contact rates unchanged compared to the pre-Covid-19 era.

4   Hybrid Acquisition Processes

The two categories of P-SBOAs, namely P-SAEAs and P-SDAs, are attractive for different budgets or landscapes as shown in [1]. In this section, we attempt to retain the best of both categories by investigating the design of hybrid APs. The generation of new candidates is envisioned *via* both IC optimization and reproduction operators.

Two APs are combined into two novel optimization algorithms. The first AP is a *cl-mean* with the voting committee EC *com-spf* and the GP_RBF surrogate model. The second AP is inspired by P-SAEA, where a BNN_MCD surrogate is only used as a filter to discard unpromising candidates (SaaF). Both APs are the most adequate for each framework on the Covid-19 problem as identified by a preliminary grid-search considering multiple surrogate models (Kriging, Bayesian Linear Regressor, GP_RBF, BNN_MCD), definitions of promisingness (Expected Improvement, Lower Confidence Bound *etc.*) and surrogate-optimizer couplings (notably Kriging Believer, *cl-max* and *cl-min*) [13].

The first new hybrid method is named HCAP for "Hybrid Concurrent Acquisition Process" and is presented in Algorithm 1. The two aforementioned APs are executed concurrently at each cycle to propose new candidates that are subsequently simulated in parallel. The algorithm starts by a search space sampling *via* LHS and the evaluation of the initial candidates (line 1). The surrogates are created and the population is initialized (lines 2 to 4). At the beginning of a cycle, the first AP generates $q_1 = 9$ new promising candidates (line 6). Thence, parents are selected from the population and reproduced to create a batch $\mathcal{P}_c$ of $n_{chld} = 288$ children (lines 7 and 8). From $\mathcal{P}_c$, the $q_2 = 63$ more promising candidates are retained and the remaining $n_{disc} = 225$ candidates are discarded (line 9). A total of $q_1 + q_2 = 72$ new candidates are simulated in parallel at each cycle (lines 10 and 11). Thereafter, the surrogates are updated (lines 13 and 14) and a new population is formed by elitist replacement (line 15).

Algorithm 1 Framework of HCAP.

Input

   simulator: real objective function
   budget: computational budget for the search
   GP_RBF: surrogate model for AP1
   com-spf: evolution control for AP1
   $q_1 = 9$: number of candidates to simulate per cycle for AP1
   $n_{pop1} = 50$: population size for AP1
   $n_{gen} = 100$: number of generations for AP1
   BNN_MCD: surrogate model for AP2
   dyn-df-incl: evolution control for AP2
   $n_{pop2} = 72$: population size for AP2
   $n_{chld} = 288$: number of new candidates issued per cycle for AP2
   $q_2 = 63$: number of candidates to simulate per cycle for AP2
   $n_{disc} = 225$: number of discarding per cycle for AP2

1: *database* ← LHS+parallel_simulations(*simulator*, n_{pop2})
2: GP_RBF ← training(*database*)
3: BNN_MCD ← training(*database*)
4: $\mathcal{P}$ ← *database*   ▷ initial population
5: **while** *budget*≠ 0 **do**
6:    $\mathcal{B}_{sim1}$ ← Constant_Liar_AP(*database*, *com-spf*, GP_RBF, q_1, n_{pop1}, n_{gen})
7:    $\mathcal{P}_p$ ← selection($\mathcal{P}$, n_{chld})   ▷ population of parents
8:    $\mathcal{P}_c$ ← reproduction($\mathcal{P}_p$, n_{chld})   ▷ population of children
9:    $\mathcal{B}_{sim2}$ ← filtering($\mathcal{P}_c$, *dyn-df-incl*, BNN_MCD, q_2, n_{disc})
10:   $\mathcal{B}_{sim}$ ← $\mathcal{B}_{sim1} \cup \mathcal{B}_{sim2}$
11:   parallel_simulation(*simulator*, $\mathcal{B}_{sim}$)
12:   *database* ← *database* ∪ $\mathcal{B}_{sim}$
13:   GP_RBF ← training(*database*, 72)
14:   BNN_MCD ← training(*database*, all)
15:   $\mathcal{P}$ ← elitist_replacement($\mathcal{P}$, $\mathcal{B}_{sim}$, n_{pop2})
16:   *budget* ← get_remaining_budget(*budget*, elapsed_time)
17: **end while**
18: (x_{min}, y_{min}) ← get_best_cost(*database*)
19: **return** x_{min}, y_{min}

The analysis led in [1] indicates that P-SDAs are relevant for few objective function evaluations and P-SAEAs to deal with moderately expensive problems. This conclusion appeals to design another hybrid method that would execute successively an AP based on IC optimization and an AP relying on evolutionary computations. The novel method is referred to as HSAP for "Hybrid Successive Acquisition Processes" and is detailed in Algorithm 2. The first stage consists of running 6 cycles of q-EGO *cl-mean* with GP_RBF and *com-spf* for $q = 18$ thus corresponding to 108 simulations (lines 2 to 11). Afterwards, P-SAEA is run with reproduction operators informed by BNN_MCD through the *dyn-df-incl* EC until the budget is totally consumed (lines 12 to 24). The population is initialized by taking a special care of balancing between exploration and exploitation. To foster exploitation, the 10 best candidates identified so far are included in the initial population (line 12). To boost exploration, a K-Means algorithm [24,

25] partitions the set of decision vectors from the database into 62 groups and one randomly-selected solution per cluster is added to the initial population (line 13).

To test the new HCAP and HSAP, the SMBO+EA from [2] is reproduced by considering $n_{cores} = 18$ as allowed by the computational budget further described in the next section. Moreover, the GP_RBF surrogate model replaces the Kriging model originally employed in [2] as this latter has not been relevant in the preliminary grid search. In SMBO+EA, a cycle consists in running three APs in parallel. The first AP, executed on one computing core, maximizes the Expected Improvement IC [12] to produce a new candidate. The second AP, also running on one computing core, minimizes $\hat{f}$ to output one new solution. The third AP generates $q = 16$ new candidates via reproduction of 16 parents extracted from the current population. The 18 new candidates are simulated in parallel on the 18 cores. After the simulation step, the database, the surrogate and the population are updated and the cycle is repeated until the computational budget is wasted.

In SMBO+EA, no EC is used in the AP based on the reproduction operators whereas a dynamic ensemble of ECs helps to discard unpromising candidates in HCAP and HSAP. Relying on an EC at this step gives more opportunity to the reproduction operators to generate good candidates. The objective pointed out in [2] for future works is to improve the performance of the method when n_{cores} increases. Indeed, in the experiments reported in [2], SMBO+EA performs similarly to P-EA (without surrogate) for $n_{cores} \geqslant 15$. In HCAP and HSAP, the use of two surrogates from different types aims at enhancing diversification in the batch of new samples and improving the overall performance of the hybrid methods. In SMBO+EA, the three APs are performed in parallel while the two APs from HCAP are performed sequentially thus giving a slight advantage to SMBO+EA regarding idleness of computing cores.

5   Experiments

The experimental protocol consists in repeating the execution of the algorithms ten independent times to compute the statistics reflecting the performance of the stochastic methods. The ten initial databases are constituted via LHS. The experiments are supported by a parallel machine made of 18 computing cores provided in an Intel Xeon Gold 5220 CPU. The parallel machine is part of the Grid5000, a French infrastructure dedicated to parallel and distributed computing and enabled by several universities [26]. A computational budget of 30 min on 18 computing cores is granted for each search.

Algorithm 2 Framework of HSAP.

Input

　　simulator: real objective function

　　budget: computational budget for the search

　　GP_RBF: surrogate model for AP1

　　com-spf: evolution control for AP1

　　$q_1 = 18$: number of candidates to simulate per cycle for AP1

　　$n_{pop1} = 50$: population size for AP1

　　$n_{gen} = 100$: number of generations for AP1

　　BNN_MCD: surrogate model for AP2

　　dyn-df-incl: evolution control for AP2

　　$n_{pop2} = 72$: population size for AP2

　　$n_{chld} = 288$: number of new candidates issued per cycle for AP2

　　$q_2 = 72$: number of candidates to simulate per cycle for AP2

　　$n_{disc} = 216$: number of discarding per cycle for AP2

1: *database* ← LHS+parallel_simulations(*simulator*, n_{pop2})
2: GP_RBF ← training(*database*)
3: *counter*=0
4: **while** *counter*< 6 AND *budget*≠ 0 **do**
5: 　　$\mathcal{B}_{sim}$ ← Constant_Liar_AP(*database*, *com-spf*, GP_RBF, q_1, n_{pop1}, n_{gen})
6: 　　parallel_simulation(*simulator*, $\mathcal{B}_{sim}$)
7: 　　*database* ← *database* ∪ $\mathcal{B}_{sim}$
8: 　　GP_RBF ← training(*database*)
9: 　　*budget* ← get_remaining_budget(*budget*, elapsed_time)
10: 　　*counter*=*counter*+1
11: **end while**
12: $\mathcal{P}$ ← get_best(*database*, 10)　　　　　　　　　　　▷ initial population
13: $\mathcal{P}$ ← $\mathcal{P}$∪ K-Means_sampling(*database*, 62)
14: BNN_MCD ← training(*database*)
15: **while** *budget*≠ 0 **do**
16: 　　$\mathcal{P}_p$ ← selection($\mathcal{P}$, n_{chld})　　　　　　　　　▷ population of parents
17: 　　$\mathcal{P}_c$ ← reproduction($\mathcal{P}_p$, n_{chld})　　　　　　　▷ population of children
18: 　　$\mathcal{B}_{sim}$ ← filtering($\mathcal{P}_c$, *dyn-df-incl*, BNN_MCD, q_2, n_{disc})
19: 　　parallel_simulation(*simulator*, $\mathcal{B}_{sim}$)
20: 　　*database* ← *database* ∪ $\mathcal{B}_{sim}$
21: 　　BNN_MCD ← training(*database*, all)
22: 　　$\mathcal{P}$ ← elitist_replacement($\mathcal{P}$, $\mathcal{B}_{sim}$, n_{pop2})
23: 　　*budget* ← get_remaining_budget(*budget*, elapsed_time)
24: **end while**
25: (x_{min}, y_{min}) ← get_best(*database*, 1)
26: **return** x_{min}, y_{min}

The GP_RBF, implemented through GPyTorch [27], is trained on a controlled-size set in HCAP whereas the whole database is used in HSAP and SMBO+EA. The BNN_MCD is built using the Keras library [28] and is always updated thanks to all the simulations performed so far. Training BNN_MCD lasts around 7 s on sets of 72 or 256 samples while the GP_RBF training varies from 40 to 100 s on non-normalized data. Normalizing the data limits the training of

GP_RBF to 1 s due to an early stopping mechanism implemented in GPyTorch. In addition to the three hybrid methods presented in the previous section, the parallel evolutionary algorithm (P-EA) without surrogate is considered. Besides, SaaF with BNN_MCD as surrogate and *dyn-df-incl* as EC, and *cl-mean* with GP_RBF as surrogate and *com-spf* as IC are also included into the comparison. The pySBO platform is used as the software framework for implementation and experimentation [29]. The calibration of the algorithms is given in Table 1. Two versions of *cl-mean* are considered: the one where the surrogate is trained on the complete training set (CTS) made of all the solutions already simulated during the search, and the other one on the restricted training set (RTS) of the last 72 simulations. In *cl-mean*, $q = 18$ simulations are performed per cycle and the optimizer is an EA where both the selection and replacement are based on the criterion defined by *com-spf*. For this specific EA, the population size and the number of generations are set by grid-search to 50 and 100 respectively and the remaining parameters are set as in Table 1 for P-EA.

Figure 1 shows the distribution of the 10 best objective values obtained at the end of the search for each strategies. The corresponding ranking according to the average final objective value is displayed in Table 2. It can be observed that the new hybrid method HSAP significantly outperforms all its competitors. The average, median and variance of the results are all improved when employing HSAP as shown in Fig. 1. The concurrent combination of APs proposed by HCAP is also a reliable strategy as, it outperforms all the non-hybrid methods and SMBO+EA as displayed in Fig. 1 and Table 2. It can be noticed that SMBO+EA behaves as expected as it produces results similar to the P-EA without surrogate.

The convergence profiles are displayed in Fig. 2. Expectedly, HSAP and the P-SDAs exhibit a similar very steep curve for less than 108 simulations. After the AP switch in HSAP, the improvement is slowed down but a continuous progress is noted until around 600 simulations where the convergence is almost reached. Figure 3 displays a zoom that highlights the benefit from using HSAP over *cl-mean* with GP_RBF trained on a reduced training set (RTS) from 300 simulations. Firstly, HSAP allows one to perform more simulations than *cl-mean* as indicates the length of the curves in Fig. 2. Secondly, the use of the surrogate to inform the reproduction operators enables a continuous improvement as soon as the IC-based AP has reached steady state. An attractive enhancement of HSAP would be to automatically detect the flatness in the convergence curve and trigger the AP switch. Such a mechanism is not trivial to design, particularly because user-defined parameters must be avoided. However, exploiting the gradient of the curve is a potential lead that we plan to investigate in the future. HCAP outperforms SMBO+EA and SaaF in Fig. 2 while SaaF overtakes SMBO+EA from 260 simulations. The bad performances of P-EA stressed by Fig. 2 demonstrate again the profit brought by surrogate models for both moderately and very expensive problems.

Table 1. Calibration of the algorithms.

Symbol	Name	Value	Calibration method
Calibration of BNN_MCD			
n_{sub}	Number of sub-networks	5	Grid search
n_{hl}	Number of fully-connected Hidden layers	1	Grid search
m_u	Number of units per layer	1024	Grid search
λ_{decay}	Weight decay coefficient	10^{-1}	Grid search
l	Normal standard deviation For weights initialization	10^{-2}	Grid search
p_{drop}	Dropout probability	0.1	Grid search
$h()$	Activation function	Relu	[30]
ξ	Adam initial learning rate	0.001	[30]
Calibration of P-EA			
n_{pop}	Population size	72	Grid search
p_c	Cross-over probability	0.9	Grid search
η_c	Cross-over distribution index	10	Set from [5]
p_m	Mutation probability	$\frac{1}{d}$	Set from [31]
η_m	Mutation distribution index	50	Set from [5]
n_t	Tournament size	2	Set from [32]
Calibration of SaaF			
n_{chld}	Children per cycle	288	Grid search
q	Simulations per cycle	$72 = 0.25 * n_{chld}$	[33,34]
n_{disc}	Discardings per cycle	216	$n_{chld} - q$
(δ_{ES}, n_{ES})	BNN_MCD early stopping	$(10^{-8}, 32)$	Grid search
	2-fold cross-validation	Yes	Grid search

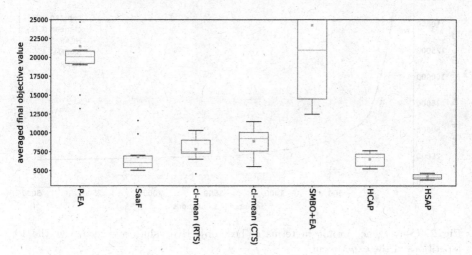

Fig. 1. Distribution of the best objective values from the 10 runs of the experiment. Averaged values are depicted by red squares, median values by red dashes and variance information is given by the length of the boxes. (Color figure online)

Table 2. Ranking of the best strategies according to the final objective value averaged over 10 runs. Ordering according to ascending average from top to bottom.

Strategy	Average
HSAP	4,178
HCAP	6,487
SaaF	6,854
cl-mean (RTS)	7,824
cl-mean (CTS)	8,897
P-EA	21,483
SMBO+EA	24,262

The length of the curves in Fig. 2 yields indications about the computational cost of the methods. Among the hybrid methods, SMBO+EA is the more computationally costly as the surrogate is trained on the entire database and IC optimizations are run at each cycle. By reducing the training set size as in HCAP, more simulations are enabled and by reducing the computational effort dedicated to IC optimization as in HSAP, the number of simulations gets closer to the one of SaaF. A possible way to relieve the computational cost of HCAP would be to execute both APs in parallel.

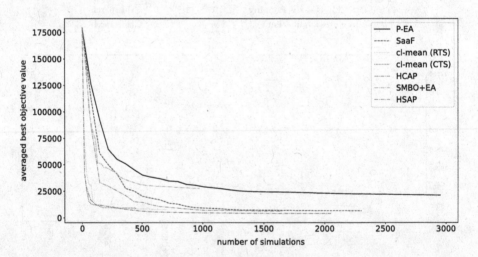

Fig. 2. Convergence profile in terms of best objective values averaged over the 10 repetitions of the experiment.

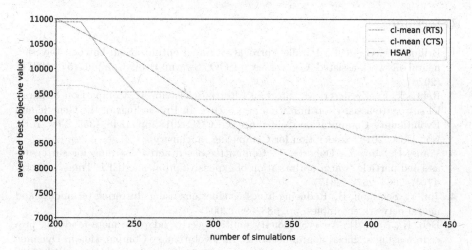

Fig. 3. Convergence profile in terms of best objective values averaged over the 10 runs of the experiment.

6 Conclusion

In this paper, the hybridization of IC optimization and informed reproduction operators is investigated to propose new candidate solutions in P-SBOAs with the aim of bringing robustness with respect to the computational budgets. The Hybrid Successive Acquisition Processes (HSAP) we propose outperforms state-of-the-art methods on a simulation-based problem of Covid-19 contact reduction with a significant number of computing cores. The new strategy consists in relying on IC optimizations during the early stages of the search and in employing informed reproduction operators at the latter stages. For tight computational budgets, only the AP inherited from P-SDAs is employed thus providing fast improvement. For larger budgets, the AP extracted from P-SAEAs is added, therefore further enhancing the search quality. The use of ensembles of ECs favors diversification in the set of newly proposed candidates consequently allowing the efficient use of multiple computing cores. Future works will consider to extend the numerical comparisons by further increasing the number of computing cores and by tackling a larger amount of benchmark problems. Moreover, the HSAP will be improved by designing a mechanism to automatically switch from one AP to another.

Acknowledgment. We thank Romain Ragonnet and the Department of Public Health and Preventive Medicine at Monash University in Melbourne, Australia, for helping us to set the AuTuMN simulator.

Experiments presented in this paper were carried out using the Grid'5000 testbed, supported by a scientific interest group hosted by Inria and including CNRS, RENATER and several Universities as well as other organizations (see https://www.grid5000.fr).

References

1. Briffoteaux, G., et al.: Parallel surrogate-assisted optimization: batched Bayesian neural network-assisted GA versus q-EGO. Swarm Evol. Comput. **57**, 100717 (2020)
2. Rehback, F., Zaefferer, M., Stork, J., Bartz-Beielstein, T.: Comparison of parallel surrogate-assisted optimization approaches. In Proceedings of the Genetic and Evolutionary Computation Conference, GECCO 2018, pp. 1348–1355, New York, NY, USA, 2018. Association for Computing Machinery
3. Wang, H., Jin, Y., Doherty, J.: Committee-based active learning for surrogate-assisted particle swarm optimization of expensive problems. IEEE Trans. Cybern. **47**(9), 2664–2677 (2017)
4. Jin, Y., Sendhoff, B.: Reducing fitness evaluations using clustering techniques and neural network ensembles, pp. 688–699 (2004)
5. Deb, K., Nain, P.: An evolutionary multi-objective adaptive meta-modeling procedure using artificial neural networks. In: Evolutionary Computation in Dynamic and Uncertain Environments, vol. 51, pp. 297–322 (2007). https://doi.org/10.1007/978-3-540-49774-5_13
6. Regis, R., Shoemaker, C.: A stochastic radial basis function method for the global optimization of expensive functions. INF. J. Comput. **19**, 497–509 (2007)
7. Emmerich, M.T.M., Giannakoglou, K.C., Naujoks, B.: Single- and multiobjective evolutionary optimization assisted by gaussian random field metamodels. IEEE Trans. Evol. Comput. **10**(4), 421–439 (2006)
8. Liu, J., Song, W., Han, Z., Zhang, Y.: Efficient aerodynamic shape optimization of transonic wings using a parallel infilling strategy and surrogate models. Struct. Multidiscip. Optim. **55**, 03 (2017)
9. Rasmussen, C.E.: Gaussian processes for machine learning. MIT Press (2006)
10. Gal, Y.: Uncertainty in Deep Learning, Ph. D. thesis, University of Cambridge (2016)
11. Jin, Y.: Surrogate-assisted evolutionary computation: recent advances and future challenges. Swarm Evolut. Comput. **1**(2), 61 – 70 (2011). https://doi.org/10.1016/j.swevo.2011.05.001
12. Jones, D.R., Schonlau, M., Welch, W.J.: Efficient global optimization of expensive black-box functions. J. Global Optim. **13**(4), 455–492 (1998)
13. Briffoteaux, G.: Parallel surrogate-based algorithms for solving expensive optimization problems, Ph. D. thesis, Université de Mons, Université de Lille (2022)
14. Ginsbourger, D., Le Riche, R., Carraro, L.: Kriging is well-suited to parallelize optimization. In: Tenne, Y., Goh, C.-K. (eds.) Computational Intelligence in Expensive Optimization Problems. ALO, vol. 2, pp. 131–162. Springer, Heidelberg (2010). https://doi.org/10.1007/978-3-642-10701-6_6
15. Briffoteaux, G., Ragonnet, R., Mezmaz, M., Melab, N., Tuyttens, D.: Evolution Control Ensemble Models for Surrogate-Assisted Evolutionary Algorithms. In: High Performance Computing and Simulation 2020, Barcelona, Spain, March 2021
16. Tian, J., Tan, Y., Zeng, J., Sun, C., Jin, Y.: Multiobjective infill criterion driven gaussian process-assisted particle swarm optimization of high-dimensional expensive problems. IEEE Trans. Evolut. Comput. **23**(3), 459–472 (2019)
17. Claeson, M., Hanson, S.: Covid-19 and the swedish enigma. Lancet **397**(10271), 259–261 (2021)
18. Medicalxpress. Weaker virus? herd immunity? omicron sparks cautious hopes. https://medicalxpress.com/news/2022-01-weaker-virus-herd-immunity-omicron.html (2022)

19. Michalewicz, Z., Dasgupta, D., Le Riche, R.G., Schoenauer, M.: Evolutionary algorithms for constrained engineering problems. Comput. Ind. Eng. **30**(4), 851–870 (1996)
20. Trauer, J.M.C., Ragonnet, R., Doan, T.N., McBryde, E.S.: Modular programming for tuberculosis control, the "autumn" platform. BMC Infect. Dis. **17**(1), 546 (2017)
21. Caldwell, J.M., et al. Modelling covid-19 in the philippines: technical description of the model. Technical report, Monash University, 2020
22. Ragonnet, R., et al.: Optimising social mixing strategies achieving COVID-19 herd immunity while minimising mortality in six European countries. medRxiv (2020)
23. Prem, K., Cook, A.R., Jit, M.: Projecting social contact matrices in 152 countries using contact surveys and demographic data. PLOS Comput. Biol. **13**(9), 1–21 (2017)
24. Sculley, D.: Web-scale k-means clustering. In: Proceedings of the 19th International Conference on World Wide Web, WWW 2010, pp. 1177–1178, New York, NY, USA, 2010. Association for Computing Machinery
25. Arthur, D., Vassilvitskii, S.: k-means++: the advantages of careful seeding. In: Proceedings of the symposium on Discrete algorithms, pp. 1027–1035 (2007)
26. Cappello, F., et al.: Grid'5000: a large scale and highly reconfigurable grid experimental testbed. In: The 6th IEEE/ACM International Workshop on Grid Computing (2005)
27. Gardner, J.R., Pleiss, G., Bindel, D., Weinberger, K.Q., Wilson, A.G.: GpyTorch: blackbox matrix-matrix gaussian process inference with GPU acceleration. In: Advances in Neural Information Processing Systems (2018)
28. Chollet, F.: Keras. https://keras.io (2015)
29. Briffoteaux, G.: pysbo: python framework for surrogate-based optimization. https://pysbo.readthedocs.io/ (2021)
30. Goodfellow, I., Bengio, Y., Courville, A.: Deep Learning. MIT Press (2016). http://www.deeplearningbook.org
31. Talbi, E.G.: Metaheuristics: from design to implementation. Wiley, Wiley Series on Parallel and Distributed Computing (2009)
32. Deb, K., Pratap, A., Agarwal, S., Meyarivan, T.: A fast and elitist multiobjective genetic algorithm: Nsga-II. IEEE Trans. Evol. Comput. **6**(2), 182–197 (2002)
33. Jin, Y., Olhofer, M., Sendhoff, B.: Managing approximate models in evolutionary aerodynamic design optimization. In: Proceedings of the 2001 Congress on Evolutionary Computation, vol. 1, pp. 592–599 (2001)
34. Buche, D., Schraudolph, N.N., Koumoutsakos, P.: Accelerating evolutionary algorithms with gaussian process fitness function models. IEEE Trans. Syst. Man Cybern. Part C (Appl. Rev.) **35**(2), 183–194 (2005)

Investigating the Impact of Independent Rule Fitnesses in a Learning Classifier System

Michael Heider[✉][iD], Helena Stegherr[iD], Jonathan Wurth[iD], Roman Sraj, and Jörg Hähner[iD]

Universiät Augsburg, Am Technologiezentrum 8, Augsburg, Germany
{michael.heider,helena.stegherr,jonathan.wurth,
roman.sraj,jorg.hahner}@uni-a.de

Abstract. Achieving at least some level of explainability requires complex analyses for many machine learning systems, such as common blackbox models. We recently proposed a new rule-based learning system, SupRB, to construct compact, interpretable and transparent models by utilizing separate optimizers for the model selection tasks concerning rule discovery and rule set composition. This allows users to specifically tailor their model structure to fulfil use-case specific explainability requirements. From an optimization perspective, this allows us to define clearer goals and we find that—in contrast to many state of the art systems—this allows us to keep rule fitnesses independent. In this paper we investigate this system's performance thoroughly on a set of regression problems and compare it against XCSF, a prominent rule-based learning system. We find the overall results of SupRB's evaluation comparable to XCSF's while allowing easier control of model structure and showing a substantially smaller sensitivity to random seeds and data splits. This increased control can aid in subsequently providing explanations for both training and final structure of the model.

Keywords: Rule-based learning · Learning classifier systems · Evolutionary machine learning · Interpretable models · Explainable AI

1 Introduction

The applicability of decision making agents utilizing machine learning methods in real-world scenarios depends not only on the accuracy of the models, but equally on the degree to which explanations of the decisions can be provided to the human stakeholders. For example, in an industrial setting, experienced machine operators often rather rely on their own knowledge instead of on—in their eyes—unsubstantiated recommendations of the model going against that knowledge. This problem is exacerbated as it is inevitable that the model is not perfect in every detail, especially when the learning task is complex and the available training data limited.

M. Mernik et al. (Eds.): BIOMA 2022, LNCS 13627, pp. 142–156, 2022.
https://doi.org/10.1007/978-3-031-21094-5_11

To still make use of the advantages of recommendations made by digital agents, increasing the trust of stakeholders in the predictions is essential. It includes providing explanations of the processes involved to produce these, as well as of the entire model. This can get to a point where easily explainable models are preferred over better performance with higher complexity. Rule-based learners such as Learning Classifier Systems (LCSs) are well suited in these settings as they facilitate extensive explanations [10].

LCSs [25] are inherently transparent and interpretable rule-based learners that make use of a finite set of if-then rules to compose their models. Each rule contains a simpler, more comprehensible submodel, related to specific areas of the feature space. The conditions under which rules apply are optimized during the training process, commonly by an evolutionary algorithm. There are two main styles of LCSs: Pittsburgh-style systems, which evolve a population of sets of rules with combined fitnesses (one per set), and Michigan-style systems, which adapt a single set of rules over time with individual fitnesses (one per rule). Therefore, optimization by the evolutionary algorithm is performed differently in the two styles, but always aimed at finding an "accurate and maximally general" [23] set of rules. Explainability requisites are commonly not directly included as optimization targets for the much more frequent Michigan-style systems, though it is to some extent represented under the concept of generality. In Pittsburgh-style systems, the evolutionary algorithm does typically include error and rule set size as targets but it has to optimize the positioning and also the selection of rules. Therefore, each iteration is comprised of several changes to rules in the set which leads to common situations where beneficial changes to a rule are not reflected in a corresponding change to the fitness of the set and might therefore be discarded for the next generation. While the suboptimal positioning of rules might not even decrease the system's performance, it is, however, a problem when explanations concerning the rule conditions or the training process should be given. Michigan-style systems, on the other hand, often generate and keep a large set of both good and suboptimal rules, in total, far more than required for the given problem. Therefore, they need additional procedures after training, especially *compaction* techniques, to reduce the population to the most important rules and therefore to enhance explainability [16,20].

The first description of a new LCS algorithm, in which the optimization of rule conditions is separated from the composition of rules to form a problem solution, was provided in [11]. This way, rule fitnesses are kept independent from other influences than their direct changes, increasing the *locality*. It also improves the explainability of these quality parameters. Additionally, explainability is improved through the direct control over population sizes and whether good rules should be optimized to be more specific or more general. In this paper, we extend the initial examinations of SupRB, as described in Sect. 3, by evaluating against a modern version of XCSF [19,27], one of the most developed and advanced LCSs, on a variety of different regression datasets (cf. Sect. 4). We find that, as intended, SupRB performs competitively based on hypothesis testing on error distributions as well as Bayesian comparison [4] across datasets, while producing more compact models directly.

2   Related Work

The XCS Classifier System (XCS) is a prominent representative of LCSs. Its many derivatives and extensions are capable of solving all three major learning tasks [25]. In the context of this paper, the most notable extensions are those concerned with applicability to real-valued problem domains and supervised function approximation. In terms of real-valued problem domains, this means replacing binary matching function with interval-based ones [26]. For supervised function approximation, XCSF was designed [27]. It replaces the constant predicted payoff with a linear function. To further enhance the performance, more complex variants were introduced to replace linear models and interval-based matching functions [6,13], however, at the cost of overall model transparency.

LCSs are commonly considered as transparent or interpretable by design, as are other rule-based learning systems, and naturally relate to human behaviour. In contrast, other systems require extensive post-hoc methods, such as visualisation or model transformation, to reach explainability. Even though LCS can be seen as inherently transparent, there can be factors that reduce these capabilities. They may arise through the encodings used, the number of rules in general and the complexity introduced by using complex matching functions or submodels in the individual rules [3].

Controlling these limitations in LCSs is typically done by design but can incorporate designated post-hoc methods. Post-hoc methods, especially visualisation techniques for classifiers, can improve the interpretability of the model [15,17,24]. However, they have to be devised or adapted to the specific needs of the problem at hand and the model itself, which requires time and expertise. Controlling transparency by design can therefore be beneficial in some cases. While some factors, for example problem-dependent complex variables/features, restrict interpretability and can hardly be influenced, other factors can compensate for these issues. This means the design must consider understandable matching functions and predictive submodels, without foregoing an adequate predictive power.

Another aspect strongly related to the interpretability of LCS models is the size of the resulting rule sets, e.g. smaller sets facilitate direct visual inspection and require less subsequent analysis. Controlling this size is handled differently in Pittsburgh-style and Michigan-style systems. Pittsburgh-style LCSs utilize the fitness function of the optimization algorithm, which often incorporates different objectives, i.e. accuracy and number of rules. A prominent example is GAssist [2], where accuracy and minimum description length form a combined objective and an additional penalty is given if the rule set size gets too small. Michigan-style systems, on the other hand, do not control the rule set size by means of the fitness function, as large populations are often beneficial for the training process. During the training, subsumption can be performed to merge two rules where one fully encompasses the other. Compaction is a post-hoc method to reduce the size of the rule set after training by removing redundant rules without decreasing the prediction accuracy [16,28]. However, most compaction methods are purely designed for classification.

3   The Supervised Rule-Based Learning System

We recently proposed [11] a new type of LCS with interchanging phases of rule discovery and solution composition, the Supervised Rule-based Learning System (SupRB). The first phase optimizes rule conditions independently of other rules, discovering a diverse pool of well proportioned rules. Subsequently, in the second phase, another optimization process selects a subset of all available rules to compose a good (accurate yet small) solution to the learning task. In contrast to other LCSs, we thus separate the *model selection* objectives of finding multiple well positioned rules (with a tradeoff between local prediction error and matched volume) and selecting a set of these rules for our final model. That allows us to predict arbitrary inputs with minimal error while the set of rules is as small as possible to keep transparency and interpretability high. As it can be difficult to determine how many rules would need to be generated before a good solution can be composed from them, the two phases are alternated until some termination criterion, e.g. a certain number of iterations, is reached (cf. Algorithm 1). Note that, in contrast to Pittsburgh-style systems, rules added to the pool remain unchanged and will not be removed throughout the training process. An advantage of alternating phases is the ability to steer subsequent rule discoveries towards exploring regions where no or ill-placed rules are found, based on information from the solution composition phase.

Algorithm 1. SupRB's main loop

1: pool ← ∅
2: elitist ← ∅
3: **for** $i \leftarrow 1, \text{n\_iter}$ **do**
4:     pool ← pool ∪ DISCOVER RULES(elitist)
5:     elitist ← COMPOSE SOLUTION(pool, elitist)
6: **end for**
7: **return** elitist

Insights into decisions are a central aspect of SupRB, therefore, its model is kept as simple and interpretable as possible [11]:

1. Rules' conditions use an interval based matching: A rule k applies for example x iff $x_i \in [l_{k,i}, u_{k,i}] \forall i$ with l being the lower and u the upper bounds.
2. Rules' submodels $f_k(x)$ are linear. They are fit using linear least squares with a l2-norm regularization (Ridge Regression) on the subsample matched by the respective rule.
3. When mixing multiple rules to make a prediction, a rule's experience (the number of examples matched during training and therefore included in fitting the submodel) and in-sample error are used in a weighted sum.

In general, a large variety of methods can be used to discover new rules, but for this paper, we utilize an evolution strategy (ES). The overall process

Algorithm 2. SupRB's Rule Discovery

1: **procedure** DISCOVER RULES(elitist)
2:     rules ← ∅
3:     **for** $i ← 1, \text{n\_rules}$ **do**   ▷ $(1, \lambda)$-ES for each new rule
4:         candidate, proponent ← INIT RULE(elitist)
5:         **repeat**
6:             children ← ∅
7:             **for** $k ← 1, \lambda$ **do**
8:                 children ← children ∪ MUTATE(proponent)
9:             **end for**
10:             proponent ← child with highest fitness
11:             **if** candidate's fitness < proponent's fitness **then**
12:                 candidate ← proponent
13:                 $j ← 0$
14:             **else**
15:                 $j ← j + 1$
16:             **end if**
17:         **until** $j = \delta$
18:         rules ← rules ∪ candidate
19:     **end for**
20:     **return** rules
21: **end procedure**

---

is displayed in Algorithm 2. While during a rule discovery phase typically multiple rules are discovered and added, this happens independently (and can be parallelized) in multiple $(1, \lambda)$-ES runs. The initial candidate and parent rule is placed around a roulette-wheel selected training example, assigning higher probabilities to examples whose prediction showed a high in-sample error in the current (intermediate) solution (or *elitist*). The non-adaptive mutation operator samples a halfnormal distribution twice per dimension to move the parent's upper and lower bounds further from the center by the respective values. This is repeated to create λ children. From these, the fittest individual is selected based on its in-sample error and the matched feature space volume as the new parent. If it displays a higher fitness than the candidate it becomes the new candidate. Specifically, the fitness is calculated as

$$F(o_1, o_2) = \frac{(1 + \alpha^2) \cdot o_1 \cdot o_2}{\alpha^2 \cdot o_1 + o_2}, \tag{1}$$

with

$$o_1 = \text{PACC} = \exp(-\text{MSE} \cdot \beta), \tag{2}$$

and

$$o_2 = V = \prod_i \frac{u_i - l_i}{\min_{x \in \mathcal{X}} x_i - \max_{x \in \mathcal{X}} x_i}. \tag{3}$$

The base form (cf. Eq. (1)) was adapted from [29], where it was combining two objectives in a feature selection context. The Pseudo-Accuracy (PACC), Eq. (2), squashes the Mean Squared Error (MSE) of a rule's prediction into a $(0, 1]$ range, while the volume share $V \in [0, 1]$ (cf. Eq. 3) of its bounds is used as a generality measure. The parameter β controls the slope of the PACC and α weighs the importance of o_1 against o_2. We tested multiple values for β and found $\beta = 2$ to be a suitable default. For α, 0.05 can be used in many problems (hyperparameter tuning for the datasets in this paper selected it in 3 out of 4 cases) but, ultimately, the value should always depend on the model size requirements, which are task dependent. If the candidate has not changed for δ generations, the optimization process is stopped and this specific elitist is added to the pool. This process of discovering a new rule and adding it to the pool of rules is repeated until the set number of rules has been found. We want to stress that this optimizer is not meant to find a single globally optimal rule as in typical optimization problems, but rather find optimally placed rules so that for all inputs a prediction can be made that is more accurate than a trivial model, i.e. simply returning the mean of all data. Therefore, independent evolution is advantageous.

Algorithm 3. SupRB's Solution Composition

1: **procedure** COMPOSE SOLUTION(pool, elitist)
2:     population ← elitist
3:     **for** $i \leftarrow 1, \text{pop\_size}$ **do**
4:         population ← population ∪ INIT SOLUTION()
5:     **end for**
6:     **for** $i \leftarrow 1, \text{generations}$ **do**
7:         elitists ← SELECT ELITISTS(population)
8:         parents ← TOURNAMENT SELECTION(population)
9:         children ← CROSSOVER(parents)   ▷ 90% probability n-point
10:        population ← MUTATE(children)   ▷ probabilistic bitflip
11:        population ← population ∪ elitists
12:    **end for**
13:    **return** best solution from population
14: **end procedure**

In the solution composition phase, a genetic algorithm (GA) selects a subset of rules from the pool to form a new solution. As with the rule discovery, many optimizers could be used and a few have already been tested in [30], finding that the GA is a suitable choice. Solutions are represented as bit strings, signalling whether a rule from the pool is part of the solution. The GA uses tournament selection to select groups of two solutions and combines two parents by using n-point crossover with a default crossover probability of 90%. Then, mutation is applied to the children, flipping each bit with a probability determined by the mutation rate. The children and some of the fittest parents (*elitism*) form the new population. The number of elitists depends on the population size of the GA, but in our experiments, we found 5 or 6 to work best with a population size of 32. Solution fitness is also based on Eq. (1). Here, the solution's in-sample mean squared error and its *complexity*, i.e. the number of rules selected, are used as first and second objective, respectively. Note that each individual in the GA always corresponds to a subset of the pool. Rules that are not part of the pool can not be part of a solution candidate and rules remain unchanged by the GA's operations.

SupRB is conceptualised and designed as a regressor. This is reflected in both the description above and the evaluation in the following section. However, we want to propose how the system could be adapted easily towards solving classification problems: The linear submodels would need to be replaced with an appropriate classifier, either simply a constant model, logistic regression or a more complex model if the explainability requirements allowed that. Additionally, the fitness functions would need to use accuracy (or an appropriate scoring for imbalanced data) instead of PACC and MSE.

4   Evaluation

For our evaluation of the proposed system, we compare SupRB to a recent XCSF[1] [19,27] with hyperrectangular conditions and linear submodels (with recursive least squares updates [14]), as they closely correspond to the conditions and submodels used in SupRB. We acknowledge that some better performing conditions, e.g. hyperellipsoids [7], have been proposed for XCSF, however, we consider them less interpretable in high dimensional space for the average user.

4.1   Experiment Design

SupRB is implemented[2] in Python 3.9, adhering to *scikit-learn* [18] conventions. Input features are transformed into the range $[-1, 1]$, while the target is standardized. Both transformations are reversible but improve SupRB's training process as they help preventing rules to be placed in regions where no sample could be matched and remove the need to tune error coefficients in fitness calculations, respectively. Based on our assumptions about the number of rules needed, 32 cycles of alternating rule discovery and solution composition are performed, generating four rules in each cycle for a total of 128 rules. For the ES we selected a λ of 20. Additionally, the GA is configured to perform 32 iterations with a population size of 32. To tune some of the more sensitive parameters, we performed a hyperparameter search using a Tree-structured Parzen Estimator in the Optuna framework [1] that optimizes average solution fitness on 4-fold cross validation. We tuned datasets independently for 256 iterations per tuning process. For XCSF we followed the same process, selecting typical default values[3] [19] and tuning the remaining parameters independently on the four datasets using the same setup as before. The final evaluation, for which we report results in Sect. 4.2, uses 8-split Monte Carlo cross-validation, each with 25 % of samples

---

[1] https://github.com/rpreen/xcsf, https://doi.org/10.5281/zenodo.5806708.

[2] https://github.com/heidmic/suprb, https://doi.org/10.5281/zenodo.6460701.

[3] https://github.com/rpreen/xcsf/wiki/Python-Library-Usage.

reserved as a validation set. Each learning algorithm is evaluated with 8 different random seeds for each 8-split cross-validation, resulting in a total of 64 runs.

We evaluate on four datasets part of the UCI Machine Learning Repository [9]. The Combined Cycle Power Plant (CCPP) [12,22] dataset shows an almost linear relation between features and targets and can be acceptably accurately predicted using a single rule. Airfoil Self-Noise (ASN) [5] and Concrete Strength (CS) [31] are both highly non-linear and will likely need more rules to predict the target sufficiently. The CS dataset has more input features than ASN but is easier to predict overall. Energy Efficiency Cooling (EEC) [21] is another rather linear dataset, but has a much higher input features to samples ratio compared to CCPP. It should similarly be possible to model it using only few rules.

4.2   Results

In our experiments we find that XCSF and SupRB achieve comparable results. Table 1 presents the dataset-specific performance in detail. All entries are calculated on 64 runs per dataset (cf. Sect. 4.1). As both systems were trained for standardized targets, we denote the results for the mean (across runs) mean squared errors (MSE) and their standard deviation (STD) as MSE_σ and STD_σ, respectively. Standardized targets allow better comparison between the datasets as results are on a more similar scale. Additionally, as many real world datasets are normally distributed, this should lighten the need to carefully hand tune the balance between solution complexity and error. Note that predictions of both models can always be retransformed into the original domain. Subsequently, MSE_{orig} references the mean MSE in units of the original dataset-specific target domain. Although this column is less helpful for cross dataset performance interpretations, it allows comparison to other works on the same data. We found that, on two datasets (CCPP and ASN), XCSF shows a better performance, albeit only slightly for CCPP, that can be confirmed through hypothesis testing (Wilcoxon signed-rank test using a confidence level of 5%). Contrastingly, for the CS dataset, the hypothesis could not be rejected. Thus, although SupRB shows a slightly lower mean MSE, this is not statistically significant. For the EEC dataset SupRB outperforms XCSF.

We found that SupRB's runs had a similar (to each other) performance much more consistently than XCSF's. This is shown by STD_σ (cf. Table 1) and specifically illustrated in Fig. 1, which shows the distribution of test errors across all 64 runs. For three of the four datasets, XCSF shows some strong outliers that go against its remaining performances. Additionally, the majority of runs is also further distributed around the mean and median values. We assume that this is largely due to the stochastic iterative nature of training in XCSF. For the CCPP dataset (Fig. 1a) no outliers were produced by XCSF and overall performance is quite similar across runs. This is especially noticeable when comparing the distribution to those on the other datasets. In fact, the runs are so similar (even across models) that it is hard to make any analysis on this scale. Although, XCSF slightly outperformed SupRB on average on CCPP, as confirmed by statistical

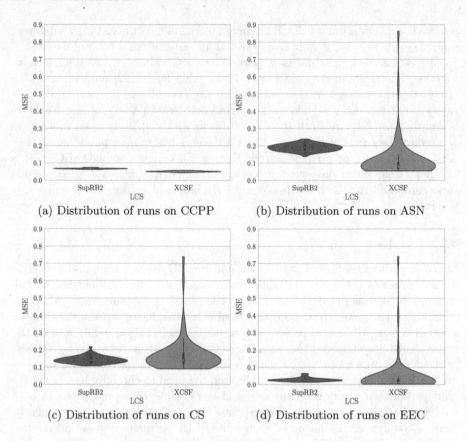

(a) Distribution of runs on CCPP   (b) Distribution of runs on ASN

(c) Distribution of runs on CS   (d) Distribution of runs on EEC

Fig. 1. Distribution of runs' errors on an equal scale.

testing, we can assume that this advantage is likely not practically significant. From a graphical perspective (cf. Fig. 1c), SupRB seems to produce more desirable models on CS, even if the hypothesis testing remained ambiguous. On EEC, XCSF achieves a slightly better median MSE performance (Median$_{XCSF}$: 0.014; Median$_{SupRB}$: 0.026), however, its mean MSE is poorer due to badly performing runs. Regardless, the overall performance can be viewed as rather close, although both sets of runs are clearly not following the same distribution. As SupRB's and XCSF's models were trained on the same random seeds and cross-validation splits, we can conclude that SupRB is overall more reliable even if not necessarily better.

For SupRB we directly control the size (number of rules; *complexity*) of the global solution via the corresponding fitness function used in the GA. Table 2 shows the complexities of the 64 runs per dataset. Note that the highest theoretical complexity is 128, as we did only add 128 rules to the pool. We find that, although theoretically a single rule is able to predict CCPP well, the optimizer prefers to use at least two but at most four rules, achieving slightly better errors

Table 1. Overview of the experimental test data results of 64 runs per dataset rounded to four decimals. MSE_{orig} and MSE_σ give the means of the mean squared errors (MSE) in the dataset's original or a standardised target space, respectively. Similarly, STD_σ displays the standard deviation of MSEs in standardised space. Highlighted in bold are the models where a 5% significance Wilcoxon signed-rank test rejected the null hypothesis of equivalent distributions and the mean was better.

	CCPP			ASN		
	MSE_{orig}	MSE_σ	STD_σ	MSE_{orig}	MSE_σ	STD_σ
XCSF	**0.8745**	**0.0512**	**0.0028**	**0.7930**	**0.1150**	**0.1195**
SupRB	1.1433	0.0669	0.0027	1.3079	0.1896	0.0199
	CS			EEC		
	MSE_{orig}	MSE_σ	STD_σ	MSE_{orig}	MSE_σ	STD_σ
XCSF	2.8291	0.1694	0.1043	0.3660	0.0385	0.1032
SupRB	2.3779	0.1424	0.0199	**0.2776**	**0.0292**	**0.0107**

Table 2. Overview of the solution complexities (number of rules in the solution proposed by SupRB or the final macro-classifier count in an XCSF population, respectively) across 64 runs per dataset.

	SupRB				XCSF			
	CCPP	ASN	CS	EEC	CCPP	ASN	CS	EEC
Mean	2.65	26.42	22.31	12.81	2253.28	962.03	562.81	1028.78
St. dev	0.62	2.47	2.60	1.71	24.70	9.17	11.73	14.90
Median	3	27	22	13	2250	962	562	1026
Min	2	19	17	9	2202	934	530	994
Max	4	30	30	17	2301	980	593	1068

than with a singular linear model. As expected, the solutions to the two highly non-linear datasets (ASN and CS) do feature considerably more rules. EEC again was solved with fewer rules, speaking to its more linear nature, although with more than CCPP, for which a linear solution exists. Standard deviations of complexities increase as the mean increases and the median stays close to the mean.

XCSF seems to have fallen into a cover-delete-cycle where rules did not stay part of the population for long. Covering is a rule generation mechanism that creates a new rule whenever there were too few matching rules. The deletion mechanism removes rules when the population is too full, as there exists a hyperparameter-imposed maximum population size. In our tuning, we did tune both the number of training steps and the maximum population size (among the many other parameters of XCSF) and find that post-training populations are at or around the maximum population size. XCSF's hyperparameter tuning opted for much larger populations than the typical rule of thumb of using ten times as many rules as would be expected (from domain knowledge or prior modelling experience) for a good problem solution [23]. Additionally, upon deeper inspection, we found that the rules were typically introduced late in the training process, however, the system error did not change in a meaningful manner long before that point. Note that we did utilize subsumption in the EA. This mechanism prevents the addition of a newly produced rule to the population when it is fully engulfed by a parent rule and instead increases the parents numerosity parameter. A rule with numerosity n counts as n rules with numerosity 1 towards the maximum population size limit. Subsumption thus theoretically decreases the actual number of classifiers in our population. However, in our experiments the cover-delete-cycle seems to have rendered this mechanism useless.

It is reasonably possible that SupRB's performance would improve in some cases if the pressure to evolve smaller rule sets was lower. However, as explainability suffers with large rule sets, we think that the presented solutions strike an acceptable balance. Afterall, XCSF's solutions were substantially larger even after applying a simple compaction technique of removing rules with an experience of 0 from the final population. This compaction method removed on average about 10% of rules from the run's populations. Table 2 reports the complexity results after compaction. However, we acknowledge that a variety of compaction techniques exists for classification problems [16] that could in some cases potentially be adjusted for the use within regression tasks. Likely, SupRB and XCSF find themselves at different points on the Pareto front between error and complexity. However, in SupRB we do not need to rely on additional post-processing but can solve this optimization problem directly and, importantly, balance the tradeoff of prediction error and rule set complexity against user needs, whereas compaction mechanisms are typically designed to decrease complexity only in a way as to not increase the LCS's system error [16].

Beyond dataset-specific performances, we would like to find a more general answer to the question whether the newly proposed SupRB does perform similarly to the well established XCSF. This would indicate that we can find a good LCS model even without the niching mechanisms employed by XCSF's rule fitness assignment. To find an initial answer based on the performed experiments we use a Bayesian model comparison approach [4] using a hierarchical model [8] that jointly analyses the cross-validation results across multiple random seeds and all four datasets. We assume a region of practical equivalence of $0.01 \cdot \sigma_{\text{dataset}}$.

$$p(\text{SupRB} \ll \text{XCSF}) \approx 63.4\,\%$$
$$p(\text{SupRB} \equiv \text{XCSF}) \approx 8.5\,\%$$
$$p(\text{SupRB} \gg \text{XCSF}) \approx 28.1\,\%$$

where:

- $p(\text{SupRB} \ll \text{XCSF})$ denotes the probability that SupRB performs worse (achieving a higher MSE on test data),
- $p(\text{SupRB} \equiv \text{XCSF})$ denotes the probability that both systems achieve practically equivalent results and

– $p(\text{SupRB} \gg \text{XCSF})$ denotes the probability that SupRB performs better (achieving a lower MSE on test data).

From these results we clearly can not make definitive assessments that XCSF is stronger than SupRB. While it might outperform SupRB in less than two thirds of cases, it also will be outperformed in almost a third of cases. [4] suggest thresholds of 0.95, 0.9 or 0.8 for probabilities to make automated decisions. The specific value needs to be chosen according to the given context. We did not perform the same analysis for the rule set sizes as the results are quite clear with SupRB being the system very likely producing much smaller rule sets. Overall, we can conclude that no clear decision can be made and that the newly developed (and to be improved in the future) SupRB should be considered an equal to the well established XCSF.

Table 3. Exemplary rule generated by SupRB on CS dataset. The target is the concrete compressive strength. The original space intervals denote the area matched by the rule in terms of the original variable scales, while the intervals in feature spaces are scaled into $[-1, 1]$ and help perceiving rule generality at a glance. Coefficients denote the weight vector used for the linear model.

		Original Space	Feature Space σ	
Input variable		Interval	Interval	Coefficient
Cement [kg/m^3]		$[104.72, 516.78]$	$[-0.99, 0.89]$	2.38
Blast Furnace Slag [kg/m^3]		$[0, 359.40]$	$[-1.00, 1.00]$	2.29
Fly Ash [kg/m^3]		$[13.45, 200]$	$[-0.87, 1.00]$	0.68
Water [kg/m^3]		$[122.64, 244.80]$	$[-0.99, 0.96]$	-1.26
Superplasticizer [kg/m^3]		$[6.02, 24.80]$	$[-0.63, 0.54]$	-0.67
Coarse Aggregate [kg/m^3]		$[950.16, 1145]$	$[-0.13, 1.00]$	0.71
Fine Aggregate [kg/m^3]		$[756.14, 992.60]$	$[-0.19, 1.00]$	0.60
Age [days]		$[18.36, 365]$	$[-0.90, 1.00]$	2.07
			intercept$_\sigma$ = 3.9160	
In-sample MSE$_{\text{orig}}$	1.5310	In-sample MSE$_\sigma$ 0.0917	Experience	84

Table 3 presents a rule trained for the CS dataset. It has an experience (number of matched examples during training) of 84 and matched another 31 examples during testing. It is part of a model consisting of 23 rules with experiences of 7 to 240 with a mean experience of 54.17 ± 55.63. The rules were, thus, either rather general or rather specific with this rule being on the more general side. Upon closer inspection, for 5 of the 8 dimensions of CS the rule matches most of the available inputs (being maximally general on the "Blast Furnace Slag" input variable). For the transformed input space (feature space) that is scaled to an interval of $[-1, 1]$ this can easily be seen without any knowledge about the datasets structure, although it is likely that users of the model will have enough domain knowledge to be able to derive this directly from the intervals in the original space. It can also be assumed that these users will generally prefer to inspect the rule in that representation. High concentrations of "Water" and "Superplasticizer" have negative effects on the compressive strength of the concrete for the aforementioned value ranges, while higher concentrations of "Cement", "Blash Furnace Slag" and "Age" of the mixture positively influence its compressive strength. The other three input variables have positive but less pronounced effects. Overall, rule inspection offers some critical insights into the decision making process and can be done fairly easily based on the rule design and the low number of rules per solution.

5 Conclusion

In this paper, we expanded the view on the Supervised Rule-based Learning System (SupRB) with an optimization perspective. We highlighted the advantages of individual rule fitnesses compared to the fitness-sharing approaches typical for other Learning Classifier Systems (LCSs) and discussed our approach to perform LCS model selection using two separated optimizers from that perspective.

To evaluate the system we compared it to XCSF, a well known LCS with a long research history, on four real world regression datasets with different dimensionalities and problem complexities. As one of the greatest advantages of LCS compared to other learning systems is their inherent interpretability and transparency, we limited our study to the use of hyperrectangular conditions and linear models for both systems. After hyperparameter searches for the more sensitive parameters (256 evaluations with 4-fold cross validation), we performed a total of 64 (8 random seeds and 8-fold cross validation with 25% test data) runs of each system on every dataset. We found that, in general, performance is relatively similar. While XCSF showed a statistically better mean test error on two datasets, it was outperformed on one and no statistically significant decision could be made on the fourth dataset. We performed a Bayesian model comparison approach using a hierarchical model and found that no clearly better model can be determined on errors. Solution sizes of SupRB were better than XCSF's even when applying some form of compaction. Additionally, SupRB was more consistent in its performance across runs. Thus, we conclude that, for now and with future research pending, both systems produce similarly performing models.

References

1. Akiba, T., Sano, S., Yanase, T., Ohta, T., Koyama, M.: Optuna: a next-generation hyperparameter optimization framework. In: Proceedings of the 25th ACM SIGKDD International Conference on Knowledge Discovery & Data Mining, KDD 2019, pp. 2623–2631. Association for Computing Machinery, New York (2019). https://doi.org/10/gf7mzz
2. Bacardit, J.: Pittsburgh genetics-based machine learning in the data mining era: representations, generalization, and run-time. Ph.D. thesis, PhD thesis, Ramon Llull University, Barcelona (2004)
3. Barredo Arrieta, A., et al.: Explainable Artificial Intelligence (XAI): concepts, taxonomies, opportunities and challenges toward responsible AI. Inf. Fusion **58**, 82–115 (2020). https://doi.org/10.1016/j.inffus.2019.12.012

4. Benavoli, A., Corani, G., Demšar, J., Zaffalon, M.: Time for a change: a tutorial for comparing multiple classifiers through Bayesian analysis. J. Mach. Learn. Res. **18**(1), 2653–2688 (2017)

5. Brooks, T., Pope, D., Marcolini, M.: Airfoil self-noise and prediction (1989)

6. Bull, L., O'Hara, T.: Accuracy-based neuro and neuro-fuzzy classifier systems. In: Proceedings of the 4th Annual Conference on Genetic and Evolutionary Computation, GECCO 2002, pp. 905–911. Morgan Kaufmann Publishers Inc., San Francisco (2002)

7. Butz, M.V.: Kernel-based, ellipsoidal conditions in the real-valued XCS classifier system. In: Proceedings of the 7th Annual Conference on Genetic and Evolutionary Computation, GECCO 2005, pp. 1835–1842. Association for Computing Machinery, New York (2005). https://doi.org/10.1145/1068009.1068320

8. Corani, G., Benavoli, A., Demšar, J., Mangili, F., Zaffalon, M.: Statistical comparison of classifiers through Bayesian hierarchical modelling. Mach. Learn. **106**(11), 1817–1837 (2017). https://doi.org/10.1007/s10994-017-5641-9

9. Dua, D., Graff, C.: UCI machine learning repository (2017). https://archive.ics.uci.edu/ml

10. Heider, M., Nordsieck, R., Hähner, J.: Learning classifier systems for self-explaining socio-technical-systems. In: Stein, A., Tomforde, S., Botev, J., Lewis, P. (eds.) Proceedings of LIFELIKE 2021 Co-located with 2021 Conference on Artificial Life (ALIFE 2021) (2021). https://ceur-ws.org/Vol-3007/

11. Heider, M., Stegherr, H., Wurth, J., Sraj, R., Hähner, J.: Separating rule discovery and global solution composition in a learning classifier system. In: Genetic and Evolutionary Computation Conference Companion (GECCO 2022 Companion) (2022). https://doi.org/10.1145/3520304.3529014

12. Kaya, H., Tüfekci, P.: Local and global learning methods for predicting power of a combined gas & steam turbine (2012)

13. Lanzi, P.L., Loiacono, D.: XCSF with neural prediction. In: 2006 IEEE International Conference on Evolutionary Computation, pp. 2270–2276 (2006). https://doi.org/10.1109/CEC.2006.1688588

14. Lanzi, P.L., Loiacono, D., Wilson, S.W., Goldberg, D.E.: Prediction update algorithms for XCSF: RLS, Kalman filter, and gain adaptation. In: Proceedings of the 8th Annual Conference on Genetic and Evolutionary Computation, GECCO 2006, pp. 1505–1512. Association for Computing Machinery, New York (2006). https://doi.org/10.1145/1143997.1144243

15. Liu, Y., Browne, W.N., Xue, B.: Absumption to complement subsumption in learning classifier systems. In: Proceedings of the Genetic and Evolutionary Computation Conference, GECCO 2019, pp. 410–418. Association for Computing Machinery, New York (2019). https://doi.org/10.1145/3321707.3321719

16. Liu, Y., Browne, W.N., Xue, B.: A comparison of learning classifier systems' rule compaction algorithms for knowledge visualization. ACM Trans. Evolut. Learn. Optim. **1**(3), 10:1–10:38 (2021). https://doi.org/10/gn8gjt

17. Liu, Y., Browne, W.N., Xue, B.: Visualizations for rule-based machine learning. Nat. Comput. (11), 1–22 (2021). https://doi.org/10.1007/s11047-020-09840-0

18. Pedregosa, F., et al.: Scikit-learn: machine learning in Python. J. Mach. Learn. Res. **12**, 2825–2830 (2011)

19. Preen, R.J., Pätzel, D.: XCSF (2021). https://doi.org/10.5281/zenodo.5806708. https://github.com/rpreen/xcsf

20. Tan, J., Moore, J., Urbanowicz, R.: Rapid rule compaction strategies for global knowledge discovery in a supervised learning classifier system. In: ECAL 2013: The Twelfth European Conference on Artificial Life, pp. 110–117. MIT Press (2013). https://doi.org/10.7551/978-0-262-31709-2-ch017
21. Tsanas, A., Xifara, A.: Accurate quantitative estimation of energy performance of residential buildings using statistical machine learning tools. Energy Build. **49**, 560–567 (2012). https://doi.org/10/gg5vzx
22. Tüfekci, P.: Prediction of full load electrical power output of a base load operated combined cycle power plant using machine learning methods. Int. J. Electri. Power Energy Syst. **60**, 126–140 (2014). https://doi.org/10/gn9s2h
23. Urbanowicz, R.J., Browne, W.N.: Applying LCSs. In: Introduction to Learning Classifier Systems. SIS, pp. 103–123. Springer, Heidelberg (2017). https://doi.org/10.1007/978-3-662-55007-6_5
24. Urbanowicz, R.J., Granizo-Mackenzie, A., Moore, J.H.: An analysis pipeline with statistical and visualization-guided knowledge discovery for Michigan-style learning classifier systems. IEEE Comput. Intell. Mag. **7**(4), 35–45 (2012). https://doi.org/10.1109/MCI.2012.2215124
25. Urbanowicz, R.J., Moore, J.H.: Learning classifier systems: a complete introduction, review, and roadmap. J. Artif. Evolut. Appl. (2009)
26. Wilson, S.W.: Get Real! XCS with continuous-valued inputs. In: Lanzi, P.L., Stolzmann, W., Wilson, S.W. (eds.) IWLCS 1999. LNCS (LNAI), vol. 1813, pp. 209–219. Springer, Heidelberg (2000). https://doi.org/10.1007/3-540-45027-0_11
27. Wilson, S.W.: Classifiers that approximate functions. Nat. Comput. **1**(2/3), 211–234 (2002). https://doi.org/10.1023/a:1016535925043
28. Wilson, S.W.: Compact rulesets from XCSI. In: Lanzi, P.L., Stolzmann, W., Wilson, S.W. (eds.) IWLCS 2001. LNCS (LNAI), vol. 2321, pp. 197–208. Springer, Heidelberg (2002). https://doi.org/10.1007/3-540-48104-4_12
29. Wu, Q., Ma, Z., Fan, J., Xu, G., Shen, Y.: A feature selection method based on hybrid improved binary quantum particle swarm optimization. IEEE Access **7**, 80588–80601 (2019). https://doi.org/10/gnxcfb
30. Wurth, J., Heider, M., Stegherr, H., Sraj, R., Hähner, J.: Comparing different metaheuristics for model selection in a supervised learning classifier system. In: Genetic and Evolutionary Computation Conference Companion (GECCO 2022 Companion) (2022). https://doi.org/10.1145/3520304.3529015
31. Yeh, I.C.: Modeling of strength of high-performance concrete using artificial neural networks. Cem. Concr. Res. **28**(12), 1797–1808 (1998). https://doi.org/10/dxm5c2

Modified Football Game Algorithm for Multimodal Optimization of Test Task Scheduling Problems Using Normalized Factor Random Key Encoding Scheme

Elyas Fadakar[✉]

Beihang University (BUAA), Beijing, China
elyas@buaa.edu.cn

Abstract. Test Task Scheduling Problems (TTSPs) are a type of scheduling problems that are very important in many big and complicated test systems in automotive industries where the reliability of the final product is fundamentally dependent on those tests while the time, workload, and agility of the production is dependent on the optimal scheduling. Scheduling problems are highly multimodal problems with high number of local and global optimum solutions. Availability of different promising solutions provide more freedom for engineering and management decisions based on other criteria which cannot be modeled easily in a design software. In this study we reviewed important researches in the field of multimodal optimization in general and specifically for scheduling problems and then proposed a major modification on Football Game Algorithm (FGA). Then we applied modified FGA for solving some practical TTSPs using a new proposed encoding scheme called Normalized Factor Random Key (NF-RK). The experimental study shows that the modified FGA in combination with NF-RK is promising in solving continuous multimodal TTSPs and shows a notable outperformance in comparison to the best-known reported results in the literature.

Keywords: Test task scheduling problem · Multimodal optimization · Niching algorithm · Encoding scheme

1 Introduction

The Test Task Scheduling Problem (TTSP) is a type of combinatorial problems that are very important in many big and complicated test systems such as automotive industries where the reliability of the final product is fundamentally dependent on those tests while the time, workload, and agility of the production is dependent on the optimal scheduling. In TTSP not only the permutation of the task matters but there are potentially one or more schemes of allocation of the tasks on the available instruments. In TTSPs when a particular scheme of a given task requires the employment of more than one instruments, those instruments must be applied on the task concurrently. Obviously when an instrument is assigned to a task for a period of time it cannot be used for any other tasks simultaneously. These characteristics of the TTSPs could be considered as a constraint on other general form of scheduling problems like Flexible Job Shop Scheduling Problems (FJSPs) and Unrelated Parallel Machines Scheduling Problems (UPMSPs). The most common objectives in the TTSPs are maximum test completion time (makespan) and the mean workload of the instruments [1, 2].

The number of the researches on the optimization of the TTSPs are growing but still not that diverse, especially when it comes to the multimodal optimization of the TTSPs it is even less. But as it is mentioned before there are other scheduling problems in the literature that share some similarities with TTSPs like JSSPs and PMSPs [2, 3], so the methos and strategies in those studies can be exchanged with minor modifications [3]. Therefore, we briefly review some of those studies other than TTSP as well, specially wherever multimodal optimization is the target.

Perez et al. [4, 5] represents one of the very few studies on JSSP with a focus on identifying multiple solutions. JSSPs are typically multi-modal, presenting an ideal case for applying niching methods. Their studies suggest that not only do niching methods help to locate multiple good solutions, but also to preserve the diversity more effectively than employing a standard single-optimum seeking genetic algorithm. In another recent study Pan Zou et al. [6] proposed a new algorithm by combining the k-means clustering algorithm and genetic algorithm (GA) for multimodal optimization of JSSPs. In the proposed algorithm, the k-means clustering algorithm is first utilized to cluster the individuals of every generation into different clusters based on some machine sequence-related features under the assumption that different global optima will have different features. Next, the adapted genetic operators are applied to the individuals belonging within the same cluster with the aim of independently searching for global optima within each cluster.

In Resource constrained multi-project scheduling problems (RCMPSP), multiple projects must be carried out and completed using a common pool of scarce resources. The difficulty is that one has to prioritize each project's tasks to optimize an objective function without violating both intra-project precedence constraints and inter-project resource constraints. A decision maker can benefit from choosing between different good scheduling solutions, instead of being limited to only one. In addition, it is also much faster than rescheduling. The deterministic crowding and clearing methods were adopted in [7] to find multiple optimal scheduling solutions for this problem.

One of the latest studies in this field carried out by Lu Hui et al. [8], in which they proposed a new strategy called Multi Center Variable Scale (MCVS) search algorithm based on the analysis and summary of the characteristics of Combinatorial problems to solve single and multi-objective problems. It is also noted that MCVS is not an algorithm for solving particular multimodal problems; it only utilizes the multimodal property to design an optimization strategy. MCVS mainly focuses on the searching center and the searching neighborhood. The multimodality of the scheduling problems also has been discussed and shown statistically [8]. In another study it is suggested that using local search algorithms which are functionally dependent on the relative smoothness of the search landscape are not suitable to be employed for the scheduling problems [9]. In order to estimate the continuity or the simplicity of the search landscape of the scheduling problems Fitness Distance Coefficient (FDC) has been adopted in [1]. FDC seems a good measure in continuous problems but for combinatorial problems it can also be used by employing Hamming or Edit distances. Also in the same study, a vector group encoding technique based on the characteristic of the problem is presented [1]. A Multi-strategy Fusion niching method has been proposed and used for solving several scheduling problems including Test Task Scheduling Problems (TTSP), Flexible Job-Shop Scheduling Problems (FJSP), and Parallel Machine Scheduling Problem (PMSP) [10].

In the rest of the paper and in Sect. 2, TTSP is mathematically described. The classic and modified FGA are reviewed in Sect. 3. The new proposed encoding scheme called normalized factor random key is explained in Sect. 4. In Sect. 5 the algorithm has been applied to solve 4 instances of TTSPs followed by a comparison discussion in Sect. 6. Finally, some of the possible future research directions are drawn in Sect. 7.

2 Problem Description and Mathematical Modeling

The mathematical representation of the TTSP is adopted from the research [2, 11] and will be as the following. For TTSP there is a set of n tasks $T = \{t_j\}_{j=1}^n$ that has to be tested by some instruments from the set of m resources $R = \{r_i\}_{i=1}^m$. The start time, the finish time and spend time of the task t_j on the instrument r_i are respectively represented by S_j^i, C_j^i, and P_j^i, where $C_j^i = S_j^i + P_j^i$. A task may have multiple available schemes $W_j = \left\{w_j^k\right\}_{k=1}^{k_j}$, where k_j is the number of available schemes for task t_j. Therefore, $K = \{k_j\}_{j=1}^n$ is the set of the number of the schemes for all the tasks. $P_j^k = \max P_j^i$, where $r_i \in w_j^k$ represent the test time of t_j when adopting the scheme w_j^k. A sample information of a TTSP is given in Table 1 with 4 tasks and 4 resources [3].

Table 1. 4 tasks and 4 instruments TTSP sample.

T	W_j	w_j^k	P_j^k	T	W_j	w_j^k	P_j^k
t_1	w_1^1	$r_1 r_2$	5	t_3	w_3^1	r_4	2
	w_1^2	$r_2 r_4$	3	t_4	w_4^1	$r_1 r_3$	4
t_2	w_2^1	r_1	4		w_4^2	$r_2 r_4$	3
	w_1^2	r_3	1		w_4^3	$r_2 r_3$	7

As it is mentioned in the introduction, the most popular objectives for TTSPs are the makespan and the mean workload of the instruments. For single objective optimization we use the makespan function which equals the maximum completion time of all tasks. $C_j^k = \max_{r_i \in w_j^k} C_j^i$ represents the completion time of task t_j when adopting the scheme w_j^k. Therefore, the makespan of all tasks is defined as the following.

$$f(x) = \max C_j^k \ for \ 1 \leq k \leq k_j, 1 \leq j \leq n \tag{1}$$

3 The Proposed Modified Football Game Algorithm (mFGA)

FGA is a population based global optimization algorithm capable to spot and converge to multiple global optimums simultaneously. Imitating a football team's teamwork during a game for finding the best positions to score goals under team's coach supervision makes the FGA able to balance diversification in a way that not only satisfies a better coverage over the search space but to maintain equally good distant positions and exhibit a final distributed convergence to multiple global solutions [12]. After its introduction in 2016, FGA has been applied into different studies. Researchers in [13] have adapted FGA for an asymmetric traveling salesman problem. FGA has also been implemented in capacitated vehicle routing problems [14]. A modified version of FGA has also been proposed and applied to the static state estimation problem of real-time measurement of power systems along with the design of an optimal hybrid active power filters [15]. Moreover, FGA has been classified and its performance has been studied alongside other algorithms as social mimic optimization algorithms [16], sports inspired algorithms [17], and soccer-inspired metaheuristics [18, 19]. So, first a brief review of the classic FGA is presented and then the details of modification on this algorithm are explained.

3.1 Classic FGA

After initial formation of the players in the pitch (initialization), every player moves around their last position biased toward the ball owner. Ball is passed between the players. Players in better position have more chance to take the ball. The new positions of the players depend on the simple random walks and a movement toward the ball when coaching effect is not applied is computed in Eq. 2.

$$x_i(t) = x_i(t - 1) + \alpha(t)\varepsilon + \beta(x_{ball}(t) - x_i(t - 1)) \tag{2}$$

where $\varepsilon \in [-1, 1]$ and $\beta \in [0, 1]$ are random numbers drawn from a uniform distribution and $\alpha(t) > 0$ is the decreasing step size which should be relative to the scales of the problem of interest.

Except from general movements of the players, there are two specific positioning strategies that can be applied by the coach to increase the chance of finding global optimum using local search. First attacking strategy in which the team's coach memorizes the best positions during the game and uses them to guide players and pushing them forward. Secondly, in the substitution strategy, coach can change a defender in low quality position with a fresh striker in the best position to increase the chance of scoring. Weaker players will be replaced with other players around the nearest best position according to the coach's memory. After applying these strategies, the new position of the players who are located beyond the limits of the distance and fitness will be determined by using a random walk around the nearest best elite solutions in the coach's memory with current step size $\alpha(t)$ according to Eq. 4.

$$x_{new} = x_{nearest-elite} + \alpha(t)\varepsilon \tag{3}$$

where $\varepsilon \in [-1, 1]$. .. The coach's memory (CM) is actually the memory of the algorithm to save the best positions (elites) and their corresponding fitness values. These positions will later be used as the base vectors ($X_{nearest-elite}$) to locate fresh substitutes or attackers. The size of the coach memory will be chosen as a fraction of population size as an algorithm's parameters. The following table shows the analogy between a real football game and the simulated FGA. The pseudo-code of the algorithm can be found in [12].

Table 2. Terminology in FGA

	Football Game	**Football Game Algorithm**
1	Football Players	Population members
2	Football pitch	Search space
3	Positioning	Random walk
4	Attacking	Hyper radius penalty
5	Substitution	Fitness value penalty
6	Scoring a goal	Convergence
7	Coach supervision	Elitism and the memory of the algorithm
8	Ball position	Randomness enhancement

3.2 Modified FGA

In the modified version of FGA (mFGA), the main focus is on the improvement of the algorithm's multimodal search capability by addressing basic challenges in MMOPs. In classic FGA, Coach's Memory (CM) is used as the memory of the algorithm to save the locations of the best positions during the search process. This actually comprises the algorithm's elitism in which by a greedy approach the best CM-size (a fraction of the whole population) positions found during the search process would be stored. This memory also saves the CM-size best solutions found after the termination criteria is met. So, the final result may consist of a number of distinct solutions less than or equal to the size of CM (CM-size).

In the modified FGA on the other hand, a distant CM strategy has been used to increase the stability of the found solutions up to the end of the run. For this purpose, a decreasing pairwise Euclidean distance metric between the CM positions is considered. It simply means that there would be a decreasing limitation for the distance between every 2 positions in the CM list. Using evolutionary algorithm analogy and in comparison to the greedy elitism in classic FGA we call this mechanism the Phenotypic Distributed Elitism (PDE). It helps the algorithm to have a better coverage over the search space in the exploration phase and consequently demonstrates an effective implicit basin identification in terms of the performance of the algorithm. This distance limit resembles niche radius in other methods however its functionality is different. For example in species conserving genetic algorithm (SCGA) [20] a constant niche radius is employed which distinguish species seeds in the population and allocate other individuals to each seed based on the same niche radius. But in the PDE the niche radius is decreasing and it does not have any effect on the population distribution in the next iteration. Also, Fig. 1 illustrates the difference between PDE and dynamic archive elitism in dADE [21]. For the one-dimensional Equal Maxima problem, considering the niche radius R and ε fitness threshold, dADE would accept only sample "A" into the archive and rejects all the other samples while in PDE if CM-size is at least 4 then samples "A", "B", "C", and D"" will be recorded in the CM as the elite samples. However, sample "E" is rejected by both methods.

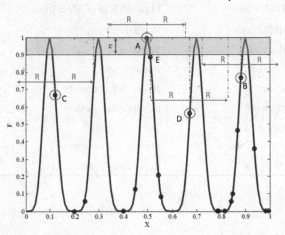

Fig. 1. Initial population for 1-D Equal Maxima problem, with R as niche radius for both dADE and PDE and ε fitness threshold for dADE

Using this mechanism allows the population to spend more searching budget on the search of promising basins with smaller size or irregular shape that are harder to be discovered in comparison to big size and smooth shape basins. Moreover, we eliminate both ball owner effect that increases instability in maintaining the found solutions and attacking strategy which elevate local search capability of FGA. Hence, the next position of the players will be reduced to a Brownian motion around their last position for general movement or around a randomly chosen CM position for fresh substitutes.

General Movement:

$$x_i(t) = x_i(t-1) + \alpha(t)\mathcal{N}(0, I_D); \forall i = 1, 2, \ldots, N_p - N_s \tag{4}$$

Substitute Players:

$$x_i(t) = CM_{position}^j + \alpha(t)\mathcal{N}(0, I_D); j \in_R [1, N_t], \forall i = (N_p - N_s) + 1, \ldots, N_p \tag{5}$$

where N_p is the number of the population, D is the problem's dimension, N_t is the number of tactics or CM size while N_s is the number of substitutions in each iteration, and $\mathcal{N}(0, I_D)$ is a D dimensional random vector drawn from the standard normal distribution. Also, j is a random index drawn from the set of N_t CM samples by fitness proportionate selection method.

Although modified FGA uses a distance metric in the way of modifying the simple FGA, this parameter is set to be a multiple of the step-size and is adaptively reduced with the step-size reduction rate as the optimization iterations go on. Therefore, for the initialization of the search process after uniform sampling of initial population using rejection sampling, the step-size for both general movement and substitution of the players will be determined by Eq. 7.

$$\alpha_0 = \left(\frac{D}{2} \frac{\Gamma(D/2)}{\pi^{\frac{D}{2}}} \frac{V_a}{N_p} \right)^{\frac{1}{D}} / \sqrt{D} \tag{6}$$

$$\alpha(t) = \alpha_0(\theta)^t \tag{7}$$

where α_0 is the initial step-size, resulted from dividing the available volume V_a by the population size N_p, considering the hyper-sphere volume formula [22] due to the fact that the local distribution of the samples follows the multivariate normal distribution, whereas the available volume V_a, without any restriction, can be found using following formula:

$$V_a = \prod_{i=1}^{D} (UB_i - LB_i) \tag{8}$$

where UB and LB representing upper and lower bound of each dimension respectively. As is can be seen in the Eq. 5, the players' positions are sampled using a multivariate normal distribution with mean $CM^j_{position}$ and covariance matrix $\Sigma = \alpha(t)^2 I_D$.

We can summarize the main modifications on FGA in the following features.

- Initialization using uniform rejection sampling
- Eliminating attacking strategy and the ball owner attraction vector
- Phenotypic Distributed Elitism (PDE) for CM
- Substitution using fitness proportionate selection of the base vector from CM

4   Normalized Factor Random Key Encoding Scheme

In order to adopt real parameter optimization algorithms researcher will choose between 2 general approaches. First those algorithms that are modified based on the discrete nature of the solution vectors as discrete version of algorithms [23, 24], and second those strategies that are equipped with a proper encoding scheme to make it possible to use the original algorithm directly in the encoded continuous domain. The disadvantages of the first group roots back to the change in the well-established mutation or random walk operators by discrete operators. Hence, the same performance as the original algorithm is not expected from the modified version of the algorithms. It is also reported in [25] that the computational overhead is often higher too. In the second approach however, employing a proper encoding scheme can resolve many existing issues with the other methods such as complexity, degradation of the original efficiency, computational overhead and generation of infeasible solutions [25, 26]. One of the most popular and simple but efficient encoding schemes that has been employed by many real-valued metaheuristic algorithms to solve permutation problems is called Random Key encoding scheme (RK) [27, 28]. In this study we take the second approach and introduce Normalized Factor Random Key (NF-RK) encoding scheme to apply modified FGA for solving the TTSPs.

In the RK encoding scheme a vector of n random values, usually between [0,1], are drown representing a permutation of n elements. Therefore, decoding of a RK vector with n elements is a mapping from the n-dimensional continues search space, $\mathbb{R}^n$, to D_n, the space of all permutations of $\{1, \ldots, n\}$. So, this mapping can be shown as RK: $[0, 1]^n \rightarrow D_n$. An example of this decoding process from a random vector x from $\mathbb{R}^n$ to a permutation array π from D_n would be as follows:

index	1	2	3	4	5	6	7	8	9	10
x	0.32	0.83	0.17	0.22	0.95	0.41	0.63	0.34	0.55	0.75

π	3	4	1	8	6	9	7	10	2	5
$sort(x)$	0.17	0.22	0.32	0.34	0.41	0.55	0.63	0.75	0.83	0.95

As it can be seen RK encoding scheme can successfully resolve feasibility and mapping challenges for the direct employment of real-valued optimization algorithms. However, in its standard version it can be used just for the permutation problems like traveling salesman problem (TSP) but not for other scheduling problems like TTSPs, FJSPs, and PMSPs because in these problems in addition to job/operation sequence determination the scheme or machine allocation is also needed to be encoded.

In order to capitalize the benefits of RK encoding scheme for TTSPs, Lu et al. proposed a new encoding scheme based on RK encoding, called Integrated Encoding Scheme (IES) [3]. The first step in IES is the same as RK encoding where the sequence of the jobs will be determined. In the second step of IES the values of the n-dimensional vector will be used to determine the scheme assignment to each job by the following equation.

$$k = mod\left(\left[x_{ij} \times 10\right], k_j\right) + 1 \tag{9}$$

where, x_{ij} is the decision variable of task t_j and k_j is the number of schemes for task t_j. The advantage of the IES is that it does not need to increase the size of the problem to more than the size of the tasks unlike other encoding schemes such as permutation with repetition [5] or 2-vector encoding [29]. So, it keeps the size of the problem as low as possible in the continuous domain. But the problem with IES is in the creation of fake multimodality in the continuous search domain.

But in this study a new encoding scheme is introduced which like the IES capitalize the benefits of the RK encoding and in addition resolve the mentioned issues with the IES. The new proposed encoding scheme is called Normalized Factor Random Key (NF-RK) encoding scheme. In the first step of NF-RK encoding, a standard RK is generated with the size of the number of tasks n. After sorting the n-dimensional vector we will have the task sequence similar to the example for the random key. In the second step in order to decide about the scheme for every task the relative position of each value in the range of its immediate higher and lower value in the sorted vector is calculated. Then this normalized factor multiplies the number of task's schemes and produces the decided scheme for the corresponding task based on the following equation. In the following a numerical example of NF-RK for a sample TTSP is provided.

$$k_i = \left[\left(\frac{s_i - s_{i-1}}{s_{i+1} - s_{i-1}} \right).m_i \right] + 1 \ \forall i \in \{1, \ldots, 10\} \qquad (10)$$

where s_i is the sorted x value at sequence i while m_i is representing available number of schemes for each task and k_i is the allocated scheme number to each task based on Eq. 10.

Task index	1	2	3	4	5	6	7	8	9	10	
m	3	4	1	2	1	3	2	5	3	4	
x	0.32	0.83	0.17	0.22	0.95	0.41	0.63	0.34	0.55	0.75	

i	0	1	2	3	4	5	6	7	8	9	10	11
Task index	-	3	4	1	8	6	9	7	10	2	5	-
m	-	1	2	3	5	3	3	2	4	4	1	-
s = sort(x)	0	0.17	0.22	0.32	0.34	0.41	0.55	0.63	0.75	0.83	0.95	1
k	-	1	2	3	2	2	2	1	3	2	1	-

So, as a result, NF-RK encoding scheme preserve the benefits of the standard RK encoding while by providing extra information exploitation from the random vector relative values enhance it for the use in the TTSPs without increasing the dimension of the problem for the scheme allocation to each task.

5   Multimodal Single-Objective Optimization of TTSP

We use modified FGA in combination with NF-RK encoding scheme in order to solve four practical TTSPs.

Four different instances of practical TTSPs are selected for the experimental study. Apart from being practical, these problems are chosen because the existing multimodal optimization results in the literature are about these problems [10] which makes the comparison study possible as well. More detailed information about these problems can be found in [2]. The other multimodal optimization algorithms in the comparison study include the multi-strategy fusion hybrids of genetic algorithm (MFGA) and multi-strategy fusion hybrids of particle swarm optimization (MFPSO) [10].

Each one of these scheduling problems are from different scales in terms of the number of the jobs (n) and the number of the machines (m). Based on the considerations in [10] the objective is to minimize the completion time of the jobs on the set of available machines. Other parameters that have to be set accordingly are the population size with 100 individuals and 100 numbers of generations. In order to have a better estimation of the overall performance of the algorithms the experiment is done 500 times for each problem. The performance and the comparison metrics based on the reference paper [10] include the Best Fitness (BF) value, the Mean Fitness (MF) value and the Mean Number (MN) of the number of found solutions in each run out of 500 runs. The statistical results are shown in Table 3.

Table 3. The performance of 3 multimodal algorithms on the TTSPs based on BF, MF, and MN measures

	TTSP					
	Instance 20 × 8			Instance 30 × 12		
	BF	MF	MN	BF	MF	MN
MFGA	29	31.91	2.83	35	37.97	2.69
MFPSO	29	31.68	2.89	36	38.53	2.86
Modified FGA	**28**	**30.99**	**6.06**	**31**	**35.37**	**6.52**
	Instance 40 × 12			Instance 50 × 15		
	BF	MF	MN	BF	MF	MN
MFGA	43	48.14	2.85	62	67.41	2.62
MFPSO	43	47.68	2.91	59	66.18	2.89
Modified FGA	**39**	**44.45**	**6.02**	**54**	**61.48**	**6.05**

6   Comparison and Discussion

Based on the presented results in Table 2, Modified FGA with NF-RK encoding scheme performs significantly better than other two multimodal algorithms. MFGA and MFPSO are the multi-strategy fusion hybrids of genetic algorithm (GA) and particle swarm optimization (PSO) respectively [10]. In the multi-strategy fusion the change in the fitness value as a measure of convergence is replaced by a more complex approach based on the fitness entropy. In terms of increasing the diversity of the population though, a distribution radius has been proposed that acts like niche radius and will be calculated by the pairwise distance between the found global solutions and the saved solution in the static archive. The algorithm also uses another measure called utility-fitness based on utility theory to manipulate the effective fitness values in the search process similar to those in fitness sharing. The proposed niching method in this work is called multi-strategy fusion that can be attached to any global search algorithms like PSO, DE, BSO, GA, and etc. [10]. In terms of the best fitness values (BF), modified FGA outperforms the other algorithms in all 4 instances and the difference gets bigger for the instances with higher number of tasks.

The average number of found solutions (MN) for modified FGA dominates the existing results in all cases with a considerable margin. When it comes to MF measure the superiority of the results for the modified FGA becomes even more clear for it

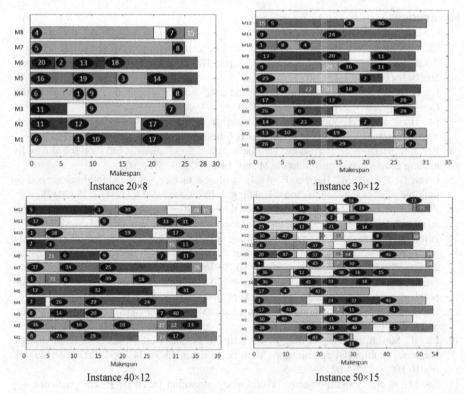

Fig. 2. Gantt Chart of sample best TTSP solutions

to be called a multimodal optimization algorithm. It easily can be concluded that the modified FGA in combination with NF-RK would have produced way better results for MN measure if reaching to the same level of MF value with other existing algorithms were desired.

Figure 2 presents the sample Gantt Charts for the best results achieved by the modified FGA for TTSP instances.

7 Conclusion and Future Works

In this study we proposed a modification on the football game algorithm (mFGA) as a dynamic niching algorithm for multimodal optimization. Also, a new encoding scheme called Normalized Factor Random Key (NF-RK) is proposed to be used as a mapping tool for mFGA to solve Test Task Scheduling Problems (TTSPs). Four instances of practical TTSPs have been solved and the results have been compared to other algorithms that solved the same problems. The statistical result shows that mFGA using NF-RK can dramatically improve the results of solving TTSPs in comparison to existing solutions in the literature. In the future studies, more diverse problems including FJSP, UMSP, and JJSP, should be solved using the current proposed methods to investigate the potential of these methods further.

References

1. Lu, H., Liu, J., Niu, R., Zhu, Z.: Fitness distance analysis for parallel genetic algorithm in the test task scheduling problem. Soft. Comput. **18**(12), 2385–2396 (2013). https://doi.org/10.1007/s00500-013-1212-6

2. Lu, H., et al.: Analysis of the similarities and differences of job-based scheduling problems. Euro. J. Oper. Res. **270**(3), 809–825 (2018)

3. Lu, H., et al.: A chaotic non-dominated sorting genetic algorithm for the multi-objective automatic test task scheduling problem. Appl. Soft Comput. **13**(5), 2790–2802 (2013)

4. Pérez, E., Herrera, F., Hernández, C.: Finding multiple solutions in job shop scheduling by niching genetic algorithms. J. Intell. Manuf. **14**(3–4), 323–339 (2003)

5. Pérez, E., Posada, M., Herrera, F.: Analysis of new niching genetic algorithms for finding multiple solutions in the job shop scheduling. J. Intell. Manuf. **23**(3), 341–356 (2012)

6. Zou, P., Rajora, M., Liang, S.Y.: Multimodal optimization of job-shop scheduling problems using a clustering-genetic algorithm based approach. Int. J. Industr. Eng. **26**(5), (2019)

7. Pérez, E., Posada, M., Lorenzana, A.: Taking advantage of solving the resource constrained multi-project scheduling problems using multi-modal genetic algorithms. Soft. Comput. **20**(5), 1879–1896 (2015). https://doi.org/10.1007/s00500-015-1610-z

8. Lu, H., et al.: Multi-center variable-scale search algorithm for combinatorial optimization problems with the multimodal property. Appl. Soft Comput. **84**, 105726 (2019)

9. Lu, H., et al.: Spatial-domain fitness landscape analysis for combinatorial optimization. Inf. Sci. **472**, 126–144 (2019)

10. Lu, H., Sun, S., Cheng, S., Shi, Y.: An adaptive niching method based on multi-strategy fusion for multimodal optimization. Memetic Comput. **13**(3), 341–357 (2021). https://doi.org/10.1007/s12293-021-00338-5

11. Lu, H., et al.: A multi-objective evolutionary algorithm based on Pareto prediction for automatic test task scheduling problems. Appl. Soft Comput. **66**, 394–412 (2018)

12. Fadakar, E., Ebrahimi, M.: A new metaheuristic football game inspired algorithm. In: 2016 1st Conference on Swarm Intelligence and Evolutionary Computation (CSIEC). IEEE (2016)

13. Raharja, F.A.: Penerapan football game algorithm untuk menyelesaikan asymmetric travelling salesman problem (2017)

14. Djunaidi, A.V., Juwono, C.P.: Football game algorithm implementation on the capacitated vehicle routing problems. Int. J. Comput. Algorithm **7**(1), 45–53 (2018)

15. Subramaniyan, S., Ramiah, J.: Improved football game optimization for state estimation and power quality enhancement. Comput. Electr. Eng. **81**, 106547 (2020)

16. Balochian, S., Baloochian, H.: Social mimic optimization algorithm and engineering applications. Expert Syst. Appl. **134**, 178–191 (2019)

17. Alatas, B.: Sports inspired computational intelligence algorithms for global optimization. Artif. Intell. Rev. **52**(3), 1579–1627 (2017). https://doi.org/10.1007/s10462-017-9587-x

18. Osaba, E., Yang, X.-S.: Soccer-inspired metaheuristics: systematic review of recent research and applications. In: Osaba, E., Yang, X.-S. (eds.) Applied Optimization and Swarm Intelligence. STNC, pp. 81–102. Springer, Singapore (2021). https://doi.org/10.1007/978-981-16-0662-5_5

19. Rashid, M.F.F.A.: Tiki-taka algorithm: a novel metaheuristic inspired by football playing style. Eng. Comput. (2020)

20. Li, J.-P., et al.: A species conserving genetic algorithm for multimodal function optimization. Evol. Comput. **10**(3), 207–234 (2002)

21. Epitropakis, M.G., Li, X., Burke, E.K.: A dynamic archive niching differential evolution algorithm for multimodal optimization. In: 2013 IEEE Congress on Evolutionary Computation. IEEE (2013)

22. Wilson, A.J.: Volume of n-dimensional ellipsoid. Sciencia Acta Xaveriana **1**(1), 101–106 (2010)
23. Baioletti, M., Milani, A., Santucci, V.: Algebraic particle swarm optimization for the permutations search space. In: 2017 IEEE Congress on Evolutionary Computation (CEC). IEEE (2017)
24. Baioletti, M., Milani, A., Santucci, V.J.I.S.: Variable neighborhood algebraic differential evolution: an application to the linear ordering problem with cumulative costs. Inf. Sci. **507**, 37–52 (2020)
25. Gao, K., et al.: A review on swarm intelligence and evolutionary algorithms for solving flexible job shop scheduling problems. IEEE/CAA J. Automat. Sinica **6**(4), 904–916 (2019)
26. Krömer, P., Uher, V., Snášel, V.J.I.T.O.E.C.: Novel random key encoding schemes for the differential evolution of permutation problems. IEEE Trans. Evol. Comput. **26**, 43–57 (2021)
27. Ponsich, A., Tapia, M.G.C., Coello, C.A.C.: Solving permutation problems with differential evolution: an application to the jobshop scheduling problem. In: 2009 Ninth International Conference on Intelligent Systems Design and Applications. IEEE (2009)
28. Tasgetiren, M.F., et al.: A particle swarm optimization algorithm for makespan and total flowtime minimization in the permutation flowshop sequencing problem. Euro. J. Oper. Res. **177**(3), 1930–1947 (2007)
29. Zhang, G.H., et al.: Solving flexible job shop scheduling problems with transportation time based on improved genetic algorithm. Math. Biosci. Eng. **16**(3), 1334–1347 (2019)

Performance Analysis of Selected Evolutionary Algorithms on Different Benchmark Functions

Jana Herzog[(✉)] [iD], Janez Brest[iD], and Borko Bošković[iD]

University of Maribor, Koroška cesta 46, 2000 Maribor, Slovenia
{jana.herzog1,janez.brest,borko.boskovic}@um.si

Abstract. This paper analyses and compares four recently published state-of-the-art evolutionary algorithms on three different sets of benchmark functions. The main intention was to show the shortcomings of the established metrics, which focus only on one variable: the obtained quality of solutions under a predetermined stopping criteria, while neglecting the runtime and the speed of a solver. Through a statistical analysis, it was established that there is no single solver which ranks as the best for each benchmark function. It is even harder to choose a solver for a specific optimization problem considering the computational complexity of the solver and the problem.

Keywords: Benchmark · Comparison · Computational complexity

1 Introduction

A thorough analysis of an algorithm's performance is helpful in designing and improving the algorithm when tackling different sets of benchmark functions. A specific optimization problem requires an algorithm, which will be able to reach the best/optimal solution in a reasonable time while spending a low number of function evaluations. However, it is difficult to deduce the criteria by which the algorithm should be evaluated while dealing with a specific problem. If the algorithm is slow (it performs a small number of function evaluations per second) and has a good convergence rate, it might be more appropriate for solving a time-consuming optimization problem than a faster algorithm with a slower convergence rate. How to deduce which algorithm is the best for a given optimization problem?

Each year, a new or an updated version of benchmark functions on the Congress of Evolutionary Computation (CEC) are proposed with the intention of comparing the newest state-of-the-art algorithms and evaluating their performance based on the solution quality they reach, while limiting the runs with a maximum number of function evaluations ($maxFEs$). This is so-called the "fixed-budget" approach [15]. The benchmark functions are treated as black-box problems and serve as a tool in comparison of the state-of-the-art evolutionary algorithms. However, there are some disadvantages to them. It is not necessary that the best algorithm reaches the best solutions on a given metric. The algorithm which is ranked as the best on one benchmark may not perform so well on other benchmark functions. Based on only one benchmark it is difficult to ascertain, how a given state-of-the-art algorithm will perform on any other types of the optimization problems or benchmark functions. Also, some aspects of the algorithm's performance are forgotten, such as the speed, defined as the number of function evaluations per second, or runtime of a solver in reaching the wanted quality of solution. The information about algorithm's speed and runtime is valuable when planning and performing an experiment or when dealing with real-world problems, which are time limited.

When comparing different evolutionary algorithms on benchmarks, the execution of the experiment plays a crucial role. Often, the results are re-used from published papers due to convenience. The role of a programming language is neglected, but for a fair comparison, the chosen algorithms should be implemented in the same programming language. The implementation in a specific programming language affects the speed of the algorithm, which is important in the runtime analysis [19]. When choosing a proper programming language, one has to keep in mind that it is better to use the language which provides compilers in contrast to languages that use interpreters. To analyze and compare the several aspects of the solvers' performances, we chose three single objective bound constrained benchmarks: CEC 2022 [1], CEC 2021 [2] and CEC 2017 [3], and four state-of-the-art solvers MadDE [5], L-SHADE [22], CS-DE [18] and j2020 [8]. The evolutionary algorithms MadDE, L-SHADE and CS-DE are implemented in Matlab, while j2020 is implemented in C++.

The main aim of this paper was to show that there is no single solver appropriate for each benchmark function or problem. The question explored was whether the state-of-the-art solvers are overfitted to a certain benchmark, which leads to a worse performance on other benchmark functions. We observed the quality of solutions reached on each benchmark function and compared the solvers based on them. The paper is structured as follows. In Sect. 2, related work is described. In Sect. 3, the experiment and analysis are provided. In Sect. 4, we summarize the results of our experiment. The Sect. 5 concludes our paper.

2   Related Work

A detailed analysis and comparison of the solvers leads to the improvement of their performance while tackling different optimization problems. Researchers focus on various aspects of the algorithm's analysis; observing the number of function evaluations, the runtime and the convergence [24]. Different performance metrics have been used for comparing the algorithms. CEC 2005 benchmark function suite [21] proposed using success rate and success performance

(SP1) to measure the runtime in terms of the number of function evaluations. The performance measures should be quantitative, well-interpretable and simple as possible, while still providing thorough knowledge about the algorithm. Considering this, the authors proposed using "the expected running time" (ERT). ERT is computed based on the expected number of function evaluations to reach the target value for the first time [16].

The analysis has evolved past this point with new approaches and methods, such as linking the performance of the algorithms with the landscape characteristics of the problem instances [17]. Recently, the studies no longer focus on using one particular algorithm on a single benchmark, but they focus on selecting the best algorithm for each problem instance. This is known as the automated algorithm selection [20]. A fair evaluation of the algorithms is no longer only dependent on the selection of the algorithms, but also on the selection of a representative benchmark [10]. Nevertheless, for the purpose of the correct analysis, one needs to apply the knowledge of statistics. In the last few years, the researchers followed statistical tests when it comes to the comparison of the solvers [11]. The statistical tests are divided into two groups, parametric and non-parametric. Parametric statistical tests are robust and quite restrictive, because the distribution of data is required. Beforehand the assumptions about the normality, homoscedasticity of variance and independence need to be checked. The non-parametric tests are less restrictive and they do not rely on the statistical distribution of the variables. For pairwise comparison of the solvers, Wilcoxon Rank Sum test is recommended, but for the comparison and ranking of several solvers, Friedman test is used [9]. Using a statistical test when comparing solvers is a common practice, but often a misuse of them can occur. Since their pitfalls, the reseachers took interest in the Bayesian trend [4]. Here, not a single probability is calculated, but a distribution of the parameter of the interest itself. The disadvantage is that this approach requires a deeper understanding of the statistics [4, 9]. To gain deeper knowledge and insight into the algorithm's performance, a deep statistical approach [13] was proposed. The main contribution of this approach is that the ranking scheme is based on the whole distribution rather on only one statistic, such as mean or median. With the intention of making a more robust performance statistics and also investigating exploration and exploitation of the optimization algorithms, a DSCTool has been proposed [12, 14]. An alternative approach for comparing and ranking the stochastic algorithms is the chess ranking system [23], which treats algorithms as chess players during a tournament.

3   Experiment

In this section, we present the experimental part and the comparison of four solvers applied to three sets of benchmark functions. The chosen benchmark functions are CEC 2022, CEC 2021 and CEC 2017. The chosen four state-of-the-art solvers, implementing the algorithms MadDE [5], which uses a multiple adaptation strategy for adapting the control parameters and an optimizer named SUBHO, L-SHADE [22], a well-known state-of-the-art solver with a linear size

Table 1. Results for CEC 2022 Benchmark for the solver MadDE.

F	Best	Worst	Median	Mean	Std	Speed	F	Best	Worst	Median	Mean	Std	Speed
1	0.00	0.00	0.00	0.00	0.00	4.2e+5	1	0.00	0.00	0.00	0.00	0.00	2.9e+5
2	0.00	0.00	0.00	0.00	0.00	5.4e+5	2	49.08	49.08	49.08	49.08	0.00	3.3e+5
3	0.00	0.00	0.00	0.00	0.00	3.9e+5	3	0.00	0.00	0.00	0.00	0.00	1.9e+5
4	3.97	9.95	6.96	6.86	1.82	4.8e+5	4	9.95	35.82	27.86	28.19	5.13	2.9e+5
5	0.00	0.00	0.00	0.00	0.00	4.4e+5	5	0.00	0.00	0.00	0.00	0.00	2.2e+5
6	0.022	0.49	0.35	0.35	0.13	4.7e+5	6	0.23	0.49	0.46	0.43	0.08	3.1e+5
7	0.00	0.01	4.3e-6	0.001	0.003	3.0e+5	7	0.02	21.06	3.31	5.86	6.13	1.3e+5
8	0.02	8.34	0.69	1.31	2.11	2.8e+5	8	17.18	20.68	20.29	20.17	0.66	1.1e+5
9	229.28	229.28	229.28	229.28	0.00	3.2e+5	9	180.78	180.78	180.78	180.78	0.00	1.3e+5
10	100.21	100.34	100.29	100.29	0.03	3.4e+5	10	100.27	100.42	100.36	100.35	0.03	1.5e+5
11	0.00	0.00	0.00	0.00	0.00	2.9e+5	11	300.00	300.00	300.00	300.00	0.00	1.8e+5
12	158.61	162.70	159.37	159.73	1.02	2.6e+5	12	228.86	232.00	231.18	231.09	0.830	9.8e+4

(a) Results for 10D.	(b) Results for 20D.

Table 2. Results for CEC 2022 Benchmark for the solver L-SHADE.

F	Best	Worst	Median	Mean	Std	Speed	F	Best	Worst	Median	Mean	Std	Speed
1	0.00	0.00	0.00	0.00	0.00	6.9e+5	1	0.00	0.00	0.00	0.00	0.00	4.2e+5
2	0.01	8.91	3.98	5.33	2.49	8.6e+5	2	44.89	49.08	49.08	48.95	0.77	4.8e+5
3	0.00	0.00	0.00	0.00	0.00	5.9e+5	3	0.00	0.00	0.00	0.00	0.00	4.8e+5
4	0.002	3.98	2.48	2.28	0.98	6.6e+5	4	0.99	6.96	3.98	4.25	1.22	4.2e+5
5	0.00	0.00	0.00	0.00	0.00	6.3e+5	5	0.00	0.00	0.00	0.00	0.00	2.4e+5
6	0.058	0.49	0.39	0.33	0.14	7.5e+5	6	0.16	1.49	0.49	0.53	0.32	4.6e+5
7	0.00	0.09	2.1e-5	0.01	0.02	4.1e+5	7	0.25	19.25	3.44	4.23	3.47	3.6e+5
8	0.25	5.43	0.65	1.19	1.34	3.6e+5	8	7.81	20.44	20.19	18.38	3.21	4.1e+5
9	229.28	248.07	248.07	248.07	0.00	4.2e+5	9	180.78	180.78	180.78	180.78	0.00	4.1e+5
10	100.18	203.72	100.22	103.67	18.89	4.6e+5	10	100.22	100.36	100.27	100.27	0.03	3.4e+5
11	0.00	0.00	0.00	0.00	0.00	3.9e+5	11	300.00	300.00	300.00	300.00	0.00	8.1e+4
12	158.61	164.93	161.21	161.21	169.93	3.5e+5	12	230.87	240.76	232.85	233.37	1.99	2.8e+4

(a) Results for 10D.	(b) Results for 20D.

reduction, CS-DE [18] with an ensemble of mutation strategies and population diversity and j2020 with two populations and a crowding mechanism [6–8]. The experiment was carried out in Matlab 2021a on Windows 11, with 16GB RAM for solvers MadDE, CSDE and L-SHADE. For the solver j2020, we used a personal computer with GNU C++ compiler version 9.3.0, Intel(R) Core(TM) i5-9400 with 3.2 GHz CPU and 6 cores under Linux Ubuntu 20.04. This section is divided into three subsections, each for the chosen benchmark. The chosen solvers are compared based on the quality of solutions (fitness value) they reach on each set of benchmark functions and analyzed with the help of statistical tests, such as Friedman test. Friedman test can detect significant differences among all solvers, but does not tell between which solvers there is a significant difference. Therefore, a post-hoc test needs to be applied, such as the Dunn's pairwise post-hoc test with Bonferroni correction [11].

3.1 CEC 2022 Single Objective Bound Constrained Numerical Optimization

For the purpose of this experiment we used CEC 2022 [1] benchmark functions which consist of 12 functions: f_1 Shifted and full Rotated Zahkarov Function, f_2 Shifted and full Rotated Rosenbrock's Function, f_3 Shifted and full Rotated Expanded Schaffer's f_6 Function, f_4 Shifted and full Rotated Non-Continuous Rastrigin's Function, f_5 Shifted and full Rotated Levy Function, $f_6 - f_8$ Hybrid Functions and $f_9 - f_{12}$ Composition Functions. For two dimensions $D = 10$ and $D = 20$, 30 runs were made for each function with each solver. The stopping criteria ($maxFEs$) were set to 200,000 and 1,000,000 for $D = 10$ and $D = 20$, respectively. We reported the best, worst, median, mean value and standard deviation for each function, while the speed is also displayed in Tables 1–4. Friedman test

Table 3. Results for CEC 2022 Benchmark for the solver CS-DE.

F	Best	Worst	Median	Mean	Std	Speed
1	0.00	0.00	0.00	0.00	0.00	4.2e+5
2	0.00	8.92	0.00	0.83	2.05	5.3e+5
3	0.003	0.005	0.008	0.001	0.001	3.9e+5
4	1.25	11.94	4.95	5.48	2.52	4.7e+5
5	0.00	0.00	0.00	0.00	0.00	4.3e+5
6	0.07	1.42	0.35	0.45	0.38	4.7e+5
7	0.64	8.30	4.59	4.32	2.19	3.0e+5
8	0.24	21.78	6.99	8.49	6.79	2.8e+5
9	229.28	229.28	229.28	229.28	0.00	4.1e+5
10	100.21	100.41	100.27	100.29	0.06	3.2e+5
11	0.00	0.00	0.00	0.00	0.00	8.1e+4
12	162.70	164.93	164.92	164.85	0.41	8.4e+4

(a) Results for 10D.

F	Best	Worst	Median	Mean	Std	Speed
1	0.00	0.00	0.00	0.00	0.00	6.5e+4
2	44.89	49.08	49.08	48.94	0.76	6.8e+4
3	0.00	0.00	0.00	0.00	0.00	2.7e+4
4	0.99	2.98	1.99	2.28	0.75	3.0e+4
5	0.00	0.00	0.00	0.00	0.00	4.6e+4
6	0.24	0.50	0.41	0.41	0.07	3.5e+4
7	0.64	20.97	5.473	7.19	5.05	2.2e+4
8	16.55	20.49	20.28	19.71	1.09	2.6e+4
9	180.78	180.78	180.78	180.78	0.00	3.2e+4
10	100.29	100.41	100.35	100.35	0.03	3.2e+4
11	300.00	300.00	300.00	300.00	0.00	4.8e+4
12	231.19	239.24	232.56	232.58	1.51	4.2e+4

(b) Results for 20D.

Table 4. Results for CEC 2022 Benchmark for the solver j2020.

F	Best	Worst	Median	Mean	Std	Speed
1	0.00	0.00	0.00	0.00	0.00	1.6e+7
2	0.00	0.00	0.00	0.00	0.00	3.3e+6
3	0.00	0.00	0.00	0.00	0.00	9.0e+6
4	2.98	12.93	5.98	6.58	2.35	1.9e+6
5	0.00	0.08	0.00	0.00	0.02	5.5e+6
6	0.12	1.35	0.43	0.63	0.41	2.6e+6
7	0.00	0.00	0.00	0.00	0.00	3.5e+6
8	0.00	0.81	0.62	0.43	0.32	1.3e+6
9	229.28	229.28	229.28	229.28	0.00	1.3e+6
10	0.00	0.25	0.12	0.09	0.07	1.0e+6
11	0.00	0.00	0.00	0.00	0.00	4.1e+6
12	158.62	164.70	161.43	161.31	1.83	6.6e+5

(a) Results for 10D.

F	Best	Worst	Median	Mean	Std	Speed
1	0.00	0.00	0.00	0.00	0.00	5.4e+6
2	0.01	28.94	0.52	2.65	6.55	1.1e+6
3	0.00	0.00	0.00	0.00	0.00	1.0e+7
4	9.95	30.02	19.11	19.59	5.33	8.4e+5
5	0.00	0.00	0.00	0.00	0.00	1.0e+7
6	3.67	56.74	19.63	22.91	14.00	1.1e+6
7	0.97	22.28	5.23	6.02	5.09	4.3e+5
8	9.92	22.17	21.69	21.06	2.21	3.6e+5
9	180.78	180.78	180.78	180.78	0.00	5.0e+5
10	0.00	0.16	0.06	0.06	0.03	4.9e+5
11	300.00	300.00	300.00	300.00	0.00	2.8e+5
12	228.86	245.35	232.84	233.74	3.48	2.6e+5

(b) Results for 20D.

Table 5. Results for CEC 2021 Benchmark for the solver MadDE.

F	Best	Worst	Median	Mean	Std	Speed	F	Best	Worst	Median	Mean	Std	Speed
1	0.00	0.00	0.00	0.00	0.00	1.5e+5	1	0.00	0.00	0.00	0.00	0.00	1.8e+5
2	0.38	129.07	10.24	21.64	32.06	1.5e+5	2	0.06	9.26	1.84	2.55	1.94	2.0e+5
3	11.25	15.60	14.15	13.99	1.13	1.3e+5	3	20.38	21.78	20.92	20.94	0.41	2.6e+5
4	0.11	0.58	0.38	0.37	0.37	1.8e+5	4	0.48	0.78	0.61	0.62	0.07	2.4e+5
5	0.00	3.40	0.42	0.96	0.85	1.9e+5	5	3.45	36.43	16.94	17.63	8.80	1.9e+5
6	0.05	0.55	0.30	0.30	0.12	2.0e+5	6	0.11	0.65	0.33	0.36	0.13	3.1e+5
7	0.01	0.87	0.06	0.17	0.21	1.9e+5	7	0.11	9.27	1.26	1.97	2.48	1.3e+5
8	0.00	100.29	100.00	91.34	26.30	1.9e+5	8	100.00	100.00	100.00	100.00	0.00	1.1e+5
9	0.00	100.00	100.00	90.00	30.00	1.5e+5	9	100.00	413.26	409.70	327.27	137.08	1.3e+5
10	397.70	397.70	397.70	397.70	0.00	1.4e+5	10	413.66	413.66	413.66	413.66	0.00	1.0e+5

(a) Results for 10D.　　　　　　　　　　　　(b) Results for 20D.

Table 6. Results for CEC 2021 Benchmark for the solver L-SHADE.

F	Best	Worst	Median	Mean	Std	Speed	F	Best	Worst	Median	Mean	Std	Speed
1	0.00	0.00	0.00	0.00	0.00	2.9e+5	1	0.00	0.00	0.00	0.00	0.00	3.9e+5
2	0.26	22.01	6.90	7.29	4.77	2.7e+5	2	0.011	6.85	6.90	1.85	1.70	4.9e+5
3	10.93	13.94	11.83	11.94	0.65	2.3e+5	3	20.68	22.52	21.52	21.48	0.47	2.6e+5
4	0.21	0.44	0.36	0.35	0.05	2.3e+5	4	0.37	0.71	0.58	0.57	0.07	3.9e+5
5	0.00	2.61	0.21	0.55	0.64	2.3e+5	5	0.10	240.27	4.14	39.79	0.64	2.9e+5
6	0.03	0.71	0.27	0.31	0.20	2.5e+5	6	0.24	0.69	0.52	0.49	0.12	4.5e+5
7	3.3e-6	0.81	0.22	0.27	0.29	2.0e+5	7	0.053	9.19	0.81	1.36	2.11	1.5e+5
8	100.00	100.00	100.00	100.00	0.00	1.8e+5	8	100.00	100.00	100.00	100.00	0.00	1.2e+5
9	100.00	335.77	330.31	295.88	81.57	1.6e+5	9	398.54	403.62	401.35	401.44	1.24	1.5e+5
10	397.74	443.36	398.01	410.05	20.42	1.5e+5	10	413.66	413.68	413.66	413.66	0.007	1.8e+5

(a) Results for 10D.　　　　　　　　　　　　(b) Results for 20D.

(level of significance $\alpha = 0.05$) was applied, since we are dealing with multiple comparisons among all methods. The analysis of their runtime and speed follows.

We applied the Friedman test, which ranks the solvers based on their mean values. The best performing solver should have the lowest rank, while the worst performing solver should have the highest rank. Note that this is correct in the case of minimization problems. As shown in Table 13, MadDE has the best rank. Friedman test indicates there is no significant difference between solvers ($p = 0.227$) for $D = 10$.

We are aware of the fact that the solvers are implemented in different programming languages which may affect their performance, however we can still statistically compare their performance. The Friedman test was also carried out to rank the four solvers according to the speed. There was found to be a significant difference between the results. When analyzing the speed of the solvers, the bigger number means that the solver is faster. The highest rank means the fastest solver. In this case, j2020 is the fastest solver.

The Dunn-Bonferroni post hoc test indicated that there were significant differences between MadDE and L-SHADE ($p = 0.027$), MadDE and j2020 ($p = 0.00$), CS-DE and L-SHADE ($p = 0.027$) and CS-DE and j2020 ($p = 0.00$).

Table 7. Results for CEC 2021 Benchmark for the solver CS-DE.

F	Best	Worst	Median	Mean	Std	Speed
1	0.00	0.00	0.00	0.00	0.00	4.0e+4
2	0.30	26.95	6.89	8.76	8.17	4.6e+4
3	11.14	13.95	12.05	12.11	0.82	4.8e+4
4	0.28	0.43	0.33	0.35	0.05	4.9e+4
5	0.42	2.61	1.20	1.37	0.69	4.5e+4
6	0.09	0.56	0.36	0.33	0.17	4.6e+4
7	0.01	0.35	0.18	0.17	0.13	5.0e+4
8	100.00	100.00	100.00	100.00	0.00	4.1e+4
9	100.00	328.16	326.95	281.58	95.69	4.3e+4
10	397.74	443.33	397.88	402.41	14.38	3.9e+4

(a) Results for 10D.

F	Best	Worst	Median	Mean	Std	Speed
1	0.00	0.00	0.00	0.00	0.00	4.2e+5
2	0.08	3.61	2.63	2.16	1.53	5.4e+5
3	21.51	22.53	22.00	22.04	0.35	3.9e+5
4	0.40	0.68	0.55	0.55	0.07	4.8e+5
5	0.00	34.65	12.19	15.59	11.16	4.4e+5
6	0.39	0.65	0.57	0.55	0.09	4.7e+5
7	0.40	8.97	5.36	4.51	2.96	3.0e+5
8	100.00	100.00	100.00	100.00	0.00	2.8e+5
9	392.70	399.78	396.23	396.37	2.64	3.2e+5
10	413.66	413.66	413.66	413.66	0.006	3.4e+5

(b) Results for 20D.

Table 8. Results for CEC 2021 Benchmark for the solver j2020.

F	Best	Worst	Median	Mean	Std	Speed
1	0.00	0.00	0.00	0.00	0.00	4.9e+7
2	0.25	75.59	18.59	23.06	21.19	3.7e+6
3	5.69	16.53	12.38	12.23	2.16	3.4e+6
4	0.01	2.04	0.75	0.66	0.43	1.7e+6
5	0.21	22.55	1.90	5.04	5.74	9.3e+6
6	0.05	1.47	0.57	0.68	0.36	1.9e+6
7	0.002	0.70	0.32	0.31	0.22	1.9e+6
8	0.00	100.55	11.56	17.48	25.47	7.1e+6
9	0.00	100.00	100.00	96.67	18.26	9.0e+6
10	100.02	398.16	397.74	269.13	142.87	9.1e+5

(a) Results for 10D.

F	Best	Worst	Median	Mean	Std	Speed
1	0.00	0.00	0.00	0.00	0.00	3.8e+5
2	0.06	5.36	2.07	2.05	1.59	1.4e+7
3	20.39	21.39	20.80	20.75	0.32	7.0e+6
4	0.75	1.52	1.24	1.22	0.17	6.3e+5
5	113.51	507.41	235.21	241.76	87.12	8.7e+5
6	0.18	0.82	0.41	0.39	0.15	7.2e+5
7	0.80	201.69	39.04	48.11	46.07	7.4e+5
8	100.00	100.00	100.00	100.00	0.00	9.0e+5
9	100.00	421.87	411.53	357.87	115.52	7.7e+5
10	399.05	413.66	400.00	403.44	5.47	3.1e+5

(b) Results for 20D.

There were no significant differences between MadDE and CS-DE, and L-SHADE and j2020.

We also repeated the procedure for $D = 20$. Rankings of the Friedman test ($\alpha = 0.05$) are shown in Table 13. There is small difference between ranks and this indicates that there is no significant difference, which is proved by the $p = 0.227 > 0.05$. We also compared the solvers for $D = 20$ based on their speed shown in Table 13. The Dunn-Bonferoni post hoc test were carried out and there were significant differences between CS-DE and MadDE ($p = 0.009$), CS-DE and j2020 ($p = 0.00$), MadDE and j2020 ($p = 0.009$).

3.2 CEC 2021 Single Objective Bound Constrained Optimization

This experiment was carried out on CEC 2021 Single Objective Bound Constrained benchmark functions [2]. The benchmark contains 10 functions: f_1 Shifted and Rotated Bent Cigar Function, f_2 Shifted and Rotated Schwefel's Function, f_3 Shifted and Rotated Lunacek bi-Rastrigin Function, f_4 Expanded Rosenbrock's plus Griewangks's Function, Hybrid Functions from f_5 to f_7, and Composition Functions from f_8 to f_{10}. Each functions undergoes 5 configura-

Table 9. Results for CEC 2017 Benchmark for the solver MadDE.

F	Best	Worst	Median	Mean	Std	Speed
1	0.00	0.00	0.00	0.00	0.00	1.5e+5
3	0.00	0.00	0.00	0.00	0.00	1.5e+5
4	0.00	0.00	0.00	0.00	0.00	1.9e+5
5	1.99	5.97	3.98	3.81	1.05	1.6e+5
6	0.00	0.00	20.00	0.00	0.00	1.4e+5
7	11.73	16.72	14.43	14.42	1.24	1.1e+5
8	1.99	7.96	4.97	4.99	1.34	4.9e+4
9	0.00	0.00	0.00	0.00	0.00	2.4e+4
10	0.37	237.77	125.46	100.35	67.29	1.3e+5
11	0.33	3.055	1.42	1.42	0.662	1.5e+5
12	0.21	119.95	0.42	21.83	45.65	1.4e+5
13	2.46e-5	6.78	2.39	2.95	2.31	1.3e+5
14	0.001	1.99	0.53	0.59	0.53	1.4e+5
15	0.036	1.18	0.00	0.28	0.22	1.4e+5
16	0.03	0.88	0.49	0.49	0.19	1.6e+5
17	0.010	1.08	0.32	0.27	0.25	1.5e+5
18	0.001	1.01	0.26	0.26	0.22	1.8e+5
19	2.5e-5	0.063	0.025	0.03	0.01	7.9e+4
20	0.00	0.00	0.00	0.00	0.00	1.5e+5
21	0.00	101.81	100.00	98.14	14.02	1.3e+5
22	27.96	100.34	100.00	889.25	22.54	1.2e+5
23	2.5e-9	308.118	304.96	2.81	82.93	1.2e+5
24	6.3e-10	100.00	100.00	98.03	14.00	1.2e+5
25	397.74	398.04	397.74	397.75	0.04	1.4e+5
26	0.00	300.00	200.00	154.90	1.49	1.2e+5
27	386.89	389.01	389.01	388.53	0.73	1.1e+5
28	0.00	300.00	300.00	282.35	71.29	1.2e+5

(a) Results for 10D.

F	Best	Worst	Median	Mean	Std	Speed
1	1175.99	3446.06	1963.12	2008.49	446.86	1.7e+5
3	510.22	48442.9	33029.3	3098.7	1022.5	1.8e+5
4	73.18	112.96	87.72	89.96	14.55	1.6e+5
5	55.35	108.78	78.92	78.47	9.61	1.8e+5
6	0.05	0.19	0.10	0.109	0.04	1.5e+5
7	79.46	130.05	109.04	107.20	12.09	1.0e+5
8	42.71	86.23	72.47	72.19	8.49	1.5e+5
9	3.39	88.78	16.23	21.76	14.95	1.6e+5
10	1924.60	3489.36	2829.21	2856.18	333.59	1.6e+5
11	37.71	102.87	77.22	73.98	16.71	1.3e+5
12	66327.3	6.8e+5	3.7e+5	3.7e+5	1.4e+5	1.6e+5
13	2637.0	22499.9	13253.2	13547.6	4271.9	1.5e+5
14	54.11	133.35	83.64	84.83	17.56	1.7e+5
15	106.89	1126.97	220.81	303.59	223.97	1.4e+5
16	126.28	741.99	450.74	468.73	132.35	1.7e+5
17	45.78	127.27	73.63	74.36	14.80	1.5e+5
18	1123.7	1.4e+5	5.73e+4	6.3e+4	2.9e+4	1.2e+5
19	27.16	1600.57	103.08	237.81	338.68	1.6e+5
20	38.44	200.29	78.17	99.49	49.076	4.4e+4
21	113.53	284.48	151.81	193.16	68.23	1.1e+5
22	100.00	100.00	100.00	100.00	1.71e-4	9.0e+4
23	390.85	434.15	415.22	413.89	10.75	8.2e+4
24	461.42	508.73	486.86	486.37	9.36	7.3e+4
25	383.40	387.13	386.95	383.40	0.83	6.8e+4
26	200.00	300.01	300.00	268.63	46.86	7.3e+4
27	503.21	520.84	514.27	513.49	3.63	6.0e+4
28	392.09	407.56	397.25	397.97	3.55	5.3e+4

(b) Results for 30D.

tions (000, 010, 110, 011 and 111) [2], but we analyzed the configuration (111 – Shift, Rotation, Translation). For each function, 30 independent runs were done. We observed the quality of solutions reached and the speed of the solvers as shown in Tables 5–8.

The Friedman test, as shown in Table 14, did not detect a significant difference between the quality of solutions of solvers for $D = 10$ and $D = 20$, but it detected a significant difference between the solvers' speed for both dimensions. We applied Dunn-Bonferroni test to show that the significant difference was detected between CS-DE and L-SHADE ($p = 0.006$), CS-DE and j2020 ($p = 0.000$), MadDE and j2020 ($p = 0.006$) for $D = 10$. For $D = 20$, the significant difference was detected between CS-DE and L-SHADE ($p = 0.002$), CS-DE and j2020 ($p = 0.000$), MadDE and j2020 ($p = 0.006$) as shown in Table 14.

3.3 CEC 2017 Single Objective Bound Constrained Optimization

The experiment was carried out on the CEC 2017 [3] test suite, which consists of 29 functions: f_1 Shifted and Rotated Bent Cigar Function, f_2 Shifted and Rotated Zakharov Function, f_3 Shifted and Rotated Rosenbrock's Function, f_4

Table 10. Results for CEC 2017 Benchmark for the solver L-SHADE.

F	Best	Worst	Median	Mean	Std	Speed	F	Best	Worst	Median	Mean	Std	Speed
1	0.00	0.00	0.00	0.00	0.00	1.7e+5	1	0.00	0.00	0.00	0.00	0.00	6.3e+4
3	0.00	0.00	0.00	0.00	0.00	2.1e+5	3	0.00	0.00	0.00	0.00	0.00	1.7e+5
4	0.00	0.00	0.00	0.00	0.00	1.9e+5	4	58.56	58.56	58.56	58.56	0.00	1.9e+5
5	0.99	4.98	2.98	2.48	0.89	1.2e+5	5	2.36	9.06	6.18	6.32	1.32	1.6e+5
6	0.00	0.00	0.00	0.00	0.00	1.3e+5	6	0.00	0.00	0.00	0.00	0.00	9.8e+4
7	10.44	14.65	12.15	12.13	0.73	1.2e+5	7	34.77	41.52	37.17	37.35	1.34	1.6e+5
8	0.99	3.98	2.98	2.46	0.85	1.1e+5	8	3.08	10.14	7.04	6.89	1.62	1.6e+5
9	0.00	0.00	0.00	0.00	0.00	1.8e+5	9	0.00	0.00	0.00	0.00	0.00	1.7e+5
10	0.34	131.88	10.49	23.32	35.36	1.2e+5	10	576.95	1845.45	1504.99	1456.39	226.88	1.3e+5
11	0.00	1.72	0.00	0.31	0.61	1.3e+5	11	0.99	68.58	7.97	21.11	25.48	1.9e+5
12	0.00	130.91	0.21	19.40	44.32	1.5e+5	12	168.49	1886.34	997.51	1028.16	363.30	1.7e+5
13	0.00	5.39	4.84	4.079	1.92	1.5e+5	13	1.01	29.90	16.92	1.47	6.99	1.9e+7
14	0.00	1.99	0.011	0.35	0.57	1.2e+5	14	20.00	25.99	21.94	21.87	1.21	1.5e+5
15	4.9e-7	0.49	0.019	0.14	0.19	1.1e+5	15	0.29	2.98	2.95	2.94	1.73	2.0e+5
16	0.04	0.894	0.31	0.34	0.19	1.5e+5	16	16.42	252.88	35.29	50.42	5.0.15	1.7e+5
17	0.0075	1.016	0.04	0.14	0.19	1.7e+5	17	17.52	44.74	33.06	32.67	5.96	1.1e+5
18	5.7e-5	0.50	0.26	0.26	0.22	5.5e+4	18	20.41	23.49	0.26	0.26	0.86	1.7e+5
19	0.00	0.04	0.002	0.009	0.02	1.7e+5	19	2.67	9.94	5.088	5.35	1.66	3.6e+4
20	0.00	0.044	0.00	0.006	0.04	1.5e+5	20	11.49	43.99	31.87	30.89	6.05	1.0e+5
21	100.00	205.52	109.52	146.43	50.59	1.3e+5	21	203.81	209.85	207.38	207.19	1.23	8.2e+4
22	0.00	100.29	100.00	97.09	15.47	1.5e+5	22	100.00	100.00	100.00	100.00	0.00	7.4e+4
23	300.00	3054.14	303.25	3.30	1.62	1.7e+5	23	342.95	356.26	349.76	350.12	314.85	6.2e+4
24	100.00	332.97	329.83	312.22	62.54	1.0e+5	24	421.12	430.35	425.81	425.78	1.39	5.7e+4
25	397.74	443.37	398.01	409.54	19.97	1.6e+5	25	386.69	386.79	386.74	386.74	0.02	6.9e+4
26	300.00	300.00	300.00	300.00	0.00	1.7e+5	26	837.21	1030.94	925.11	930.09	40.27	5.3e+4
27	389.01	389.52	389.52	389.46	0.17	1.5e+5	27	493.28	512.02	503.05	503.16	4.22	4.6e+4
28	0.00	611.82	300.00	330.30	110.45	1.8e+5	28	300.00	413.98	300.00	325.77	46.99	5.3e+4

(a) Results for 10D. (b) Results for 30D.

Shifted and Rotated Rastrigin's Function, f_5 Shifted and Rotated Expanded Scaffer's f_6 Function, f_6 Shifted and Rotated Lunacek bi-Rastrigin Function, f_7 Shifted and Rotated Non-Continuous Rastrigin's Function, f_8 Shifted and Rotated Levy Function, f_9 Shifted and Rotated Schwefel's Function, Hybrid Functions from f_{10} to f_{19} and Composition Functions from f_{20} to f_{29}. 30 independent runs were made for each function with each chosen solver. The stopping criteria ($maxFEs$) was set as 100,000 for $D = 10$ and 300,000 for $D = 30$. We observed the solvers based on the quality of solutions they reached and their average speed for each run on each function as shown in Tables 9 to 12.

By using the Friedman test, we see that there is no significant difference between the solvers ($p = 0.145$) for the $D = 10$ as shown in Table 15. We compared them based on the speed. It was established that there was significant difference between solvers ($p < 0.05$) with the Friedman test. Dunn-Bonferroni post-hoc test shows there is a significant difference between CS-DE and MadDE ($p = 0.007$), L-SHADE and j2020 ($p = 0.000$), and L-SHADE and MadDE ($p = 0.000$).

Table 11. Results for CEC 2017 Benchmark for the solver CS-DE.

F	Best	Worst	Median	Mean	Std	Speed	F	Best	Worst	Median	Mean	Std	Speed
1	0.00	0.00	0.00	0.00	0.00	4.7e+4	1	0.00	0.00	0.00	0.00	0.00	5.4+4
3	0.00	0.00	0.00	0.00	0.00	5.2e+4	3	0.00	0.00	0.00	0.00	0.00	5.2e+4
4	0.00	0.00	0.00	0.00	0.00	4.9e+4	4	0.00	64.12	58.56	54.16	15.69	5.6e+4
5	0.99	3.98	1.99	2.36	0.77	3.3e+4	5	0.00	0.00	0.00	0.00	0.00	4.1e+4
6	0.00	0.00	0.00	0.00	0.00	4.8e+4	6	0.00	0.00	0.00	0.00	0.00	6.6e+4
7	10.668	13.79	12.25	12.19	0.65	3.5e+4	7	0.00	0.00	0.00	0.00	0.00	5.6e+4
8	0.99	3.98	1.99	2.36	0.82	3.9e+4	8	7.05	15.019	11.45	11.31	1.88	2.9e+4
9	0.00	0.00	0.00	0.00	0.00	5.0e+4	9	0.00	0.00	0.00	0.00	0.00	2.0e+4
10	0.28	241.29	13.72	60.36	67.20	3.3e+4	10	1073.8	2003.8	1620.6	1617.1	207.4	4.2e+4
11	0.00	2.56	1.66	1.45	0.75	3.3e+4	11	1.51	67.66	4.03	7.18	12.19	3.3e+4
12	0.00	120.15	0.21	9.63	32.41	4.4e+4	12	307.49	1269.18	792.26	794.52	247.88	2.8e+4
13	0.00	6.69	0.00	1.42	2.24	5.0e+4	13	0.99	20.43	15.44	12.63	6.11	2.8e+4
14	0.00	2.29	0.00	0.33	0.51	3.3e+4	14	20.03	25.98	23.19	23.24	1.35	2.8e+4
15	0.008	0.50	0.04	0.13	0.18	5.0e+4	15	0.27	5.48	2.41	2.668	1.29	2.4e+4
16	0.06	0.94	0.56	0.52	0.23	3.3e+4	16	10.11	373.51	240.782	201.79	91.12	3.3e+4
17	0.0051	1.069	0.09	0.23	0.29	3.3e+4	17	17.73	47.74	34.06	33.80	5.80	3.0e+4
18	0.0010	0.50	0.22	0.22	0.20	5.0e+4	18	20.29	24.40	21.39	21.48	0.93	2.9e+4
19	0.00	0.12	0.02	0.018	0.019	3.3e+4	19	1.81	11.88	6.21	6.11	1.84	2.0e+4
20	0.00	0.00	0.00	0.00	0.00	3.3e+4	20	28.99	59.62	41.29	41.44	7.45	3.4e+4
21	100.00	205.16	107.38	141.03	49.18	3.3e+4	21	206.69	215.50	210.99	211.18	1.82	1.4e+4
22	100.00	100.00	100.00	100.00	0.00	5.0e+4	22	100.00	100.00	100.00	100.00	0.00	3.8e+4
23	300.00	305.05	302.89	302.11	1.69	1.0e+5	23	341.47	353.72	347.55	347.63	3.15	3.6e+4
24	96.58	331.99	327.16	282.45	90.67	1.0e+5	24	413.19	425.83	422.08	421.79	2.79	3.6e+4
25	397.74	443.37	398.01	407.73	18.86	5.0e+4	25	386.69	386.79	386.75	386.75	0.016	3.6e+4
26	300.00	300.00	300.00	300.00	0.00	5.0e+4	26	794.15	1023.96	939.29	933.23	51.84	3.4e+4
27	389.24	394.23	393.82	393.19	1.50	5.0e+4	27	483.71	510.24	501.99	501.31	5.99	3.1e+4
28	300.00	396.57	300.00	301.89	13.52	5.0e+4	28	300.00	413.98	300.00	317.04	39.96	3.2e+4

(a) Results for 10D.	(b) Results for 30D.

There was a significant difference between solvers for the $D = 30$ as shown in Table 15. The p-value was 0.001. Dunn's Bonferroni test showed that the significant difference was between MadDE and L-SHADE ($p = 0.00$), CS-DE and MadDE ($p = 0.007$), j2020 and MadDE ($p = 0.007$). There was no significant difference detected between other pairs.

Note that, when analyzing speed of the solvers (see Table 15), the highest rank means the fastest solver. There was a significant difference detected between the solvers by the Friedman's test. It was deduced that the significant difference is between CS-DE and MadDE ($p = 0.000$), CS-DE and L-SHADE ($p = 0.000$), CS-DE and j2020 ($p = 0.000$), L-SHADE and j2020 ($p = 0.000$) and MadDE and j2020 ($p = 0.001$).

4   Discussion

In this section, we will summarize the results of the experiment. On the CEC 2022 benchmark functions, the best performance according to the quality of

Table 12. Results for CEC 2017 Benchmark for the solver j2020.

F	Best	Worst	Median	Mean	Std	Speed	F	Best	Worst	Median	Mean	Std	Speed
1	0.00	0.00	0.00	0.00	0.00	5.6e+6	1	0.00	0.00	0.00	0.00	0.00	7.5e+5
3	0.00	0.00	0.00	0.00	0.00	1.8e+6	3	27.69	15775.6	315.7	1512.3	2919.7	5.7e+5
4	0.00	0.46	0.020	0.049	0.08	7.3e+6	4	0.01	114.90	79.67	72.14	35.12	5.5e+5
5	0.00	7.96	2.62	2.57	1.54	2.5e+6	5	39.90	83.66	56.85	58.12	10.06	3.9e+5
6	0.00	0.00	0.00	0.00	0.00	1.8e+6	6	0.00	0.00	0.00	0.00	0.00	9.3e+5
7	9.68	23.25	14.65	15.02	2.6	3.6e+6	7	74.11	123.96	94.82	96.08	13.14	4.9e+5
8	0.00	7.96	3.02	3.47	1.73	2.0e+6	8	36.84	84.22	60.03	58.21	10.39	4.4e+5
9	0.00	0.00	0.00	0.00	0.00	1.7e+6	9	0.00	0.00	0.00	0.00	0.00	1.7e+6
10	6.83	375.37	104.15	119.18	102.36	6.5e+6	10	2635.4	4835.3	3590.5	3582.1	440.3	3.5e+5
11	0.00	4.98	1.22	1.83	1.39	1.3e+6	11	7.46	89.25	38.88	44.36	26.09	5.5e+5
12	0.03	158.0	11.72	43.40	56.84	2.2e+6	12	4213.16	45935	13616	15374	8945.1	4.6e+5
13	0.00	11.57	6.07	5.57	2.88	1.8e+6	13	152.88	10278.0	734.7	1394.8	2011.9	5.4e+5
14	0.00	1.99	0.00	0.44	0.60	2.0e+6	14	38.02	75.03	52.19	53.28	7.93	4.3e+5
15	0.00	3.12	0.99	0.72	0.69	1.9e+6	15	18.45	109.03	33.85	43.48	22.90	5.4e+5
16	0.02	11.93	0.53	1.63	3.36	1.9e+6	16	205.33	708.77	458.09	449.96	122.56	4.5e+5
17	0.00	1.31	0.31	0.29	0.27	1.7e+6	17	53.95	148.09	99.29	99.09	22.07	3.1e+5
18	0.001	2.20	1.00	0.78	0.63	1.2e+6	18	50.76	1096.41	86.40	135.69	160.43	5.1e+5
19	0.00	0.059	0.019	0.02	0.01	1.9e+6	19	14.85	36.59	24.03	24.91	4.67	9.4e+4
20	0.00	0.31	0.00	0.01	0.06	3.3e+5	20	36.75	227.75	93.44	104.77	56.58	3.1e+5
21	100.00	109.92	100.00	100.63	1.68	2.7e+6	21	172.06	282.27	262.66	260.66	16.29	2.6e+5
22	0.00	100.39	32.83	38.19	28.88	1.1e+6	22	100.00	100.00	100.00	100.00	0.00	1.9e+5
23	302.80	313.50	307.15	307.05	2.64	7.4e+5	23	379.73	434.24	410.21	409.94	12.53	1.7e+5
24	0.00	200.90	100.00	98.64	22.37	7.5e+5	24	455.23	513.31	481.66	481.18	10.99	1.6e+5
25	100.02	399.59	397.74	310.49	129.20	7.9e+5	25	383.41	387.34	386.91	386.79	0.70	1.7e+5
26	0.00	300.00	0.00	84.17	110.15	9.3e+5	26	906.40	1824.4	1562.8	1529.3	190.44	1.1e+5
27	386.89	400.93	390.76	391.69	3.05	7.1e+5	27	492.80	522.58	511.52	510.96	5.76	1.1e+5
28	0.00	300.00	300.00	203.35	138.89	5.9e+5	28	300.00	453.96	387.89	362.53	57.22	1.3e+5

(a) Results for 10D.   (b) Results for 30D.

solutions was reached by the solver MadDE ($rank = 2.13$) for $D = 10$ and by
L-SHADE ($rank = 2.25$) for $D = 20$. But according to the speed, the best
performance was obtained by j2020 ($rank = 4.00$) for both dimensions.

On the CEC 2021 benchmark functions, the best performance according
to the quality of the solutions was reached by MadDE ($rank = 2.35$ and

Table 13. Rankings of the solvers according to mean values and speed values for
dimensions 10 and 20 for the CEC 2022 benchmark functions.

Solver	Quality of Solutions		Speed	
	$D = 10$	$D = 20$	$D = 10$	$D = 20$
MadDE	**2.13**	2.58	1.50	2.33
L-SHADE	2.58	**2.25**	3.00	2.67
j2020	2.29	2.75	**4.00**	**4.00**
CS-DE	3.00	2.42	1.50	1.00

Table 14. Rankings of the solvers according to mean values and speed values for dimensions 10 and 20 for the CEC 2021 benchmark functions for the configuration 111.

Solver	Quality of Solutions		Speed	
	$D = 10$	$D = 20$	$D = 10$	$D = 20$
MadDE	**2.35**	**2.15**	2.10	2.00
L-SHADE	2.50	2.90	2.90	3.10
j2020	2.55	2.50	**4.00**	**3.90**
CS-DE	2.60	2.45	1.00	1.00

Table 15. Rankings of the solvers and their speed for dimensions 10 and 30 for the CEC 2017 benchmark functions.

Solver	Quality of Solutions		Speed	
	$D = 10$	$D = 30$	$D = 10$	$D = 30$
MadDE	2.50	3.39	2.32	2.61
L-SHADE	2.46	**1.48**	2.64	2.43
j2020	2.89	2.91	**4.00**	**3.96**
CS-DE	**2.15**	2.22	1.04	1.00

$rank = 2.15$) for both dimensions. But according to the speed of the solvers, j2020 ($rank = 4.00$) was the fastest for both dimensions.

Finally, on the CEC 2017 benchmark functions, CS-DE ranked the highest for $D = 10$ ($rank = 2.15$) and L-SHADE for $D = 30$ ($rank = 1.48$). The solver j2020 was ranked as the first according to the speed.

Some algorithms obtain good ranks on a specific benchmark, but they do not show this trend on the other benchmarks. The reason for this might be overfitting of an algorithm to a specific benchmark. Therefore, to mitigate this problem, we compare the four solvers on all benchmark functions, considering all dimensions, shown in Table 16, L-SHADE is ranked as the best according to the quality of solutions ($rank = 2.20$), but according to the speed, j2020 is the fastest ($rank = 3.96$). From the observed results, we can conclude that there is

Table 16. Rankings of the solvers based on the quality of solutions and their speed for all dimensions on all benchmark functions.

Solver	Quality of Solutions	Speed
MadDE	2.72	2.15
L-SHADE	**2.20**	2.64
j2020	2.64	**3.96**
CS-DE	2.44	1.29

no solver which prevails on every chosen benchmark. The reason for the success of j2020 according to speed is also the programming language in which the solver was implemented (C++). In contrast to j2020, all other solvers are implemented in Matlab. It is difficult to choose the best solver for all benchmarks. It is even harder, when the speed of the solvers is taken into the account as shown in Table 16.

5   Conclusion

This paper explores the analysis and comparison of different state-of-the-art evolutionary algorithms applied to three sets of benchmark functions with the intention of showing that the established comparison metrics have some short-comings. The observed variables in each run were the quality of solutions reached under a predetermined stopping criteria and the speed, defined as the number of function evaluations per second of each solver for a chosen benchmark. The main aim was to compare the state-of-the-art solvers based on both variables. Often happens that the runtime and the speed of the solver get neglected in favor of analyzing the quality of solutions and establishing the best solver only based on that. We show that observing only the quality of the solutions is not enough. Depending on the type of an optimization problem, an information about the runtime and the speed of the solver can be helpful, either in planning an exper-iment or while dealing with optimization problems. Through statistical analysis with post-hoc tests, we showed that it is not enough to observe only the quality of solutions, but it is also necessary to consider the speed of the solver. By com-paring the state-of-the-art solvers on different benchmarks, we show that there is no solver which is appropriate for all benchmarks. We also show that the choice of a programming language might be important for the solver's perfor-mance. The comparison of the solvers should include the analysis of the number of function evaluations, runtime and speed of the solvers, while considering the computational complexity of an optimization problem.

Acknowledgements. This work was supported by the Slovenian Research Agency (Computer Systems, Methodologies, and Intelligent Services) under Grant P2-0041.

References

1. Mohamed, A.W., Hadi, A.A., Mohamed, A.K., Agrawal, P., Kumar, A., Suganthan, P.N.: Problem definitions and evaluation criteria for the CEC 2022 special session and competition on single objective bound constrained numerical optimization. Technical report, Nanyang Technological University, Singapore (2021). https:// github.com/P-N-Suganthan/2022-SO-BO
2. Mohamed, A.W., Hadi, A.A., Mohamed, A.K., Agrawal, P., Kumar, A., Suganthan, P.N.: Problem definitions and evaluation criteria for the CEC 2021 special session and competition on single objective bound constrained numerical optimization. Technical report, Nanyang Technological University, Singapore (2020). https:// github.com/P-N-Suganthan/2021-SO-BCO

3. Awad, N.H., Ali, M.Z., Liang, J.J., Qu, B.Y., Suganthan, P.N.: Problem definitions and evaluation criteria for the CEC 2017 special session and competition on
   single objective bound constrained real-parameter numerical optimization. Technical report, Nanyang Technological University, Singapore (2016). https://www.ntu.
   edu.sg/home/epnsugan/
4. Benavoli, A., Corani, G., Demšar, J., Zaffalon, M.: Time for a change: a tutorial
   for comparing multiple classifiers through Bayesian analysis. J. Mach. Learn. Res.
   18(1), 2653–2688 (2017)
5. Biswas, S., Saha, D., De, S., Cobb, A.D., Das, S., Jalaian, B.A.: Improving differential evolution through Bayesian hyperparameter optimization. In: 2021 IEEE
   Congress on Evolutionary Computation (CEC), pp. 832–840. IEEE (2021)
6. Brest, J., Zamuda, A., Fister, I., Boskovic, B.: Some improvements of the self-
   adaptive JDE algorithm. In: 2014 IEEE Symposium on Differential Evolution
   (SDE), pp. 1–8 (2014). https://doi.org/10.1109/SDE.2014.7031537
7. Brest, J., Maučec, M.S., Bošković, B.: Single objective real-parameter optimization:
   algorithm JSO. In: 2017 IEEE Congress on Evolutionary Computation (CEC), pp.
   1311–1318. IEEE (2017)
8. Brest, J., Maučec, M.S., Bošković, B.: Differential evolution algorithm for single objective bound-constrained optimization: algorithm J2020. In: 2020 IEEE
   Congress on Evolutionary Computation (CEC), pp. 1–8. IEEE (2020)
9. Carrasco, J., García, S., Rueda, M., Das, S., Herrera, F.: Recent trends in the use
   of statistical tests for comparing swarm and evolutionary computing algorithms:
   practical guidelines and a critical review. Swarm Evol. Comput. **54**, 100665 (2020)
10. Cenikj, G., Lang, R.D., Engelbrecht, A.P., Doerr, C., Korošec, P., Eftimov, T.:
    Selector: selecting a representative benchmark suite for reproducible statistical
    cómparison. arXiv preprint arXiv:2204.11527 (2022)
11. Derrac, J., García, S., Molina, D., Herrera, F.: A practical tutorial on the use of
    nonparametric statistical tests as a methodology for comparing evolutionary and
    swarm intelligence algorithms. Swarm Evol. Comput. **1**(1), 3–18 (2011)
12. Eftimov, T., Korošec, P., Koroušić Seljak, B.: Data-driven preference-based deep
    statistical ranking for comparing multi-objective optimization algorithms. In:
    Korošec, P., Melab, N., Talbi, E.-G. (eds.) BIOMA 2018. LNCS, vol. 10835, pp.
    138–150. Springer, Cham (2018). https://doi.org/10.1007/978-3-319-91641-5_12
13. Eftimov, T., Korošec, P., Seljak, B.K.: A novel approach to statistical comparison
    of meta-heuristic stochastic optimization algorithms using deep statistics. Inf. Sci.
    417, 186–215 (2017)
14. Eftimov, T., et al.: Deep statistics: more robust performance statistics for single-
    objective optimization benchmarking. In: Proceedings of the 2020 Genetic and
    Evolutionary Computation Conference Companion, pp. 5–6 (2020)
15. Hansen, N., Auger, A., Ros, R., Mersmann, O., Tušar, T., Brockhoff, D.: COCO: a
    platform for comparing continuous optimizers in a black-box setting. Optim. Meth-
    ods Softw. **36**, 114–144 (2021). https://doi.org/10.1080/10556788.2020.1808977
16. Hansen, N., Auger, A., Finck, S., Ros, R.: Real-parameter black-box optimization
    benchmarking 2010: experimental setup. Ph.D. thesis, INRIA (2010)
17. Kostovska, A., Vermetten, D., Džeroski, S., Doerr, C., Korosec, P., Eftimov, T.: The
    importance of landscape features for performance prediction of modular CMA-ES
    variants. In: Proceedings of the Genetic and Evolutionary Computation Confer-
    ence, pp. 648–656 (2022)
18. Meng, Z., Zhong, Y., Yang, C.: CS-DE: Cooperative strategy based differential
    evolution with population diversity enhancement. Inf. Sci. **577**, 663–696 (2021)

19. Ravber, M., Moravec, M., Mernik, M.: Primerjava evolucijskih algoritmov imple-
    mentiranih v različnih programskih jezikih. Elektrotehniski Vestnik **89**(1/2), 46–52
    (2022). (In Slovene)
20. Škvorc, U., Eftimov, T., Korošec, P.: Transfer learning analysis of multi-class clas-
    sification for landscape-aware algorithm selection. Mathematics **10**(3), 432 (2022)
21. Suganthan, P.N., et al.: Problem definitions and evaluation criteria for the
    CEC 2005 special session on real-parameter optimization. KanGAL Rep.
    2005005(2005), 2005 (2005)
22. Tanabe, R., Fukunaga, A.S.: Improving the search performance of SHADE using
    linear population size reduction. In: 2014 IEEE Congress on Evolutionary Compu-
    tation (CEC), pp. 1658–1665. IEEE (2014)
23. Veček, N., Mernik, M., Črepinšek, M.: A chess rating system for evolutionary algo-
    rithms: a new method for the comparison and ranking of evolutionary algorithms.
    Inf. Sci. **277**, 656–679 (2014)
24. Yang, X.-S.: Metaheuristic optimization: algorithm analysis and open problems.
    In: Pardalos, P.M., Rebennack, S. (eds.) SEA 2011. LNCS, vol. 6630, pp. 21–32.
    Springer, Heidelberg (2011). https://doi.org/10.1007/978-3-642-20662-7_2

Refining Mutation Variants in Cartesian Genetic Programming

Henning Cui[1]([✉])[iD], Andreas Margraf[2][iD], and Jörg Hähner[1][iD]

[1] University of Augsburg, 86159 Augsburg, Germany
{henning.cui,joerg.haehner}@uni-a.de
[2] Fraunhofer IGCV, Institution for Casting, Composite and Processing Technology,
86159 Augsburg, Germany
andreas.margraf@igcv.fraunhofer.de

Abstract. In this work, we improve upon two frequently used mutation algorithms and therefore introduce three refined mutation strategies for Cartesian Genetic Programming. At first, we take the probabilistic concept of a mutation rate and split it into two mutation rates, one for active and inactive nodes respectively. Afterwards, the mutation method *Single* is taken and extended. *Single* mutates nodes until an active node is hit. Here, our extension mutates nodes until more than one but still predefined number n of active nodes are hit. At last, this concept is taken and a decay rate for n is introduced. Thus, we decrease the required number of active nodes hit per mutation step during CGP's training process. We show empirically on different classification, regression and boolean regression benchmarks that all methods lead to better fitness values. This is then further supported by probabilistic comparison methods such as the Bayesian comparison of classifiers and the Mann-Whitney-U-Test. However, these improvements come with the cost of more mutation steps needed which in turn lengthens the training time. The third variant, in which n is decreased, does not differ from the second mutation strategy listed.

Keywords: Cartesian genetic programming · Genetic programming · Evolutionary algorithm · Mutation strategy

1 Introduction

Cartesian genetic programming (CGP) is a form of genetic programming and a nature inspired search heuristic. It can be used to automatically generate programs and was first introduced by Miller [20] in 1999. Since its introduction it has been used for a multitude of applications like evolving electronic circuits [19], image processing [16] or evaluation of sensor data [3]. CGP employs a directed, acyclic graph-based representation. This means that these graphs consists of nodes which are arranged in a two-dimensional grid. These nodes can in turn be active or inactive, meaning that they do or do not contribute to an output.

While some versions of CGP utilize or suggest the usage of a crossover operator, as is seen in Kalkreuth et al. [13] or Wilson et al. [25], they remain unused in standard CGP as it does not universally profit from this operation [11,18]. Thus, mutation is oftentimes the only genetic operation used. In the literature, there are two different mutation strategies which protrude and are frequently applied: a probabilistic approach or the *Single* mutation by Goldman and Punch [6]. As for the first one, every node has a chance of mutation while in the latter one, nodes are mutated until an active one is hit.

While the mutation function and the mutation rate, if needed, are very important for the success of CGP, there is little research performed trying to improve upon these two operations. Because of that, this work focuses on improving the probabilistic and *Single* mutation approach. At first, the probabilistic approach is further discussed. Normally, only a single mutation rate is used. However, we hypothesize that utilizing a different mutation rate for both active and inactive genes may lead to faster convergence. As for one *Single* mutation step, nodes are mutated until one active node is altered. This notion is suspended now and we allow the mutation of nodes to take place until n active nodes are mutated. To evaluate this concept, we experiment with different n values. We also investigate the effect of decreasing the mutation rate and number of active nodes changed over time.

We show on multiple datasets that our extensions achieve statistically significantly better fitness values on these datasets compared to the standard ones for the cost of a slower convergence rate.

We follow this introduction with Sect. 2, which gives a brief overview of related work. Afterwards, we present an introduction into CGP in Sect. 3. In Sect. 4, the refined mutation concepts are introduced. Furthermore, a short theoretical explanation is given as to why these concepts should lead to better fitness values. Afterwards, the experimental design is introduced in Sect. 5 and Sect. 6 presents and discusses our experimental results. Finally, we conclude our work in Sect. 7.

2   Related Work

In this work, the focus lies on extending the probabilistic mutation strategy as well as *Single*.

Previously, there has been other mutation strategies employed as well. Goldman and Punch [6] created other mutation algorithms alongside *Single*. Both algorithm extends the probabilistc mutation and the first one skips redundant evaluations, which leads to less computational time needed. The second algorithm is more complex and not recommended by the authors, as it does not improve upon existing strategies.

Another work done by Kalkreuth [12] introduced two different mutation operators. He directly altered the type of a node and changed it from active to inactive or vice versa. As a consequence thereof, the whole phenotype might be changed to maintain CGP's restrictions and rules. In his experiments, these two operators lead to improvements in the search performance regarding boolean regression benchmarks. However, more tests should be conducted before a definite statement can be made.

The authors of [10] focused on a new mutation operator for the evolutionary design of combinational circuits. They propose a semantically-oriented mutation algorithm, which reduces the computational complexity as well as improves search performance for this problem domain.

There is also the possibility to ulitze concepts from reinforcement learning to extend the mutation algorithm, as is done in [22]. They applied their new operator to multiple logic circuit benchmarks and show improved performances for selected problems.

In other studies, there has been some other work done investigating the effect of mutating inactive genes. The authors Turner and Miller [23] did an in depth investigation about genetic drift and genetic redundancy. Their findings, included but are not limited to, are that performance significantly worsens when there is no mutation of inactive genes. Furthermore, they hypothesized that a single mutation rate may be inferior to utilizing two mutation rates, for active and inactive genes respectively. By utilizing a mutation rate for inactive genes of up to 100%, they believe that this may enhance genetic drift and therefore CGP's performance.

Kaufmann and Kalkreuth [14] examined, among other things, the effects of *Single* mutation but mutated every inactive node while doing one mutation step. By doing so, they found an improvement and other, better suited mutation methods as well. One of these improvements would be to turn off the mutation of function genes when boolean regression benchmarks are used. Thus, they laid out the first steps for this work.

3   Cartesian Genetic Programming

This section gives a brief overview of CGP and its mutation algorithm.

3.1   Introduction to Cartesian Genetic Programming

CGP is traditionally represented as a directed, acyclic and feed-forward graph. It consists of *nodes* which are arranged in a $n_c \times n_r$ grid, wherein n_c declares the grid's number of columns and n_r indicates its number of rows. It takes one or multiple program inputs and feeds it forward through partially connected nodes before writing final values to output nodes.

Each node consists of a number of genes, namely a function gene, two connection genes and a parameter gene. The function gene addresses the encoded computational function of the node. If this function depends on a parameter, its value is taken from the parameter gene. The required input is taken from its connection genes, as they indicate where the node gets its data from. This can either be a program input or the output of a previous node. However, a node cannot get its input from an arbitrary former node, as the hyperparameter *levels-back l* restricts the connectivity of a node. Furthermore, the nodes are partitioned into two groups: active and inactive nodes. Active nodes contribute to a program output, while inactive nodes do not. The parameter l defines the number of columns to the nodes left it can receive its input from. Oftentimes, l is equal to n_c meaning that every node receives its input from every prior node.

In our work, a slightly modified version of CGP inspired by Harding et al. [8] and Leitner et al. [15] is used. Here, the handling of input and output are different from regular CGP and are adopted from *Self-Modifying CGP* [9]. Traditionally, if a program has n_i many inputs and requires n_o outputs, CGP contains $n_i + n_o$ additional genes. Each additional gene represents an address to its respective program input or output node. In our work though, no further input and output genes are used. The function set used is extended by four special functions taken from Harding et al. [9], indicating which node serves as an input or output.

An illustrative example of a genotype can be seen in Fig. 1 as it shows a graph with $n_r = 1$ and $n_c = 7$. The function of the first two nodes are 'INPUT', indicating that the next program input is to be read. Afterwards, both input values are added in the third node. However, it does not link to a node with an output function, rendering it inactive. In the fourth node, the same inputs are subtracted and then added in node six with an additional program input taken from node five. At last, node seven indicates a program output.

Fig. 1. An example genotype of CGP.

As is the case for many CGP algorithms, the standard $(1 + \lambda)$ Evolutionary Algorithm (EA) is used where the individual with the highest fitness is chosen as the next parent. This parent individual is taken to evolve λ offsprings. Additionally, neutral search is performed. This means that, when an offspring and the parent have the same fitness score, the offspring is always chosen as the next parent, even in the rare case that both offspring and parent are identical. This allows to generate better offsprings [23, 26].

As is described by Miller [21], inactive nodes are part of *non coding* genes which are not used to provide an output. Such nodes can be exemplified by the third node in Fig. 1 as it is part of the genotype but not contributing to the genotypes output.

By having non coding genes, genetic drift is allowed to occur as mutating them does not affect the fitness of the phenotype. These inactive nodes have the possibility to later be changed to active nodes when connection genes are mutated [7]. The works of Turner and Miller [23] and Yu and Miller [26] show that such neutral genes are highly beneficial for the evolutionary process of

Evolutionary Algorithms and CGP as they help escape local optima and the evolutionary search of CGP. Albeit this theory is doubted by Goldman and Punch [5].

3.2 Mutation Algorithm

The most common mutation algorithms are of probabilistic nature [21] and *Single* [6].

As for the first one, the nodes are mutated according to a predefined mutation probability p. Furthermore, there is a distinction between *point mutation* and *probabilistic mutation*. For the point mutation, p percent of all nodes are randomly chosen and mutated. Concerning the probabilistic mutation, we iterate through every node with each node mutating with a probability p. However, with a probabilistic mutation strategy, it is possible that only inactive genes are mutated. Thus, it does not change the fitness value and nothing can be said about the quality of the changes.

Single, on the other hand, takes random nodes and mutates them until an active node is mutated. This offers the benefit to generate good results without setting a mutation rate. Goldman and Punch [6] evaluated *Single* on four boolean benchmarks and found that *Single* is preferred when the optimal mutation rate is unknown. Another benefit is that there is a guaranteed change in the phenotype. This avoids wasted evolutions where the phenotype stays the same. However, When the mutation rate can be optimized, a probabilistic mutation method is able to outperform *Single*.

4   Further Changes in the Mutation Algorithm

The following section gives an overview about the refined mutation strategies utilized in this work. Afterwards, a short theoretical explanation is given to motivate and reason their respective effectiveness.

4.1   Probabilistic Mutation

A caveat of the probabilistic mutation is that it does not differentiate between active and inactive nodes. We hypothesize that this could lead to a slower convergence and/or worse fitness. The main motivation is established by the findings of Turner and Miller [23]. This work addresses neutral drift in the context of CGP. They found that, among other things, not mutating inactive genes leads to a worse evolutionary search and in turn worse results overall. The authors Miller and Turner [23] also hypothesize that it could be favorable to explicitly use a higher mutation rate for inactive genes; and perhaps change it as high as up to 100%, meaning that every inactive node in the genotype is mutated after a single evaluation step. As having neutral drift is highly beneficial for the evolution of the CGP phenotype, such high mutation rates for inactive genes could lead to lower convergence rates.

To test this hypothesis, we split the mutation rate into two. A user defined probability p_i for the mutation of inactive nodes is introduced as well as a mutation rate p_a for active nodes. We compare different combinations of p_i and p_a for different classification and regression datasets as well as some selected boolean regression benchmarks.

4.2 Single and Multiple Mutation

Fig. 2. An example of a changed phenotype as compared by Fig. 1. By changing the connection gene of the sixth node, it is possible to alter the phenotype greatly. The computation does not rely on the second and fourth node anymore.

Single randomly changes inactive genes until one active gene is hit. Thus, only incremental changes are possible. However, these changes may affect the phenotype greatly by changing a connection gene as is exemplified in Fig. 2. The connection gene of the sixth node is mutated and now merely depends on the first and third input. Compared to the previous phenotype version in Fig 1, the computation of the program output does not rely on the second and fourth node anymore. Nevertheless, the single incremental change may, on the other hand, not affect the phenotype at all or only slightly. As is shown by Kaufmann and Kalkreuth [14], changing connection genes can improve the fitness at most and is the most meaningful mutation according to their estimation. Looking at the CGP implementation used in this work, every node has four possible genes which can be mutated: two connection genes, one function and one parameter gene. However, only the first connection gene guarantees a change in the phenotype. The second connection gene is only used if the corresponding function requires two inputs, which is needed by 6 out of 30 functions in our case. This leads to a chance of mutating a meaningful connection gene at about 29%. Albeit one could argue to exclusively mutate connection genes, only selected problems benefit from solely changing the in-going connections rendering this procedure not viable for every problem [14].

As is hypothesized by Goldman et al. [6], the incremental change of *Single* may perform worse on problems where larger changes per evaluation step are necessary. Thus, we introduce a modified version of *Single*: *Multi-n* and *Decreasing Multi-n* (*DMulti-n*).

Multi-n takes the concept of *Single* but mutates the genotype until n active nodes are mutated. It should be noted that, when an active node is hit, it is possible that the phenotype is completely changed. Thus, the list of active nodes may not stay the same during mutation and it is important to update the new active nodes before mutation is continued.

We speculate that *Multi-n* could lead to faster convergence due to several reasons. At first, more inactive genes are mutated per evolution step which leads to more genetic drift. The original authors of *Single* report an average of $\frac{t-a}{a+1} + 1$ expected mutations, where t is the number of total genes and a being the number of active parent genes. With *Multi-n*, we should expect an average of $\prod_{i=1}^{n} \left(\frac{t-a_i}{a_i+1} + 1 \right)$ mutations with a_i being the number of active nodes after i active nodes have been hit. As is shown by Miller and Smith [17], typically there are more inactive genes than active ones. Considering *Multi-n*, this leads to even more mutated inactive genes per evolution cycle which should also introduce more genetic drift. Another reasoning in favor of *Multi-n* is a higher chance of a more impactful mutation in the phenotype, as more active nodes are mutated. A higher n leads to a higher probability of mutating a meaningful connection gene while also maintaining the chance to mutate function or parameter genes. With $n = 2$, the probability rises to 49% or even 74% for $n = 4$.

While *Multi-n* could lead to a faster convergence to possible solution spaces and local optima, it may also introduce a lot of changes per evolution step. It looses the ability to make small and incremental changes. *DMulti-n* introduces a slight modification to *Multi-n*. It can be imagined as *Multi-n*, but n decreases over time. Here, we employ a simple stepped decay rate:

$$n_{current} = n_{start} - \left\lfloor i_{current} \cdot \frac{n_{start}}{i_{total}} \right\rfloor \tag{1}$$

with $n_{current}$ being the current n value, n_{start} the initial n value, $i_{current}$ the current evaluation step and i_{total} the total evaluation steps to perform. We have, for example, the following two starting values: $i_{total} = 100$ and $n_{start} = 5$. This means that, in the beginning, we start mutating nodes until five active nodes are hit for each mutation step. After the 20th evaluation step, $n_{current}$ reduces to 4. Now the nodes are mutated until only four active ones are hit per mutation step. This repeats until a lower limit of $n_{current} = 1$ is reached and the training is finished with the equivalent of *Single* mutation.

5 Preliminaries

This section is now dedicated to the experiments conducted to explore the previously defined mutation concepts.

5.1 Experiment Description

As far as hyperparameters are concerned, default values found in the literature [21] are used. This means that $n_r = 1$ and $l = n_c$ are adopted. For the genotype, a length of $n_c = 100$ is employed. The number of maximum evaluations are $100,000$. If the parameter gene is needed, its value is randomly changed to a value in the range of $[-10, 10]$. Each experiment is repeated for a total of 15 times. The program inputs of the real-world datasets are standardized and standard k-fold cross validation with $k = 5$ is used when no train/test split is defined by the dataset.

For classification problems, the fitness function used in this work is subject to the Matthews Correlation Coefficient (MCC) [1]. Its score is in range $[-1, 1]$, with 1 and -1 indicating that every sample is correctly classified (but inverted, in case of -1). A value of 0 indicates only falsely classified samples. Thus, the fitness value is defined as:

$$fitness = 1 - |MCC| \tag{2}$$

For regression problems, the mean squared error is used.

As optimization algorithm, a $(1 + \lambda)$ with $\lambda = 4$ selection algorithm as mentioned in Sect. 3.1 is utilized.

Furthermore as is discussed in the later Sect. 5.2, two different kinds of datasets are used to evaluate the mutation operators. We employed real-world classification and regression datasets as well as symbolic regression benchmarks. Both utilize different function sets to accommodate to their respective problem. The function sets are specified in Table 1.

Table 1. Functions used in this work. The number of required inputs is given by arity.

Function Name	Arity	Description
Functions for Real-World Datasets		
INP, INPP, SKIP	0	Special input functions taken from [9]
OUTPUT	1	Special output functions taken from [9]
Add, Sub, Mul, Div	2	Standard mathematical functions
Addc, Subc, Mulc, Divc	1	Mathematical function, the operation takes one input i and the nodes' parameter as a constant c to calculate $i \circ c$
Sin, Cos, Tan, Tanh, Log, Log1p, Sqrt, Abs, Ceil, Floor	1	Standard mathematical functions
Max, Min	1	Compares the input with the nodes' parameter. Returns the bigger / lower value
Const	0	Returns the nodes' parameter
Negate	1	Returns the negated value
Functions for Symbolic regression datasets		
INP, INPP, SKIP	0	Special input functions taken from [9]
OUTPUT	1	Special output functions taken from [9]
Add, Sub, Mul, Div	2	Standard mathematical functions
Sin, Cos, Log	1	Standard mathematical functions

5.2   Datasets

We used a mixture of classification, regression and symbolic regression bench-marks to assess our mutation algorithms for multiple problem definitions. These datasets are chosen according to White et al. [24] as they surveyed and recommended multiple benchmarks for genetic programming. For classification and regression, we use Abalone, Breast Cancer, Credit, Forest Fire, Page Block and Spect, downloaded from the UCI repository [4]. We also employ symbolic regression benchmarks, namely Nguyen-7, Pagie-1 and Vlad-4 [24].

6   Experiments

In this work, several experiments are conducted to empirically test the different mutation strategies. We state the average fitness value as well as the average number of evaluations it takes until CGP converges, i.e. the number of evaluations until the best fitness result is achieved. Both values are then compared to their standard equivalent mutation strategy[1].

Furthermore, we utilize the Bayesian comparison of classifiers introduced by Benavoli et al. [2] to compare our models trained on classification datasets. The advantage here is that no null hypothesis is needed. Hence the results are presented as a triplet $(p_{default}, p_{equal}, p_{extend})$. These values indicate different probability values with $p_{default}$ stating the probability that the standard classifier is better; p_{equal} expresses the probability that the differences are within the region of practical equivalence and p_{extend} presents the probability that the modified mutation strategy is better. However, if the Bayesian comparison is not applicable since this only works for classification models, we utilize the Mann-Whitney-U-Test at $\alpha = 0.05$ between the default and the extended model. Again, we report our results as a triplet (U, Z, p) with U being the Mann-Whitney-U value, Z the z-statistic and p the p-value.

6.1   Impact of Different Probabilistic Mutation Strategies

To test the impact of splitting a single mutation rate p into two mutation rates p_a and p_i for the respective active and inactive nodes, we ran multiple experiments with different mutation rates and combinations thereof. For comparison, we utilize $p \in \{0.03, 0.1, 0.15, 0.2, 0.25\}$. As for utilizing two mutation rates, we ran experiments for every combination of the following probability values: $p_i \in \{0.1, 0.25, 0.5, 0.75, 1.0\}$ and $p_a \in \{0.03, 0.1, 0.15, 0.2, 0.25\}$. These mutation rates should cover a wide range of varieties while not being too finegrained or too fuzzy. For space-savings sake, the p and p_a values are averaged into p_{avg} or $p_{a,avg}$ respectively.

The averaged fitness values can be seen in Table 2. Here, even with averaged values there is a clear trend noticeable towards utilizing two different mutation

---

[1] The code as well as its datasets preprocessing can be found in the following GitHub repository: https://github.com/CuiHen/Refining-Mutation-in-CGP.git.

rates instead of one as a split mutation rate almost always yields higher fitness values. Additionally, a higher mutation rate for inactive nodes oftentimes generate better results. This is in accordance to the same findings of other works, such as Kaufmann and Kalkreuth [14] who tested the influence of mutating every inactive node per mutation step. Still, it is not recommended to always set $p_i = 1.0$ as a more optimal p_i value is oftentimes lower than 1.0.

This finding is supported by the probabilistic evaluation in Table 3. Oftentimes, the comparison lies in favor of a split mutation rate or both methods being equal. However, only *Page Block* shows a trend towards the single mutation rate when a higher p_i value is used. Additionally, the p-values for *Forest Fire* are higher than 0.05, which means that those are not statistically significant.

Table 4 shows the convergence speed. Here, the results are averaged as is seen in Table 2. Interestingly, two different mutation rates oftentimes leads to more mutation steps needed for the model to converge. Moreover, the convergence speed for the best fitness results is oftentimes the lowest, too.

Table 2. The average fitness value achieved with a single and two mutation rates. The p and p_a values are averaged into p_{avg} or $p_{a,avg}$ respectively. Lower values are better. Bold symbols indicate the best values in the current row.

Dataset	p_{avg}	$p_{a,avg}$, $p_i = 0.1$	$p_{a,avg}$, $p_i = 0.25$	$p_{a,avg}$, $p_i = 0.50$	$p_{a,avg}$, $p_i = 0.75$	$p_{a,avg}$, $p_i = 1.0$
Abalone	6.59	5.96	5.84	5.99	5.84	**5.77**
Breast Cancer	0.074	0.054	0.049	0.042	0.049	**0.048**
Credit	0.229	0.230	0.241	**0.223**	0.223	0.229
Forest Fire	1.78	1.79	1.75	**1.72**	1.73	**1.72**
Heart Disease	0.763	0.743	0.744	0.736	**0.734**	0.742
Page Block	0.253	**0.212**	0.298	0.246	0.310	0.274
Spect	0.445	0.423	0.423	0.408	0.414	**0.406**
Nguyen-7	$1.42 \cdot 10^{-2}$	$5.44 \cdot 10^{-3}$	$6.16 \cdot 10^{-3}$	$5.94 \cdot 10^{-3}$	$\mathbf{5.30 \cdot 10^{-3}}$	$6.67 \cdot 10^{-3}$
Pagie-1	0.175	**0.093**	0.103	0.100	0.109	0.115
Vlad-4	0.037	**0.034**	**0.034**	**0.034**	0.035	0.035

6.2   Impact of *Multi-n* and DMulti-n

As for the other mutation strategy tested in this work, *Single* is compared to our *Multi-n* mutation strategy with $n \in \{2, 3, 4, 5\}$ at first.

Our results are shown in Table 5, 7 and 6. Here, we can see a clear trend towards higher n values, as they generally deliver better results. Interestingly, the highest n value of 5 seldom leads to the best results. It is possible that there are too many changes in the phenotype per mutation step so it becomes impossible to grasp more optimal results.

As the probability values in Table 6 suggest, the trend towards higher n values lead to better results, too. This reinforces the hypothesis that *Multi-n* can be better than *Single*. However, a caveat is that the datasets evaluated on the Mann-Whitney-U-Test oftentimes only significantly differ for $n > 3$.

Table 3. The probability of one classifier being better than the other, evaluated on a single mutation rate compared to a split one. The p and p_a values are averaged into p_{avg} or $p_{a,avg}$ respectively. Datasets where the Bayesian comparison is used are marked with (B); the Mann-Whitney-U-Tests are marked with (MW). Every $(p_{a,avg}, p_i)$ value is compared to p_{avg}.

Dataset	$p_{a,avg},$ $p_i = 0.1$	$p_{a,avg},$ $p_i = 0.25$	$p_{a,avg},$ $p_i = 0.50$	$p_{a,avg},$ $p_i = 0.75$	$p_{a,avg},$ $p_i = 1.0$
Abalone (MW)	$(450, -2.6, 0.01)$	$(460, -2.9, 0.00)$	$(470, -3.0, 0.00)$	$(460, -2.9, 0.00)$	$(500, -3.6, 0.00)$
Breast Cancer (B)	$(0.00, 0.12, 0.88)$	$(0.00, 0.03, 0.97)$	$(0.00, 0.00, 1.0)$	$(0.00, 0.05, 0.95)$	$(0.00, 0.02, 0.98)$
Credit (B)	$(0.21, 0.63, 0.16)$	$(0.54, 0.26, 0.2)$	$(0.14, 0.49, 0.37)$	$(0.08, 0.61, 0.31)$	$(0.22, 0.58, 0.2)$
Forest Fire (MW)	$(310, -0.02, 0.98)$	$(370, -1.1, 0.25)$	$(370, -1.0, 0.30)$	$(360, -0.81, 0.42)$	$(390, -1.5, 0.14)$
Heart Disease (B)	$(0.00, 0.17, 0.83)$	$(0.00, 0.13, 0.87)$	$(0.00, 0.04, 0.96)$	$(0.00, 0.04, 0.96)$	$(0.00, 0.07, 0.93)$
Page Block (B)	$(0.08, 0.11, 0.81)$	$(0.87, 0.09, 0.04)$	$(0.32, 0.21, 0.47)$	$(0.82, 0.09, 0.09)$	$(0.59, 0.14, 0.27)$
Spect (B)	$(0.10, 0.21, 0.69)$	$(0.08, 0.22, 0.70)$	$(0.00, 0.02, 0.98)$	$(0.00, 0.05, 0.95)$	$(0.00, 0.03, 0.97)$
Nguyen-7 (MW)	$(490, -3.5, 0.00)$	$(470, -3.1, 0.00)$	$(480, -3.2, 0.00)$	$(470, -3.0, 0.00)$	$(460, -2.9, 0.00)$
Pagie-1 (MW)	$(510, -3.7, 0.00)$	$(490, -3.5, 0.00)$	$(510, -3.8, 0.00)$	$(480, -3.2, 0.00)$	$(510, -3.8, 0.00)$
Vlad-4 (MW)	$(520, -4.1, 0.00)$	$(500, -3.7, 0.00)$	$(500, -3.6, 0.00)$	$(510, -3.8, 0.00)$	$(500, -3.5, 0.00)$

Table 4. The average number of evaluations until convergence is achieved; given for a single and two mutation rates. The p and p_a values are averaged into p_{avg} or $p_{a,avg}$ respectively. Lower values are better. The number of evaluations for the best model (i.e. the best fitness, cf. Table 2) is underlined; the lowest number indicates the fastest convergence is in bold.

Dataset	p_{avg}	$p_{a,avg},$ $p_i = 0.1$	$p_{a,avg},$ $p_i = 0.25$	$p_{a,avg},$ $p_i = 0.50$	$p_{a,avg},$ $p_i = 0.75$	$p_{a,avg},$ $p_i = 1.0$
Abalone	**663**	2034	2531	2280	2352	<u>2654</u>
Breast Cancer	**962**	1406	1275	1412	1286	<u>1724</u>
Credit	**759**	1086	1108	<u>1412</u>	<u>1653</u>	847
Forest Fire	**866**	2077	1978	<u>1826</u>	2473	<u>2490</u>
Heart Disease	**759**	1440	1426	1702	<u>2185</u>	1364
Page Block	2142	<u>2549</u>	2210	2787	**1871**	2818
Spect	**834**	1244	1461	1398	1007	<u>1689</u>
Nguyen-7	276	316	282	300	<u>468</u>	**266**
Pagie-1	**700**	<u>2330</u>	2002	2021	1901	2433
Vlad-4	**819**	<u>1553</u>	<u>1616</u>	<u>2213</u>	2094	1959

Table 5. The average fitness value for *Single* and *Multi-n*. Lower values are better. Bold symbols indicate the best values in the current row.

Dataset	*Single*	*n* = 2	*n* = 3	*n* = 4	*n* = 5
Abalone	6.07	5.84	5.71	**5.60**	5.64
Breast Cancer	0.059	0.042	0.039	**0.035**	0.034
Credit	0.389	0.258	**0.222**	0.239	0.228
Forest Fire	1.75	1.72	1.71	**1.67**	1.69
Heart Disease	0.756	0.752	**0.736**	0.739	0.742
Page Block	0.273	0.282	0.255	**0.238**	0.239
Spect	0.489	0.433	0.445	**0.391**	0.408
Nguyen-7	$5.60 \cdot 10^{-3}$	$\mathbf{4.15 \cdot 10^{-3}}$	$5.19 \cdot 10^{-3}$	$6.15 \cdot 10^{-3}$	$7.04 \cdot 10^{-3}$
Pagie-1	0.075	0.056	0.044	0.042	**0.036**
Vlad-4	0.034	0.035	0.034	**0.031**	0.032

Table 6. The probability of one classifier being better than the other, evaluated on *Single* compared to *Multi-n*. Datasets where the Bayesian comparison is used are marked with (B); the Mann-Whitney-U-Tests are marked with (MW). Every *n* value is compared to *Single*.

Dataset	*n* = 2	*n* = 3	*n* = 4	*n* = 5
Abalone (MW)	(130, −0.54, 0.59)	(150, −1.6, 0.11)	(170, −2.3, 0.02)	(170, −2.3, 0.02)
Breast Cancer (B)	(0.01, 0.25, 0.73)	(0.01, 0.20, 0.78)	(0.00, 0.08, 0.92)	(0.01, 0.11, 0.88)
Credit (B)	(0.04, 0.02, 0.94)	(0.01, 0.01, 0.99)	(0.02, 0.02, 0.96)	(0.01, 0.01, 0.98)
Forest Fire (MW)	(130, −0.58, 0.56)	(140, −1.2, 0.21)	(150, −1.6, 0.11)	(120, −0.41, 0.68)
Heart Disease (B)	(0.23, 0.41, 0.36)	(0.02, 0.22, 0.76)	(0.04, 0.29, 0.67)	(0.06, 0.35, 0.59)
Page Block (B)	(0.49, 0.19, 0.32)	(0.26, 0.17, 0.57)	(0.25, 0.10, 0.64)	(0.12, 0.14, 0.74)
Spect (B)	(0.01, 0.04, 0.95)	(0.02, 0.07, 0.91)	(0.00, 0.00, 1.00)	(0.00, 0.00, 1.00)
Nguyen-7 (MW)	(150, −1.50, 0.15)	(120, −0.37, 0.71)	(99, −0.54, 0.59)	(79, −1.4, 0.17)
Pagie-1 (MW)	(140, −1.10, 0.28)	(160, −1.80, 0.07)	(160, −1.9, 0.06)	(170, −2.4, 0.02)
Vlad-4 (MW)	(120, −0.25, 0.80)	(110, 0.00, 1.00)	(210, −3.9, 0.00)	(170, −2.4, 0.02)

Table 7. The average number of evaluations until CGP converges for *Single* and *Multi-n*. The lower the better. The number of evaluations for the best model (i.e. the best fitness, cf. Table 5) is underlined; the lowest number indicates the fastest convergence is in bold.

Dataset	*Single*	*n* = 2	*n* = 3	*n* = 4	*n* = 5
Abalone	2084	**2043**	2244	<u>2239</u>	2230
Breast Cancer	1497	**1319**	1632	<u>2085</u>	1561
Credit	1553	2113	<u>1734</u>	2175	**1440**
Forest Fire	**1816**	1946	2437	<u>2462</u>	2264
Heart Disease	2174	**2042**	<u>2802</u>	3048	2805
Page Block	**846**	957	1044	<u>1194</u>	1041
Spect	**461**	891	724	<u>1353</u>	1508
Nguyen-7	108	<u>**21**</u>	64	38	20
Pagie-1	1866	2141	**1849**	2481	<u>1997</u>
Vlad-4	2149	**2052**	2494	<u>3309</u>	2527

Nonetheless, the convergence speed shows the same behavior as for the probabilistic approaches. This performance boost comes with a higher training time as most often CGP converges faster with a lower n value.

DMulti-n Afterwards, *DMulti-n* is utilized and n is decreased over time until $n = 1$. Interestingly, this does in fact not lead to better results because its fitness values and training times are very identical to the results in Table 5. This is why we did not list the results separately. Moreover, this implies that the later decreases of n do not improve the fitness values as all models converge before n is decreased. Hence it can be said that there is no need for smaller and more incremental changes in this setting. This leads to the assumption that there is no gain derived from decreasing n over time.

However, these findings are highly counter intuitive to the findings of *Multi-n* and its slight decrease in performance between $n = 4$ and $n = 5$ or $n = 3$ and $n = 5$. One may assume that, by decreasing n, CGP should be able to counter balance the negative effects of a starting value of $n = 5$. The reason is that there is a big portion of mutation steps utilizing lower n values. Thus, it should be able to balance the best results of *Multi-n* seen in Table 5. This may, in theory, come with the cost of having a lower convergence rate. Nevertheless, this is not the case here. Our theory is that CGP seems to get stuck in a local optimum after the first iterations. In later stages of training, decreasing the hyperparameter n apparently cannot significantly help to increase performance. Further research should be done in this direction to explore the reasoning behind this phenomenon.

7   Conclusion

In this work, we empirically evaluated the fitness values as well as the convergence speed for three new mutation algorithms. The mutation based on probability as well as *Single* was extended. At first, two different mutation rates for active and inactive nodes respectively were employed. On the utilized datasets, we found that it is favorable to utilize two mutation rates. However, while a higher rate for inactive genes is encouraged, the optimal probability differs between the problems. Albeit most of the times, it is better to utilize a mutation rate of 50% or higher.

Afterwards, we extended the *Single* Algorithm to *Multi-n* and *DMulti-n*. We found that, in these datasets, *Multi-n* always outperforms *Single* but its improvements decline for high values of n.

However, there is the caveat that all of these improvements in fitness value oftentimes comes with the downside of longer training time. Most of the time, a single mutation rate or *Single* need less evaluation steps until a solution was found.

As for future work, the introduced concepts could be merged and used inter-changeably. As Goldman et al. [6] found, *Single* performs better than the prob-abilistic approach when the best mutation rate is unknown. Both of these muta-tion strategies could be changed depending on the current training status and or fitness value.

Another possibility is to keep two mutation rates for active and inactive genes. In addition, these mutation rates could change with their position in the grid. In the work of Goldman and Punch [5], they found a high positional bias in CGP as well as challenged the status quo in terms of neutral search, bloat and the importance of genetic drift. By applying higher mutation rates for genes positioned in the back, a partial success could be achieved to reduce CGP's positional bias.

At last, it is unclear as to why *DMulti-n* does not show an advantage over *Multi-n* as well as why it does not improve when n is lowered. Further works might investigate into this phenomena, possibly leading into deeper understand-ing of CGP.

References

1. Baldi, P., Brunak, S., Chauvin, Y., Andersen, C.A.F., Nielsen, H.: Assessing the accuracy of prediction algorithms for classification: an overview. Bioinformatics **16**(5), 412–424 (2000). https://doi.org/10.1093/bioinformatics/16.5.412
2. Benavoli, A., Corani, G., Demšar, J., Zaffalon, M.: Time for a change: a tutorial for comparing multiple classifiers through Bayesian analysis. J. Mach. Learn. Res. **18**(1), 2653–2688 (2017)
3. Bentley, P.J., Lim, S.L.: Fault tolerant fusion of office sensor data using carte-sian genetic programming. In: 2017 IEEE Symposium Series on Computational Intelligence (SSCI), pp. 1–8. IEEE (2017)
4. Dua, D., Graff, C.: UCI machine learning repository (2017). https://archive.ics.uci.edu/ml
5. Goldman, B.W., Punch, W.F.: Length bias and search limitations in cartesian genetic programming. In: Proceedings of the 15th Annual Conference on Genetic and Evolutionary Computation, pp. 933–940. GECCO '13, Association for Com-puting Machinery, New York, NY, USA (2013)
6. Goldman, B.W., Punch, W.F.: Reducing Wasted Evaluations in Cartesian Genetic Programming. In: Krawiec, K., Moraglio, A., Hu, T., Etaner-Uyar, A.Ş, Hu, B. (eds.) EuroGP 2013. LNCS, vol. 7831, pp. 61–72. Springer, Heidelberg (2013). https://doi.org/10.1007/978-3-642-37207-0_6
7. Goldman, B.W., Punch, W.F.: Analysis of cartesian genetic programming's evolu-tionary mechanisms. IEEE Trans. Evol. Comput. **19**(3), 359–373 (2015)
8. Harding, S., Graziano, V., Leitner, J., Schmidhuber, J.: MT-CGP: mixed type cartesian genetic programming. In: Proceedings of the 14th Annual Conference on Genetic and Evolutionary Computation, pp. 751–758 (2012)
9. Harding, S.L., Miller, J.F., Banzhaf, W.: Self-modifying cartesian genetic program-ming. In: Miller, J. (ed.) Cartesian Genetic Programming, pp. 101–124. Springer, Heidelberg (2011). https://doi.org/10.1007/978-3-642-17310-3_4

10. Hodan, D., Mrazek, V., Vasicek, Z.: Semantically-oriented mutation operator in cartesian genetic programming for evolutionary circuit design. Genetic Programm. Evol. Mach. **22**(4), 539–572 (2021)

11. Husa, J., Kalkreuth, R.: A comparative study on crossover in cartesian genetic programming. In: Castelli, M., Sekanina, L., Zhang, M., Cagnoni, S., García-Sánchez, P. (eds.) Genetic Programming, pp. 203–219. Springer International Publishing, Cham (2018). https://doi.org/10.1007/978-3-319-77553-1_13

12. Kalkreuth, R.: Towards advanced phenotypic mutations in cartesian genetic programming. arXiv preprint arXiv:1803.06127 (2018)

13. Kalkreuth, R., Rudolph, G., Droschinsky, A.: A new subgraph crossover for cartesian genetic programming. In: McDermott, J., Castelli, M., Sekanina, L., Haasdijk, E., García-Sánchez, P. (eds.) Genetic Programming, pp. 294–310. Springer International Publishing, Cham (2017). https://doi.org/10.1007/978-3-319-55696-3_19

14. Kaufmann, P., Kalkreuth, R.: On the parameterization of cartesian genetic programming. In: 2020 IEEE Congress on Evolutionary Computation (CEC), pp. 1–8 (2020)

15. Leitner, J., Harding, S., Forster, A., Schmidhuber, J.: Mars terrain image classification using cartesian genetic programming. In: Proceedings of the 11th International Symposium on Artificial Intelligence, Robotics and Automation in Space, i-SAIRAS 2012, pp. 1–8. European Space Agency (ESA) (2012)

16. Margraf, A., Stein, A., Engstler, L., Geinitz, S., Hahner, J.: An evolutionary learning approach to self-configuring image pipelines in the context of carbon fiber fault detection. In: 2017 16th IEEE International Conference on Machine Learning and Applications (ICMLA), pp. 147–154. IEEE (2017)

17. Miller, J., Smith, S.: Redundancy and computational efficiency in cartesian genetic programming. Evol. Comput. IEEE Trans. **10**, 167–174 (2006)

18. Miller, J.F., Thomson, P.: Cartesian genetic programming. In: Poli, R., Banzhaf, W., Langdon, W.B., Miller, J., Nordin, P., Fogarty, T.C. (eds.) EuroGP 2000. LNCS, vol. 1802, pp. 121–132. Springer, Heidelberg (2000). https://doi.org/10.1007/978-3-540-46239-2_9

19. Miller, J., Thomson, P., Fogarty, T., Ntroduction, I.: Designing electronic circuits using evolutionary algorithms, arithmetic circuits: a case study. Genetic Algorithms Evol. Strateg. Eng Comput. Sci. (1999)

20. Miller, J.F.: An empirical study of the efficiency of learning Boolean functions using a cartesian genetic programming approach. In: Proceedings of the 1st Annual Conference on Genetic and Evolutionary Computation - Volume 2. pp. 1135–1142. GECCO'99, Morgan Kaufmann Publishers Inc., San Francisco, CA, USA (1999)

21. Miller, J.F.: Cartesian genetic programming: its status and future. Genetic Programm. Evol. Mach. **21**(1), 129–168 (2020)

22. Möller, F.J.D., Bernardino, H.S., Gonçalves, L.B., Soares, S.S.R.F.: A reinforcement learning based adaptive mutation for cartesian genetic programming applied to the design of combinational logic circuits. In: Cerri, R., Prati, R.C. (eds.) Intelligent Systems, pp. 18–32. Springer International Publishing, Cham (2020). https://doi.org/10.1007/978-3-030-61380-8_2

23. Turner, A.J., Miller, J.F.: Neutral genetic drift: an investigation using cartesian genetic programming. Genetic Programm. Evol. Mach. **16**(4), 531–558 (2015). https://doi.org/10.1007/s10710-015-9244-6

24. White, D., et al.: Better GP benchmarks: community survey results and proposals. Genetic Programm. Evol. Mach. **14**, 3–29 (2013)

25. Wilson, D.G., Miller, J.F., Cussat-Blanc, S., Luga, H.: Positional cartesian genetic programming. arXiv preprint arXiv:1810.04119 (2018)
26. Yu, T., Miller, J.: Neutrality and the evolvability of Boolean function landscape. In: Miller, J., Tomassini, M., Lanzi, P.L., Ryan, C., Tettamanzi, A.G.B., Langdon, W.B. (eds.) EuroGP 2001. LNCS, vol. 2038, pp. 204–217. Springer, Heidelberg (2001). https://doi.org/10.1007/3-540-45355-5_16

Slime Mould Algorithm: An Experimental Study of Nature-Inspired Optimiser

Petr Bujok$^{(\boxtimes)}$ ⓘ and Martin Lacko

University of Ostrava, 30. dubna 22, 70200 Ostrava, Czech Republic
petr.bujok@osu.cz

Abstract. In this paper, the performance of the slime mould algorithms (SMA) is studied. The original SMA algorithm is enhanced by several mechanisms to achieve better results in various problems. The Eigen transformation, linear reduction of population size, and perturbation of the solution are proposed and combined together with various settings of control parameters. All 16 newly proposed variants of SMA are compared with the original SMA and 16 various nature-inspired methods. All the algorithms are applied to 22 real-world problems called CEC 2011. Achieved results illustrate the good performance of the newly proposed SMA variants, especially compared with the original SMA algorithm.

Keywords: Global optimisation · Slime mould algorithm · Swarm-intelligence · Experiment · Real-world problems

1 Introduction

Optimisation techniques are divided into several groups based on the principle of how to evaluate individuals. One of the most frequently used groups of these methods is called swarm-intelligence (SI) algorithms inspired by the behaviour of the swarm-individuals in nature. These methods use stochastic or learning approaches, which bring a successful solution to even hard tasks in an acceptable time. There have been many representatives of SI algorithms presented over two past decades – particle swarm optimisation [9], bat-inspired algorithm [26], firefly algorithm [27], grey wolf optimiser [15], etc. Many of these techniques are successfully applied to real-world problems to optimise various systems of the human environment.

Generally, each optimisation problem is represented by the objective function $f(\boldsymbol{x})$, $\boldsymbol{x} = (x_1, x_2, \ldots, x_D) \in \mathrm{I\!R}^D$ which is defined on the search space Ω bounded by $\Omega = \prod_{j=1}^{D}[a_j, b_j]$, $a_j < b_j$, $j = 1, 2, \ldots, D$. Then the point $\boldsymbol{x}^*$ is called a global minimum if it satisfies $f(\boldsymbol{x}^*) \leq f(\boldsymbol{x}), \forall \boldsymbol{x} \in \Omega$.

The goal of the paper is to enhance the slime mould algorithm by several various approaches and settings of the control parameters. The original variant

M. Mernik et al. (Eds.): BIOMA 2022, LNCS 13627, pp. 201–215, 2022.
https://doi.org/10.1007/978-3-031-21094-5_15

of the algorithm provides interesting results on several test problems and the application of the algorithm to several real-world problems also illustrates its efficiency. Newly proposed variants of the slime mould algorithm are applied to a set of 22 real-world problems called CEC 2011 to show how the enhancing elements truly help the algorithm in practice.

The rest of the paper is organised as follows. A short description of the slime mould algorithm is in Sect. 1.1. Newly proposed variants of the slime mould algorithm are introduced in Sects. 2.1–2.4. Other optimisation methods used in the comparison are briefly assumed in Sect. 3. Settings of the experiment and results of the comparison are discussed in Sects. 4 and 5. The conclusions are made in Sect. 6.

1.1 Slime Mould Algorithm

In 2020, Li et al. proposed a new method for stochastic optimisation called the Slime Mould Algorithm (SMA) [12]. The authors modelled the behaviour and morphological changes of slime mould when collecting food. The model does not focus on other parts of the slime mould's life cycle. More precisely, the SMA represents the model of Physarum polycephalum which was first classified as slime mould [8]. The authors of MSA comprehensively described the slime mould behaviour when living and especially foraging. A previous analysis clearly shows that faster decision of slime mould about the food source means lower probability to achieve a prime source [12]. In other words, slime mould needs more appropriate settings of speed and accuracy when leaving one area and finding another food area.

The slime mould individuals allocated the food sources based on the smell in the air. The process of approaching the food of slime mould is modelled by:

$$x_i = \begin{cases} x_b + vb \cdot (W \cdot x_A - x_B), & \text{if } r < p \\ vc \cdot x_i & \text{if } r \geq p. \end{cases} \tag{1}$$

where elements of the tuning vector vb randomly oscillate in $[-aa, aa]$ (3), vc is linearly decreasing from 1 to 0, vb,vc are used for the tuning process during iterations. Further, x_b is the best individual, which is the location with the current highest concentration of the food. Position of x_i is for the current individual, W is the slime mould weight, x_A and x_B represent two randomly selected slime mould individuals, r is a uniformly sampled random number from $[0, 1]$, and the parameter p is updated by:

$$p = tanh|S_i - DF| \tag{2}$$

where S_i is the function value of ith slime mould and DF the function value of the best slime mould in the algorithm's history.

$$aa = arctanh(-(\frac{FES}{maxFES}) + 1) \tag{3}$$

FES and $maxFES$ denote the current and the total function evaluations for the run. The weight of slime mould is updated by:

$$W(SmellIndex(i)) = \begin{cases} 1 + r \cdot \log(\frac{bF-S(i)}{bF-wF} + 1), & condition \\ 1 - r \cdot \log(\frac{bF-S(i)}{bF-wF} + 1) & others. \end{cases} \quad (4)$$

where $SmellIndex = sort(S)$ is the sequence of the population function evaluations sorted in ascending, the $condition$ determines that $S(i)$ is in the first half of the sorted population and bF is the best and wF is the worst function evaluation of the population.

After approaching the food, the contracting mode is performed using:

$$X^* = \begin{cases} rand \cdot (b-a) + a, & if\ rand < z \\ x_b + vb \cdot (W \cdot x_A - x_B), & if\ r < p \\ vc \cdot x_i & if\ r \geq p. \end{cases} \quad (5)$$

W oscillates with respect to the quality of the food-smell. When the smell is higher, W is increased, otherwise decreased. It means that slime mould approaches the food faster in case the quality of food is higher and if the quality of food is smaller, slime mould approaches the food more slowly. Where r is a randomly selected number from uniform interval $[0, 1]$. The parameters of z and p enable us to reach better adaptability of SMA in different search phases. The parameter z is the only input of the SMA (besides population size). The pseudo-code of the SMA algorithm is in Algorithm 1.

Algorithm 1. Pseudo-code of the SMA

---

    initialise population $X = \{x_1, x_2, \ldots, x_N\}$
    while stopping condition not reached **do**
        update p, vb, vc
        update W by (4)
        for $i = 1, 2, \ldots, N$ **do**
            update positions x_i by (5)
            evaluate $f(y)$
        end for
        increase FES
        update x_b
    end while

1.2   Previous Works

The SMA algorithm was proposed in 2020, and many scientists proposed some real applications or enhanced variants of this optimisation method. In 2020, Kumar et al. applied the SMA to parameters of a photovoltaic system [11]. In 2021, Rizk-Allah et al. proposed a chaos-opposition-based variant of the SMA algorithm to achieve optimal energy cost for a wind turbine [19]. In 2020, Zubaidi et al. proposed hybridised neural network that cooperated with the SMA algorithm to optimise the urban water demand [31]. In 2020, Abdel-Basset et al. proposed a hybrid SMA algorithm that cooperated with the whale optimisation algorithm applied to chest x-ray image segmentation [1]. In 2021, Liang et al. proposed an enhanced variant of the SMA algorithm for the design of the IIR filter [13]. In 2022, Ornek et al. proposed a novel SMA for real-world engineering problems [16]. Here, a sine and cosine oscillation mode help to perturb the individuals' positions. The proposed ESMA provides the best results compared to the original SMA and several evolution algorithms. In 2022, Yin et al. proposed Equilibrium Optimiser SMA (EOSMA) [28]. This approach was used to solve benchmark and also engineering problems, where EOSMA achieved very promising results (regarding quality and time-complexity) compared with the original SMA and various evolutionary algorithms. In 2022, Zhu proposed SMA for modelling the load dispatch issue in an electric power system [30]. The proposed SMA with proposed changeable weights provided the best results out of five various optimisation methods. The table below lists the mentioned variants of SMA and their application.

Table 1. Review of SMA variants and application area.

SMA variants	Application area	Method
Boosting SMA [9]	For the proposal of parameters of a photovoltaic system	Nelder-Mead Simplex strategy, Chaotic map
Chaos-opposition-based [16]	For achieving optimal energy cost for a wind turbine	Chaotic mapping
Neural network model with SMA [26]	For prediction of urban stochastic water demand	Using SMA for update neuron-network model
HSMA WOA [1]	To chest x-ray image segmentation	Whale optimisation
Enhanced SMA [11]	For the design of the IIRfilter	Learning strategy, Chaotic initialisation strategy,
ESMA [16]	Four engineering problems	Oscillation of SMA position
EOSMA [28]	Nine engineering problems	Equilibrium strategy
SMA [30]	Load dispatch issue in an electric power system	Changeable weights

2   Newly Proposed Variants of SMA

In this paper, 16 newly proposed variants of the SMA algorithm are proposed using several settings of control parameters and enhancing mechanisms. These settings and mechanisms are used separately or combined with another mechanism to achieve better results.

2.1   Linear Reduction of the Population Size

A well-known mechanism to reduce the population size linearly was proposed for successful L-SHADE [21]. At the end of each generation, the proper population size is computed (6), and in the case of a lower value compared to N it is reduced:

$$N = round[(\frac{N_{\min} - N_{\text{init}}}{maxFES})FES + N_{\text{init}}], \qquad (6)$$

where FES indicates the current number of function evaluations, N_{init} is the initial population size, $N_{\min}$ represents the final population size at the end of the search process (allowed by number of $maxFES$ function evaluations). The SMA variants using this enhancing mechanism are denoted by the letter 'L' (linear). This approach enables reducing the population size by one. To achieve faster reduction to the proper population size, all excess individuals are removed from the population. This approach is labelled as 'LF' (linear fast).

2.2   Eigen Transformation

In 2014, Wang et al. introduced the Eigen transformation approach introduced for CoBiDE [25]. In the beginning of each generation, Eigenvalues (matrix D) and Eigenvectors (matrix B) are computed from the covariance matrix (C) computed from a ps part of a better individuals of population:

$$C = BD^2B^T. \qquad (7)$$

After applying the Eigen transformation, a new solution is produced in an Eigen coordinate system for several elements:

$$x_b^i = B^{-1}x_b = B^Tx_b, \qquad (8)$$

where the same approach is applied to vb, x_A, and x_B. A new position is allocated using (5), x_i^i which is transformed back into a standard coordinate system:

$$x_i = Bx_i^i. \qquad (9)$$

This approach is used for the whole generation if the randomly generated number is lower than the control parameter $peig$, which controls the frequency of the Eigen transformation. The SMA variants using this approach use label 'E' with four numbers denoting settings of ps and $peig$ (for example SMA$_{E0504}$ for $ps= 0.5$ and $peig= 0.4$).

2.3   Perturbation

Random perturbation of several elements of an individual before evaluation is proposed. When a new position is achieved, one or two randomly selected positions in the vector of solution are selected and replaced by reinitialised values from the search space. The SMA variant using this approach only in one coordinate is denoted by 'R1' (one random element), and the variant using one or two (in the case of one-dimensional problems) dimensions are labelled 'R1-2'.

2.4   Adaptation of Parameter z

The last enhancing approach uses a simple adaptation of the main control parameter used in SMA - z. The value of this parameter is set based on the cosine function from 0.05 at the beginning, then 0.01 in the middle of the run, and 0.05 at the end of the SMA run.

3   Methods Used in Experiments

In this paper, 16 various nature-inspired optimisation methods were selected to be compared with 17 variants of the SMA algorithm (including the original variant). A brief list of the involved algorithms follows.

Particle swarm optimisation (PSO) is the most popular swarm-intelligence-inspired optimiser proposed in 1995. Here, the value of the parameter balancing between the local and the global part of the updated velocity is set $c = 1.05$, and for next-generation, Velocity is computed as $v_{i,G+1} = w_{G+1} \times v_{i,G} + c\,U(0,1)\,(p_{best} - x_i) + c\,U(0,1)\,(g_{best} - x_i)$. Moreover, an advanced cooperative variant of PSO with the Firefly algorithm (called HFPSO) is used [2]. The parameters of HFPSO are set to the recommended values at $\alpha = 0.2$, $\beta_0 = 2$, $\gamma = 1$, and $c_1 = c_2 = 1.49445$.

The self-organising migration algorithm (SOMA) is a model of a pack of predators hunting the prey [29]. The parameters are set to $PathLenght = 2$, $Step = 0.11$, and $Prt = 0.1$. The strategy all-to-one is used in the comparison.

In 2014, the dispersive flies optimisation algorithm (DFO hereafter) was proposed [18]. The only control parameter is the disturbance threshold $dt = 1 \times 10^{-3}$.

The firefly algorithm (FFL) models the real fireflies [27]. The control parameters are set to $\alpha = 0.5$, $\gamma = 1$, and $\beta_0 = 1$, $\beta_{min} = 0.2$.

The grey wolf optimiser (GWO) represents the hunting and hierarchic behaviour of grey wolves [15]. The only control parameter of GWO is component a, and it decreases linearly from 2 to 0.

Monarch butterflies optimisation (MBO) [23] is applied with settings at $keep = 2$, $MaxStepSize = 1$, $period = 1.2$, and $part = 5/12$.

The tree-seed algorithm (TSA) models the relationship between the seeds and the trees in nature [10]. The only control parameter of this algorithm is set recommended value $TS = 0.1$.

The elephant herd optimisation algorithm (EHO) inspired by the hierarchical behaviour of elephants in an elephant herd [24] is used with values - elitism parameter is 2, the number of clans is 5, and $\alpha = 0.5$ and $\beta = 0.1$.

The bacterial foraging optimisation algorithm (BFO) [6] is inspired by the chemotaxis of *E. coli* bacteria. The recommended settings are used - number of chemotactic steps $N_c = 20$, swim steps $N_s = 10$, reproductive steps $N_r = 20$, probability of elimination $P_{ed} = 0.9$, and step-size $C = 0.01$.

The bat algorithm (BAT) [26] uses settings - $f_{max} = 2$, $f_{min} = 0$, $A_i = 1.2$ $\alpha = 0.9$, $r_i = 0.1$, and $\gamma = 0.9$.

The invasive weed optimisation algorithm (IWO) models the spreading technique of weeds [14]. The recommended settings are used - $s_{max} = 5$, $s_{min} = 0$, $\sigma_{init} = 0.5$, $\sigma_{fin} = 0.001$, and $\eta = 2$.

The biogeography-based optimisation algorithm (BGO) is inspired by the geographical distribution of biological organisms [20]. The control parameters of BGO were set to the recommended values - $KeepRate = 0.2$, emigration rate is a vector of length N (population size) with linearly decreasing values from 1 to 0, immigration rate is a vector of length N with linearly increasing values from 0 to 1, and the probability of mutation is 0.1.

The Vortex Search Algorithm (VSA) is inspired by the vortex flow during the stirring of the fluids [7]. The initial radius circle r is based on an inverse incomplete gamma function controlled by the recommended value $x = 0.1$.

The Sonar Inspired Optimisation algorithm (SIO) employs acoustics to allocate the solution [22]. The only input parameter of the SIO algorithm is $I_0 = 1 \times 10^{-12}$. The simplest stochastic optimisation method is random blind search (RS) [17]. This method has no control parameters.

4   Experimental Settings

In this comparison, a test suite of 22 real-world problems of the CEC 2011 competition in the Special Session on Real-Parameter Numerical optimisation is used [5]. The functions differ in complexity and dimensionality (from $D = 1$ to $D = 240$). For each algorithm and problem, 25 independent runs were performed to achieve statistical significance. The run of the algorithm is stopped when it achieves the prescribed number of function evaluations $MaxFES = 150000$. Moreover, partial results after one-third and two-thirds of $MaxFES$ are also analysed.

Each of the algorithms (except for RS) employs the parameter of the population size. The parameter for each algorithm was set to the best possible values recommended by the authors or achieved in previous experiments [3,4]. Therefore, $N = 50$ is for TSA, VSA, MBO, EHO, and IWO. $N = 30$ is for GWO, HFPSO, BFO, and SIO. $N = 90$ is for SOMA and FFL. $N = 40$ is for BAT, FFL, PSO, BGO, and DFO. The original SMA algorithm uses $N = 30$. In this experiment, two bigger population size values are used $N = 90$ and $N = 180$. The variants of SMA using the population size reduction mechanism starts with a population of size $N_{init} = 18 \times D$. This causes population size 18 for the problems with $D = 1$. Therefore, the initial population size was also increased by a simple mechanism to values from 18×10 for the problems with dimension $D < 10$ (these variants are denoted 'B' as bigger initial population size).

Moreover, the SMA uses the control parameter of $z = 0.03$. In this study, the value is also set to $z = 0.1$, $z = 0.3$, and $z = 0.9$. These variants of SMA are labelled by a combination of the population size and the value of z (with a dash, the original setting is SMA_{30-003}).

Parameters of the Eigen transformation used in the SMA algorithms are set to $ps = 0.5$ and $peig = 0.4$ or $peig = 0.1$.

Table 2. Overview of the SMA variants in comparison.

Algorithm	Description
SMA_{30-003}	Original SMA, $N = 30$, $z = 0.03$
SMA_{30-01}	Original SMA, $N = 30$, $z = 0.1$
SMA_{30-03}	Original SMA, $N = 30$, $z = 0.3$
SMA_{30-09}	Original SMA, $N = 30$, $z = 0.9$
SMA_{90-01}	Original SMA, $N = 90$, $z = 0.1$
SMA_L	Linear population size mechanism (L)
SMA_{E0504}	Eigen transformation (E), $ps = 0.5$, $peig = 0.4$
SMA_{E0501}	Eigen transformation, $ps = 0.5$, $peig = 0.1$
SMA_{EL0501}	L, E, $ps = 0.5$, $peig = 0.1$
$SMA_{ELF0501}$	L, faster decreation (F), E, $ps = 0.5$, $peig = 0.1$
$SMA_{ELBF0501}$	L, F, bigger initial pop.size, E, $ps = 0.5$, $peig = 0.1$
$SMA_{ELBF0504}$	L, F, bigger initial pop.size, E, $ps = 0.5$, $peig = 0.4$
SMA_{R1}	One element in the SMA position is changed (R)
SMA_{R1-2}	One or two elements in the SMA position are changed
$SMA_{RELF0501}$	R, L, F, E, $ps = 0.5$, $peig = 0.1$
SMA_Z	Value of z is adapted by cosine function

The remaining parameters of the algorithms are set to the values recommended by the authors. All the algorithms are implemented in Matlab 2020b, where the statistical analysis is assessed as well. All computations were carried out on a standard PC with Windows 7, Intel(R) Core(TM)i7-4700 CPU 3.0 GHz, 16 GB RAM.

5   Results

In this comparison, 17 variants of the SMA algorithm are proposed and compared with 16 various nature-inspired methods. All 33 algorithms are applied to a set of 22 CEC 2011 test problems and the results are statistically analysed using the significance level $\alpha = 0.05$. To produce a good insight into the algorithm's performance, first, the Friedman test is applied. The test is applied on medians achieved from minimum values at three stages of the search ($FES = 50,000$, $100,000$, and $150,000$). The results are presented in Table 3 where also the absolute ranks are depicted in brackets (the heading of the table is only on the left side). The null hypothesis on the equivalent performance of the algorithms is rejected with $p < 5 \times 10^{-10}$. The algorithms are ordered in the tables based on the mean ranks from the Friedman test at the end of the search (column 'stage3') and on the average ranks of the overall dimensions.

The newly proposed $SMA_{ELBF0504}$ achieves the best overall performance in the experiment regarding all 22 problems. This variant of the SMA uses the

Table 3. Mean ranks and absolute ranks of all algorithms from the Friedman tests.

Alg.	Stage1	Stage2	Stage3	Alg.	Stage1	Stage2	Stage3
$SMA_{ELBF0504}$	9.4 (7)	**8.7 (2)**	**7.0 (1)**	$SMA_{180-003}$	10.5 (10)	12.4 (13)	13.8 (17)
$SMA_{ELF0501}$	8.7 (4)	9.5 (6)	**8.2 (2)**	VSA	22.4 (24)	12.9 (16)	14.1 (18)
$SMA_{RELF0501}$	14.5 (15)	11.6 (11)	**8.4 (3)**	SOMA	**7.7 (3)**	12.5 (14)	15.5 (19)
SMA_{E0504}	11.0 (11)	**8.9 (3)**	8.9 (4)	SMA_{30-03}	21.6 (23)	19.6 (23)	15.6 (20)
$SMA_{ELBF0501}$	9.5 (8)	10.4 (8)	9.9 (5)	PSO	14.8 (16)	15.9 (19)	18.5 (21)
SMA_{E0501}	8.8 (5)	9.2 (4)	10.0 (6)	BGO	9.5 (9)	14.5 (17)	18.8 (22)
SMA_{R1-2}	16.2 (17)	12.2 (12)	10.1 (7)	TSA	18.1 (19)	19.0 (21)	19.4 (23)
SMA_Z	**6.9 (2)**	**7.0 (1)**	10.4 (8)	SMA_{30-09}	25.4 (26)	24.3 (25)	21.9 (24)
SMA_{EL0501}	12.0 (12)	11.3 (10)	11.6 (9)	MBO	20.6 (22)	23.9 (24)	24.7 (25)
SMA_{R1}	14.2 (14)	10.9 (9)	11.6 (10)	EHO	23.6 (25)	25.1 (26)	25.2 (26)
SMA_L	12.9 (13)	12.6 (15)	12.3 (11)	RS	28.4 (32)	28.5 (31)	27.3 (27)
SMA_{30-003}	9.2 (6)	10.1 (7)	12.7 (12)	DFO	26.9 (29)	27.1 (27)	27.5 (28)
SMA_{90-01}	18.4 (20)	15.9 (20)	13.0 (13)	SIO	27.3 (30)	27.5 (28)	27.5 (29)
HFPSO	**6.5 (1)**	9.4 (5)	13.0 (14)	BAT	26.7 (28)	28.5 (30)	28.8 (30)
SMA_{30-01}	16.2 (18)	14.7 (18)	13.1 (15)	FFL	25.8 (27)	28.2 (29)	29.0 (31)
GWO	19.1 (21)	19.1 (22)	13.6 (16)	BFO	28.3 (31)	29.1 (32)	29.4 (32)
				IWO	30.0 (33)	30.3 (33)	30.3 (33)

Eigen transformation with $ps = 0.5$ and $peig = 0.4$, linearly reducing population size from a bigger initial value (from 180 to $18 \times D$), and it uses a faster reduction of the population size to achieve the proper size (not only by one individual).

Moreover, it is obvious that the first 11 positions in the ordered table are occupied by the proposed variants of SMA (the 12th position is for the original SMA).

The best results after one-third of the run are provided by the SOMA algorithm, whereas after two-thirds, it is the proposed SMA variant with an adaptation of z (SMA_z).

More details of the algorithm's comparison of each real-world problem independently are provided by the Kruskal-Wallis test. The null hypothesis is rejected on the significance level 1×10^{-10}. The median values of the methods are provided in Tables 4 and 5. In these tables, the best achieved median value for each problem is printed in bold. In the case of problems $T03$ and $T08$, there is no best-performing method.

For a better overview, there are numbers of the first, second, third, and last positions for each algorithm (delimited by a slash), based on the mean ranks. We can see that the best algorithm in the comparison is HFPSO which achieves the best rank in four problems, and the second is SOMA with the two best mean ranks. SMA_{E0504}, SMA_L, and SMA_{R1} provide potentially good performance. It is obvious that the number of the best mean ranks is not the same as the number of the least median values of the algorithms.

Having compared all 33 algorithms, the 16 proposed SMA variants are compared with the original SMA_{30-003} using the Wilcoxon rank-sum test. This comparison is performed in order to show how the enhancing mechanisms increase the performance of the SMA algorithm. Because of space, the results are assumed in Table 7. In this table, the numbers of better (B), similar (S), and worse (W) results achieved for each problem independently are illustrated. Moreover, a

Table 4. Medians and numbers of the first, second, third, and last positions from the Kruskal-Wallis tests.

F	BAT	BFO	BGO	DFO	EHO	FFL	GWO
T01	29.9997	23.0459	19.5091	27.3572	25.3804	28.4112	11.6669
T02	−2.46801	−3.13836	−15.6066	−6.55608	−12.7726	−12.6795	−23.807
T03	1.15E−05	1.15E−05	1.15E−05	1.15E−05	1.15E−05	1.15E−05	1.15E−05
T04	13.9343	21.0146	20.9722	14.6533	14.3221	19.2793	13.8308
T05	−11.4317	−20.2997	−31.7313	−9.82837	−18.9533	−18.2279	**−34.1542**
T06	−7.51923	−16.4406	−19.5118	−7.03242	−10.2898	−14.7311	−23.0059
T07	2.52058	1.48588	1.13308	2.57839	1.82214	2.37051	**0.772**
T08	310.916	1444.23	230	282	220	350.555	220
T09	997647	1.67E+06	8452.52	464478	1.27E+06	3.18E+06	24326.4
T10	−7.78773	−10.2486	−15.8281	−8.83998	−8.35503	−7.59723	−12.9574
T11.1	1.91E+06	1.10E+07	65577	9.73E+07	8.56E+06	2.50E+06	474417
T11.2	1.34E+07	1.59E+07	1.11E+06	5.75E+06	5.94E+06	7.33E+06	1.12E+06
T11.3	97335.1	146284	15459.4	18083.8	15492.6	52975.7	15463.6
T11.4	19385.8	19443	19389.7	19161.6	19138.9	19243	19210.7
T11.5	2.25E+06	8.59E+06	33095.2	251810	164692	495711	32993
T11.6	150956	9.61E+07	140684	164780	1.30E+07	7.78E+07	136626
T11.7	1.05E+10	2.32E+10	1.96E+06	1.28E+10	4.68E+09	1.48E+10	2.49E+06
T11.8	7.16E+07	1.23E+08	1018940	8366530	64553100	1.47E+08	956785
T11.9	5.93E+07	1.36E+08	1.35E+06	1.25E+07	6.57E+07	1.58E+08	1.30E+07
T11.10	6.72E+07	1.24E+08	998405	1.14E+07	6.48E+07	1.48E+08	957703
T12	70.1237	59.1886	18.7061	75.8614	41.1322	46.1224	24.3415
T13	70.2957	53.7055	24.5615	58.8164	39.6229	39.8236	21.7653
#ranks	0/0/0/3	0/0/0/4	0/0/0/0	0/0/0/4	0/0/1/0	0/0/0/1	1/1/2/0

F	HFPSO	IWO	MBO	PSO	RS	SIO	SMA$_{180-003}$
T01	14.5269	24.7506	15.5475	12.8765	21.9619	23.9607	12.3061
T02	**−26.1267**	−3.02094	−13.176	−15.3444	−4.00667	−5.63485	−17.5827
T03	1.15E−05	1.15E−05	1.15E−05	1.15E−05	1.15E−05	1.15E−05	1.15E−05
T04	14.3291	14.2251	14.3291	14.3291	14.0724	14.2853	14.3291
T05	−33.6333	−18.1002	−21.5977	−31.2519	−18.4336	−18.205	−34.107
T06	−23.0059	−15.5533	−15.9741	−22.8628	−12.8843	−12.307	−23.0059
T07	0.87531	1.77031	1.80721	1.45337	1.75981	1.51637	0.97916
T08	220	262	220	220	220	238	220
T09	18356.8	2.02E+06	1.90E+06	429738	2.57E+06	2.52E+06	3272.39
T10	−20.4544	−3.18672	−11.6599	−15.917	−8.61655	−8.05445	−20.9435
T11.1	**52296.4**	3.50E+08	8.38E+06	1.47E+06	2.12E+08	2.67E+06	53243.3
T11.2	1.08E+06	1.55E+07	1.10E+07	4.78E+06	1.24E+07	7.93E+06	1.08E+06
T11.3	15480	229271	15463.4	15457.2	15744.8	15660.7	15464
T11.4	19236	826388	19346.3	**18860.7**	19409	19574.6	19222.9
T11.5	33003	1.02E+07	33220	33134.5	178465	171324	33066.3
T11.6	144160	9.51E+07	1.03E+07	140281	3.25E+06	699476	141240
T11.7	**1.95E+06**	2.50E+10	7.32E+09	2.28E+06	7.37E+09	5.52E+09	1.97E+06
T11.8	955539	1.50E+08	4.63E+07	955368	9.00E+07	8.84E+07	950468
T11.9	**1.17E+06**	1.65E+08	5.02E+07	1.44E+06	9.37E+07	9.40E+07	1.48E+06
T11.10	951985	1.67E+08	4.93E+07	1.04E+06	9.28E+07	8.76E+07	951090
T12	14.9149	54.4457	29.212	21.1637	40.235	37.9648	16.437
T13	24.5003	43.3834	37.7085	25.3219	39.9143	42.0761	20.3451
#ranks	4/0/1/0	0/0/0/9	0/0/0/0	1/0/0/0	0/0/0/0	0/0/0/0	1/0/0/0

Table 5. Medians and numbers of the first, second, third, and last positions from the Kruskal-Wallis tests.

F	SMA_{30-003}	SMA_{30-01}	SMA_{30-03}	SMA_{30-09}	SMA_{90-01}	$SMA_{ELBF0501}$
T01	11.8063	11.7076	11.7077	15.8266	11.7076	12.5458
T02	−21.5109	−21.2454	−16.513	−9.23923	−20.86	−20.0322
T03	1.15E−05	1.15E−05	1.15E−05	1.15E−05	1.15E−05	1.15E−05
T04	13.7715	13.7721	13.7743	13.9269	13.7723	13.7709
T05	−34.1075	−31.6959	−31.5647	−26.2346	−34.1071	−31.8164
T06	−26.5001	−23.0059	−23.0058	−19.5105	−26.5	−27.4297
T07	0.9595	0.95364	0.93268	1.09094	0.93718	0.898908
T08	220	220	220	220	220	220
T09	5308.43	17890.4	78319.7	242346	18738.2	3800.93
T10	−21.0726	−20.7995	−21.1297	−17.6703	−20.844	−21.1787
T11.1	53090.2	53672.6	56177.5	4.42E+06	54509.5	52771
T11.2	1.10E+06	1.12E+06	1.11E+06	1.46E+06	1.08E+06	1114470
T11.3	15465.3	15465.9	15466.8	15482.1	15464.5	15459.8
T11.4	19176.7	19136	19127.3	19339.3	19152.1	19238.5
T11.5	33007.5	32982.2	33052.1	33131.5	33026.3	33006.1
T11.6	142094	141092	141602	144746	138739	138871
T11.7	2.00E+06	1.97E+06	1.97E+06	2.80E+06	1.98E+06	1963690
T11.8	952147	950708	955232	1.04E+06	956158	**947945**
T11.9	1.48E+06	1.55E+06	1.52E+06	1.76E+06	1.49E+06	1438560
T11.10	950990	951167	954447	1002460	954262	**948665**
T12	16.5118	16.5185	17.6693	21.304	15.392	16.0985
T13	20.6651	19.4512	22.4152	22.8457	21.3559	20.6505
#ranks	0/0/0/0	0/0/0/0	0/0/0/0	0/0/0/0	0/0/1/0	0/0/0/0

SMA_{E0501}	SMA_{E0504}	SMA_{EL0501}	$SMA_{ELF0501}$	SMA_L	SMA_{R1}	$SMA_{ELBF0504}$
11.7567	11.3342	11.2072	11.2072	12.3061	**0.00069**	11.3805
−24.1084	−20.9548	−20.1813	−22.9585	−21.046	−23.0163	−22.1525
1.15E−05	1.15E−05	1.15E−05	1.15E−05	1.15E−05	1.15E−05	1.15E−05
13.7719	13.7714	13.7709	**13.7708**	13.7709	13.7749	13.771
−34.1064	−34.107	−34.1076	−34.1075	−34.1076	−34.1067	−34.1075
−27.4294	−23.0059	−27.4297	−27.4297	−27.4297	−27.4293	−23.0059
0.93762	0.92161	0.958043	0.995723	0.967852	0.985047	0.868851
220	220	220	220	220	220	220
3599.94	3400.98	35172.9	3463.97	49825	2791.99	3240.45
−21.0732	−21.1498	−21.169	**−21.2479**	−21.1566	−20.8572	−21.1738
53374.7	53072.9	681414	52727.6	510737	53841.2	53015.2
1.08E+06	**1.08E+06**	2501720	1122970	2512590	1080630	1157170
15459.3	**15451.2**	15459.8	15461.5	15469.5	15459.9	15457.8
19179.7	19202.6	19202.2	19204.2	19240.9	19257.9	19113.2
32985.4	32929.9	32991.2	32964.8	33039.7	32958.4	32987
140977	139251	140698	139627	141530	139240	138292
1.97E+06	1.96E+06	1959870	1955450	1961640	2005940	1954530
950450	950903	951100	948097	948755	950669	949000
1.54E+06	1.59E+06	1493920	1486720	1418850	1600570	1453950
948903	951417	953066	949195	949036	951877	949066
15.6797	15.3917	15.6844	16.0799	16.3311	15.1139	15.384
20.2318	19.8552	20.7516	**18.7242**	19.3124	21.6121	19.1586
0/1/1/0	1/3/1/0	0/1/0/0	0/0/0/0	1/0/3/0	1/1/0/0	0/0/0/0

number of problems where the proposed variant performs significantly better (row denoted 'sig.'), similar, and worse (based on the Wilcoxon tests) are provided, too. For a better overview, the algorithms are ordered from the top left (best algorithm) to the bottom right (worst algorithm).

We can see that the best performing method (compared with the original SMA_{30-003}) is $SMA_{RELF0501}$ which outperforms the original algorithm in 16 problems, and significantly in six problems (it is worse in four, and one significantly). Very similar performance is provided by $SMA_{ELF0501}$ (the only difference is the missing perturbation mechanism) and SMA_{E0504}. All newly proposed methods using some of the enhanced mechanisms perform rather better compared with the original SMA algorithm. Two variants performing rather worse are SMA_{EL0501} and SMA_L.

Table 6. Medians and numbers of the first, second, third, and last positions from the Kruskal-Wallis tests.

F	SMA_{R12}	$SMA_{RELF0501}$	SMA_Z	SOMA	TSA	VSA
T01	11.2089	11.3343	11.3812	11.7076	1.53821	17.9793
T02	−22.6416	−22.006	−25.4959	−18.7298	−−5.16811	−10.1061
T03	1.15E−05	1.15E−05	1.15E−05	1.15E−05	1.15E−05	1.15E−05
T04	13.7721	13.8154	13.7726	14.3291	13.9016	**13.7708**
T05	−34.107	−34.1076	−34.1073	−32.3813	−20.8928	−26.5547
T06	−27.4293	**−29.1658**	−27.4294	−27.4297	−17.6841	−21.2481
T07	0.986185	0.953269	0.978724	1.11902	1.69353	0.848128
T08	220	220	220	220	220	220
T09	2839.38	**2771.84**	3545.74	221914	142246	26221.8
T10	−21.0291	−21.0599	−21.1223	−16.8617	−18.6144	−12.3067
T11.1	54634.5	53070.4	53347.6	1273590	84746800	53998.3
T11.2	1079220	1133150	1077930	3644920	5636820	1090110
T11.3	15456.3	15456.8	15463.1	15461.2	15460.2	15462.2
T11.4	19199.2	19192.3	19253.4	19219.6	19214.7	19134.2
T11.5	32996.3	32936.9	33017.1	**32863.6**	32947	33033.7
T11.6	138085	139193	140742	**133372**	136988	136941
T11.7	1969490	1986960	1957730	1978580	3118680	1955100
T11.8	953259	949153	951727	980383	2659120	948646
T11.9	1393940	1468450	1439210	1318600	3433670	1350050
T11.10	952516	949249	949768	988279	2645350	951881
T12	15.5863	**14.5627**	15.641	18.0423	28.3102	21.2392
T13	20.3867	19.4555	19.2042	22.8158	26.1825	30.3203
#ranks	0/0/0/0	0/0/0/0	0/2/2/0	2/0/1/0	1/1/0/0	1/1/0/0

These results enable us to state that a simple linear reduction of the population size (by one individual) provides a rather worse performance.

Table 7. Number of wins, equalities, and losses from the Wilcoxon tests.

SMA_{30-003} vs	$SMA_{RELF0501}$	$SMA_{ELF0501}$	SMA_{E0504}	SMA_{R1}
B/S/W	16/2/4	15/3/4	14/2/6	10/2/10
B/S/W (sig.)	6/15/1	5/17/0	5/17/0	3/17/2
SMA_{30-003} vs	SMA_{E0501}	SMA_{R1-2}	$SMA_{ELBF0501}$	SMA_Z
B/S/W	15/2/5	12/2/8	15/2/5	14/2/6
B/S/W (sig.)	3/18/1	3/18/1	3/19/0	2/20/0
SMA_{30-003} vs	$SMA_{ELBF0504}$	$SMA_{180-003}$	SMA_{EL0501}	SMA_L
B/S/W	17/3/2	9/2/11	12/2/8	11/2/9
B/S/W (sig.)	2/19/1	2/19/1	1/18/3	1/18/3
SMA_{30-003} vs	SMA_{90-01}	SMA_{30-01}	SMA_{30-03}	SMA_{30-09}
B/S/W	8/2/12	8/2/12	6/2/14	0/2/20
B/S/W (sig.)	1/18/3	0/19/3	0/17/5	0/2/20

6   Conclusion

In this paper, 17 variants of the SMA algorithm are compared with 16 nature-inspired optimisers. All the methods are applied to 22 real-world problems to provide a real performance for the real application. The results were statistically compared by various methods. We can clearly state that most of the proposed SMA variants enable solving some of the real-world problems in the best way. Regarding the overall comparison of all the problems, all proposed SMA variants achieved the first 11 positions based on the mean ranks. SMA variants on the first three positions employed the linear reduction of the population size removing more than one individual.

Concerning each problem separately, variants $SMA_{ELF0501}$, $SMA_{ELBF0501}$, and $SMA_{RELF0501}$ provided the best median values in two or three problems. It means that the linear population size reduction with more than one individual removed and the Eigen transformation with $ps = 0.5$ and $peig = 0.1$ seem to be a good choice.

Most of the newly proposed enhanced SMA variants perform significantly better compared with the original SMA_{30-003}. More enhancing approaches will be studied in future work to achieve more promising results.

References

1. Abdel-Basset, M., Chang, V., Mohamed, R.: HSMA_WOA: a hybrid novel Slime mould algorithm with whale optimization algorithm for tackling the image segmentation problem of chest X-ray images. Appl. Soft Comput. **95**, 106642 (2020). https://doi.org/10.1016/j.asoc.2020.106642

2. Aydilek, I.B.: A hybrid firefly and particle swarm optimization algorithm for computationally expensive numerical problems. Appl. Soft Comput. **66**, 232–249 (2018)

3. Bujok, P., Tvrdík, J., Poláková, R.: Nature-inspired algorithms in real-world optimization problems. MENDEL Soft Comput. J. **23**, 7–14 (2017)

4. Bujok, P., Tvrdík, J., Poláková, R.: Comparison of nature-inspired population-based algorithms on continuous optimisation problems. Swarm Evol. Comput. **50**, 100490 (2019). https://doi.org/10.1016/j.swevo.2019.01.006

5. Das, S., Suganthan, P.N.: Problem definitions and evaluation criteria for CEC 2011 competition on testing evolutionary algorithms on real world optimization problems. Technical report, Jadavpur University, India and Nanyang Technological University, Singapore (2010)

6. Das, S., Biswas, A., Dasgupta, S., Abraham, A.: Bacterial foraging optimization algorithm: theoretical foundations, analysis, and applications. In: Abraham, A., Hassanien, A.E., Siarry, P., Engelbrecht, A. (eds.) Foundations of Computational Intelligence, vol. 203, pp. 23–55. Springer, Heidelberg (2009). https://doi.org/10.1007/978-3-642-01085-9_2

7. Dogan, B., Ölmez, T.: A new metaheuristic for numerical function optimization: Vortex search algorithm. Inf. Sci. **293**, 125–145 (2015)

8. Howard, F.L.: The life history of physarum polycephalum. Am. J. Botany **18**(2), 116–133 (1931). https://doi.org/10.2307/2435936

9. Kennedy, J., Eberhart, R.: Particle swarm optimization. In: 1995 IEEE International Conference on Neural Networks Proceedings, vol. 1–6, pp. 1942–1948. IEEE, Neural Networks Council (1995)

10. Kiran, M.S.: TSA: tree-seed algorithm for continuous optimization. Expert Syst. Appl. **42**(19), 6686–6698 (2015). https://doi.org/10.1016/j.eswa.2015.04.055

11. Kumar, C., Raj, T.D., Premkumar, M., Raj, T.D.: A new stochastic Slime mould optimization algorithm for the estimation of solar photovoltaic cell parameters. Optik 223 (2020). https://doi.org/10.1016/j.ijleo.2020.165277

12. Li, S., Chen, H., Wang, M., Heidari, A.A., Mirjalili, S.: Slime mould algorithm: a new method for stochastic optimization. Future Generation Computer Systems-The International Journal Of Escience **111**, 300–323 (2020). https://doi.org/10.1016/j.future.2020.03.055

13. Liang, X., Wu, D., Liu, Y., He, M., Sun, L.: An enhanced Slime mould algorithm and its application for digital IIR filter design. Discrete Dynamics in Nature and Society 2021 (2021). https://doi.org/10.1155/2021/5333278

14. Mehrabian, A., Lucas, C.: A novel numerical optimization algorithm inspired from weed colonization. Ecological Informat. **1**(4), 355–366 (2006)

15. Mirjalili, S., Mirjalili, S.M., Lewis, A.: Grey wolf optimizer. Adv. Eng. Softw. **69**, 46–61 (2014)

16. Ornek, B.N., Aydemir, S.B., Duzenli, T., Ozak, B.: A novel version of Slime mould algorithm for global optimization and real world engineering problems enhanced slime mould algorithm. Math. Comput. Simul. **198**, 253–288 (2022). https://doi.org/10.1016/j.matcom.2022.02.030

17. Rastrigin, L.: The convergence of random search method in extremal control of many-parameter system. Autom. Remote. Control. **24**, 1337–1342 (1963)

18. al Rifaie, M.M.: Dispersive flies optimisation. In: Federated Conference on Computer Science and Information Systemss, 2014. ACSIS-Annals of Computer Science and Information Systems, vol. 2, pp. 529–538 (2014)

19. Rizk-Allah, R.M., Hassanien, A.E., Song, D.: Chaos-opposition-enhanced Slime mould algorithm for minimizing the cost of energy for the wind turbines on high-altitude sites. ISA Trans. **121**, 191–205 (2022). https://doi.org/10.1016/j.isatra.2021.04.011

20. Simon, D.: Biogeography-based optimization. IEEE Trans. Evol. Comput. **12**(6), 702–713 (2008)

21. Tanabe, R., Fukunaga, A.S.: Improving the search performance of shade using linear population size reduction. In: IEEE Congress on Evolutionary Computation (CEC) 2014, pp. 1658–1665 (2014)

22. Tzanetos, A., Dounias, G.: A new metaheuristic method for optimization: sonar inspired optimization. In: Engineering Applications of Neural Networks (EANN), pp. 417–428 (2017)

23. Wang, G.G., Deb, S., Cui, Z.: Monarch butterfly optimization. Neural Comput. Appl., 1–20 (2015)

24. Wang, G.G., Deb, S., Gao, X.Z., Coelho, L.D.S.: A new metaheuristic optimisation algorithm motivated by elephant herding behaviour. Int. J. Bio-Inspired Comput. **8**(6), 394–409 (2017)

25. Wang, Y., Li, H.X., Huang, T., Li, L.: Differential evolution based on covariance matrix learning and bimodal distribution parameter setting. Appl. Soft Comput. **18**, 232–247 (2014)

26. Yang, X.S.: A new metaheuristic bat-inspired algorithm. In: Gonzalez, J., Pelta, D., Cruz, C., Terrazas, G., Krasnogor, N. (eds.) NICSO 2010: Nature Inspired Cooperative Strategies for Optimization. Studies in Computational Intelligence, vol. 284, pp. 65–74 (2010)

27. Yang, X.S.: Nature-Inspired Optimization Algorithms. Elsevier (2014)

28. Yin, S., Luo, Q., Zhou, Y.: EOSMA: an equilibrium optimizer Slime mould algorithm for engineering design problems. Arab. J. Sci. Eng. **7**, 1–32 (2022). https://doi.org/10.1007/s13369-021-06513-7

29. Zelinka, I., Lampinen, J.: SOMA - self organizing migrating algorithm. In: Matousek, R. (ed.) MENDEL, 6th International Conference On Soft Computing, pp. 177–187. Czech Republic, Brno (2000)

30. Zhu, Z.: An improved solution to generation scheduling problem using slime mold algorithm. Front. Ener. Res. 10 (2022). https://doi.org/10.3389/fenrg.2022.878810

31. Zubaidi, S.L., et al.: Hybridised artificial neural network model with slime mould algorithm: a novel methodology for prediction of urban stochastic water demand. Water 12(10) (2020). https://doi.org/10.3390/w12102692

SMOTE Inspired Extension
for Differential Evolution

Dražen Bajer[✉], Bruno Zorić, and Mario Dudjak

Faculty of Electrical Engineering, Computer Science and Information Technology
Osijek, Josip Juraj Strossmayer University of Osijek, Osijek, Croatia
{drazen.bajer,bruno.zoric,mario.dudjak}@ferit.hr

Abstract. Although differential evolution (DE) is a well-established optimisation method, proven on a wide variety of problems, modifications are proposed on a regular basis attempting to ever more improve its performance. Typical avenues for improvement include the introduction of new (mutation) operators or parameter control schemes. Another, less common approach, is the incorporation of additional, complementary, search mechanisms. This paper proposes one such mechanism, based on the idea of producing new solutions akin to the manner of the SMOTE algorithm producing synthetic minority instances in supervised machine learning. The conducted experimental analysis showed it to be highly competitive against comparable mechanisms on the CEC2014 benchmark suite when incorporated into standard DE, whilst being especially beneficial on simpler multimodal problems. Its incorporation into improved DE variants, although still undoubtedly bringing value on these problems, does hint at complex interactions with already integrated enhancements, suggesting that extending already enhanced algorithm variants is not simple, to say the least.

Keywords: Auxiliary search mechanism · Differential evolution · Oversampling · SMOTE

1 Introduction

Differential evolution (DE) [22] is a stochastic population-based optimisation method. It is one of the best-performing evolutionary algorithms (EAs) for numerical optimisation, but also proved an effective tool for discrete and combinatorial optimisation problems. Like other common EAs, DE uses variation operators, i.e. mutation and crossover, to sample the search space and selection to steer the search towards areas containing promising solutions. Due to the notable performance demonstrated at its inception, a myriad of modifications and improvements were proposed. Most of these are related to the mutation operation, which is the prominent distinguishing factor of DE [1], whereby scaled differences between population members are used to attain perturbations (hence the name). Accordingly, a wide range of mutation operators/strategies are available for the algorithm that differ not only in their complexity, but also in their capability to promote exploration and exploitation (see, e.g., [1,17,25]). Further, due to the sensitivity to used parameter configurations (common to virtually all bio-inspired optimisation algorithms) and inherent limitations of (offline) parameter tuning, various types of parameter control are typically integrated to facilitate the search process throughout its different phases. Diverse parameter control schemes were devised for DE that adjust the parameter values by different principles (see, e.g., [5,16]). Naturally, many high-performing DE variants incorporate both of the aforementioned modifications (see, e.g., [17,25]) since different modifications affect the algorithm behaviour in different manners.

Modifications or enhancements of the mutation strategy do not alter the search process in a groundbreaking way since these are still represented by linear combinations of population members. Additionally, as the population converges, exploration via mutation (and crossover) becomes limited [15], whilst there is only so much parameter control can achieve to compensate for this. Thus, auxiliary or complementary search mechanisms are sometimes incorporated into the algorithm. However, these are notably less common than the above-mentioned improvements to the DE algorithm. Nevertheless, several such mechanisms can be found in the literature (see, e.g., [6,15,19]) that attempt to complement the search performed by the variation operators, and hence, achieve a synergistic effect. Yet, not all are made equal and some are more successful in this endeavour than others. Ideally, such a mechanism should contribute both to exploration and exploitation since a balance is paramount [23], but this is difficult to achieve and focus on one or the other is usually inherent to such mechanisms. In the extremes, too much focus on exploration hinders convergence, whilst too much focus on exploitation leads to premature convergence or search stagnation. Accordingly, the incorporation of such mechanisms must take into account the nature of the employed variation operators i.e. whether they are aimed at exploration or exploitation in order to gain a synergistic effect. In DE, this relates mainly to the employed mutation strategy. The extent of the utilisation of such mechanisms must also be considered. From this perspective, it should always be viewed as an auxiliary search operator. This is to say the search must be primarily conducted by the variation operators and guided by the selection operator. Based on the aforementioned, the incorporation of such mechanisms into the algorithm is not a straightforward task, let alone its design. This paper represents an attempt to devise such a (complementary) search mechanism for DE by taking into account the above-mentioned issues. The proposed mechanism takes inspiration from the SMOTE (Synthetic Minority Oversampling Technique) algorithm [7] used in machine learning to handle the class imbalance problem by producing synthetic data as convex combinations of available minority samples. The proposal utilises essentially the same method for generating a new population of candidate solutions. The viability and efficiency of the proposed mechanism was assessed in comparison with several DE variants incorporating different auxiliary mechanisms for enhancing the search process. Highly competitive performance was observed on the CEC2014 benchmark functions. The merit of the proposal was additionally demonstrated by incorporation into

DE variants already enhanced by parameter control schemes and/or improved mutation strategies. However, mixed results were obtained, suggesting that interactions between different enhancements may be detrimental in some instances.

The rest of the paper is organised as follows. A concise outline of the DE and SMOTE algorithms is given in Sect. 2 as well as a literature overview of complementary search mechanisms that have been presented to enhance the performance of DE. Section 3 presents the proposed search mechanism inspired by SMOTE systematically and in detail. Section 4 describes the conducted experimental analysis and provides an overview of the obtained results alongside discussions. The drawn conclusions and possible directions for future work are presented in Sect. 5

2   Background

Generally, a numerical optimisation problem can be defined by the pair (f, S), where $f \colon S \to \mathbb{R}$ is an objective function, and $S = [\mathbf{s}^{min}, \mathbf{s}^{max}] \subseteq \mathbb{R}^d$ the search space. The optimal solution is represented by a point $\mathbf{x}^* \in S$ such that $\forall \mathbf{x} \in S \colon f(\mathbf{x}^*) \le f(\mathbf{x})$.

2.1   Differential Evolution

Developed for numerical optimisation, DE proved itself in numerous occasions and on diverse problems [8,18]. The structure of DE, as can be observed from the outline in Algorithm 1, is in line with the general structure of other typical EAs. Correspondingly, a population of candidate solutions (referred to as vectors) is subjected to the application of mutation and crossover to create new solutions that compete through selection for survival (due to the constant size of the population). This is iteratively repeated until a termination criterion is met.

The population $P(t) = (\mathbf{v}^{1,t}, \dots, \mathbf{v}^{NP,t})$, where $\mathbf{v}^{j,t} = (v_1^{j,t}, \dots, v_d^{j,t}) \in \mathbb{R}^d$ for $j = 1, \dots, NP$, of size NP is usually initialised uniformly at random at $t = 0$ inside the search space [2]. Each iteration/generation $t > 0$, and for each target vector $\mathbf{v}^{j,t}$ a donor/mutant is created as the linear combination

$$\mathbf{w}^{j,t} = \mathbf{v}^{r1,t} + F \cdot (\mathbf{v}^{r2,t} - \mathbf{v}^{r3,t}), \tag{1}$$

where $F \in \langle 0, 1]$ is the scale factor and $r1, r2$, and $r3$ are randomly chosen indices such that $j \ne r1 \ne r2 \ne r3$. The donor and target vector are crossed over to produce the trial vector

$$y_i^{j,t} = \begin{cases} w_i^{j,t} & \text{if } \mathcal{U}_{j,i} \le CR \text{ or } i = r_j \\ v_i^{j,t} & \text{else} \end{cases}, \quad i = 1, \dots, d, \tag{2}$$

where $\mathcal{U}_{j,i}$ is a uniform random number in $[0, 1]$, $CR \in [0, 1]$ is the crossover-rate, and r_j is chosen randomly from $\{1, \dots, d\}$. Eventually, the target and trial vector compete (in terms of quality) for survival into the next generation $t + 1$ i.e. selection is performed as

$$\mathbf{v}^{j,t+1} = \begin{cases} \mathbf{y}^{j,t} & \text{if } f(\mathbf{y}^{j,t}) \le f(\mathbf{v}^{j,t}) \\ \mathbf{v}^{j,t} & \text{else} \end{cases} . \tag{3}$$

It must be noted that linear combination given by Eq. (1) is not the only mutation strategy available for DE, yet is rather often used. Unsurprisingly, due to the importance of mutation, numerous other schemes exist (see, e.g., [8]). Also, mutant vectors outside S may be created, which implies that bound constraint handling is required (see, e.g., [4]). Rarely, instead of Eq. (2) a different crossover operation is applied. Also worth noting is the necessity to appropriately set the parameters NP, F, and CR to attain acceptable performance. Finally, the described algorithm represents the standard or canonical DE algorithm, typically denoted by DE/rand/1/bin.

Algorithm 1. Outline of the general DE structure

1: Set the population size NP, scale factor F, and crossover-rate CR;
2: Initialise the population $P(0)$, $t := 0$;
3: **while** *termination criterion not met* **do**
4:     **for** $j := 1, \ldots, NP$ **do**
5:         create mutant/donor vector $\mathbf{w}^{j,t}$;   % e.g. as per (1)
6:         create trial vector $\mathbf{y}^{j,t}$;   % e.g. as per (2)
7:         select between $\mathbf{v}^{j,t}$ and $\mathbf{y}^{j,t}$;   % as per (3)
8:     **end for**
9:     $t := t + 1$;
10: **end while**

2.2 Synthetic Minority Oversampling Technique (SMOTE)

The class imbalance problem is a well-known issue encountered in supervised learning. Simply put, it refers to situations in which one class is substantially less represented than the others. A frequent approach to this problem is oversampling for which various algorithms exist with SMOTE being the best-known. Although relatively simple, it proved its effectiveness on countless problems [12]. An outline of the SMOTE algorithm is provided in Algorithm 2. It generates a given number of synthetic instances from each available minority sample based on its nearest neighbours (in terms of Euclidean distance) as an attempt to better represent the minority class, and thus, make learning from the dataset easier.

For each sample of the minority class $\mathbf{x}^j \in M \subset X \subseteq \mathbb{R}^d$, a set number q of synthetic instances is generated as the convex combination

$$\mathbf{z}^{j,i} = \mathbf{x}^j + \mathcal{U}_{j,i} \cdot (\mathbf{x}^{n(j,i)} - \mathbf{x}^j), \quad i = 1, \ldots, q, \tag{4}$$

where $\mathcal{U}_{j,i}$ is a uniform random number in $[0, 1]$, and $\mathbf{x}^{n(j,i)}$ is randomly chosen from the k-neighbourhood of $\mathbf{x}^j$ (comprised of the k nearest samples in M). Notably, the algorithm takes only the minority class into account, which may become an issue when the dataset has certain unfavourable intrinsic characteristics, like overlap or small disjuncts [10]. Hence, an appropriate choice of the parameters k and q may be crucial as these control the oversampling process.

Algorithm 2. Outline of the SMOTE algorithm

1: Extract minority class samples $M \subset X$;
2: Set the neighbourhood size k, and oversampling-rate q;
3: **for all** $\mathbf{x}^j \in M$ **do**
4:     determine k-neighbourhood;
5:     **for** $i := 1 \rightarrow q$ **do**
6:         create synthetic instance $\mathbf{z}^{j,i}$;   % as per (4)
7:     **end for**
8: **end for**

2.3 Literature Overview

As mentioned earlier, several works in the literature devised auxiliary search mechanisms for DE to complement its search behaviour and, thereby, improve its performance. Essentially, their aim is to enable the discovery of better solutions or at least to reduce the computations required to find solutions of satisfactory quality. Many of these mechanisms are represented by various other techniques that are more or less related to optimisation, like data clustering, various types of random walks, and opposition-based learning (OBL).

Rahnamayan et al. [19] introduced opposition-based learning into DE. The proposed algorithm (ODE) uses opposite numbers during population initialisation and also to generate new populations during the search (termed "generation jumping"). In particular, the initial population of ODE consists of NP best solutions from the union of a randomly generated population and its opposite population. Generation jumping is performed probabilistically, where the NP fittest individuals are selected into the new population from the union of the current one and its opposite. The notion of utilising opposite numbers to accelerate the optimisation process was later used in many other algorithms (see, e.g., [14]). Additionally, Rahnamayan et al. [20] proposed a quasi-oppositional DE algorithm (QODE) that is conceptually the same as ODE, but creates quasi-oppositional points i.e. solutions instead of opposite ones. However, it was noted that population diversity is affected more since quasi-oppositional solutions are, generally, closer to current solutions than (true) opposite ones.

Further, Cai et al. [6] incorporated the k-means algorithm for data clustering into DE (CDE) that acts as several multi-parent crossover operations. It periodically performs clustering of all solutions using only one iteration of the k-means algorithm, whereby the found cluster centres represent new solutions competing to enter the population. The population is updated by selecting the k best solutions amongst the obtained cluster centres and k solutions that are randomly selected from the population itself. This clustering step is applied periodically as to avoid rapid loss in population diversity and thereby premature convergence.

Zhan and Zhang [24] proposed a DE algorithm (DERW) that incorporates a simple random walk (although, technically, not representing a true random walk). It is built into the binomial crossover operator [see, Eq. (2)] so that a trial vector component is set to a random value with a small given probability, instead of representing the usual recombination of target and mutant vector components. The probability is dynamically adjusted, so that this mechanism is applied more often at earlier stages of the search, and less so later on. On the other hand, in [15] a DE algorithm that incorporates macromutations (DEMM) was proposed. Since they represent crossover of a given population vector with a randomly generated vector, macromutations allow for more extensive exploration of search space. In DEMM, they are performed instead of the standard mutation and crossover operators with a dynamically adjusted probability.

There are significant distinctions amongst the aforementioned mechanisms, which are mostly reflected in the way new solutions are created. Given their differences and the lack of experimental comparisons between them in the literature, it is difficult to assess the most suitable one for a particular problem (class). It should also be kept in mind that all those mechanisms introduce one or more additional parameters that control their behaviour and, thereby, impact the search process. Presumably, these need to be tuned to attain the best possible performance. In some cases, the associated parameters are adjusted, mostly dynamically, to alleviate this issue. Although embedded into DE, the majority of the reviewed mechanisms are performed independent of the standard variation operators and can therefore be considered as auxiliary search operators. Hence, they may be embedded into other, improved, variants of the DE algorithm as needed, or even into other bio-inspired algorithms. It is therefore unsurprising that some of these mechanisms are popular, like OBL, and have been applied to a myriad of optimisation problems (see, e.g., [3, 21]).

3   Proposed Mechanism for Differential Evolution

The proposed auxiliary search mechanism for DE takes inspiration in the SMOTE algorithm. Simply put, a small number of solutions in the population of the DE algorithm are oversampled to generate new ones. From a high-level viewpoint this may seem straightforward since the convex combination (4) can be directly applied to the population of DE.

Remark 1. The convex combination (4) resembles whole arithmetic crossover [11], but with two important distinctions: (a) the weight is a random number and (b) the second parent is always chosen randomly from the k-neighbourhood of the first.

Although perhaps simple at first glance, several critical questions need to be answered before the incorporation of the mechanism is even attempted. These questions are as follows:

- When and how often should the oversampling be performed?
- Which solutions should be oversampled?
- How should new solutions enter the population?
- How should the parameters k and q be set?

For potential answers, at least to some of the above-mentioned questions, the previous works, like [6,19,20], can be consulted. The auxiliary mechanisms in these are employed at the end of each iteration of the DE algorithm either probabilistically [19,20] or periodically [6]. Such an implementation does not intervene with the standard algorithm elements, like e.g. the mechanism in [24] does, since there is a clear line separating them from the auxiliary mechanism. This also keeps the incorporation of the mechanism as simple as possible. The whole population is used in these mechanisms to generate new solutions, but this may be considered mechanism-specific. More important, in a general sense, is the number of newly created solutions, which is equal to the population size in [19,20], but is substantially smaller in [6]. On the one hand, a large number, especially in conjunction with a high selection pressure during the replacement may excessively bias the search. On the other hand, a small number may be insufficient in that regard. Coincidentally, the same selection procedure is used in all three mechanisms to determine which solutions enter the population and survive into the next iteration (replacement). It exhibits a high selection pressure, which is, however, not inherent to the procedure employed in DE, that exhibits a mild selection pressure and is, thus, beneficial for maintaining diversity [18].

The aforementioned considerations have been taken into account in the design of the proposed mechanism, which is outlined in Algorithm 3 as incorporated into the standard DE algorithm. It is applied at the end of each iteration with a small probability p. However, instead of the whole population, a relatively small subset is selected randomly for oversampling. The subset comprises 25% of the population and always includes the best-so-far solution as an attempt to balance exploration and exploitation to some extent. A number of new solutions equal to the population size is generated, which implies a oversampling-rate of $q = 4$. Also essential to the oversampling is the neighbourhood size that is defined across the whole population (not only to the selected solutions) and is determined as $k = 3$. Newly generated solutions compete only with the one oversampled to obtain them, which is in line with the common DE selection procedure.

Remark 2. From an evolutionary computation point of view, the proposal may be considered a simple $(\mu + \lambda)$-EA. Specifically, based on μ parents (solutions selected for oversampling), λ offspring (new solutions) are created through crossover (see Remark 1).

4   Experimental Analysis

A comprehensive experimental analysis was conducted to evaluate the proposed mechanism in terms of potential to enhance the performance of DE. The analysis is divided into two parts. The first compares the proposal with different

---

Algorithm 3. Outline of DE incorporating the proposed mechanism

---

1: Set parameters and initialise the population $P(0)$, $t := 0$;
2: $k = 3$, $q = 4$;
3: **while** *termination criterion not met* **do**
4:     **for** $j := 1, \ldots, NP$ **do**
5:         perform mutation [Eq. (1)], crossover [Eq. (2)], and selection [Eq. (3)];
6:     **end for**
7:     $t := t + 1$;
        {Proposed search mechanism}
8:     **if** $\mathcal{U}_t < p$ **then**
9:         extract $M = \{\mathbf{v}^{best}\} \cup \{\mathbf{v}^{r1}, \ldots, \mathbf{v}^{rn}\}$ from $P(t)$, $|M| = \frac{NP}{4}$;
10:        **for all** $\mathbf{v}^j \in M$ **do**
11:            determine k-neighbourhood of $\mathbf{v}^j$ in $P(t)$;
12:            **for** $i := 1, \ldots, q$ **do**
13:                create new solution $\mathbf{z}^{j,i}$;   % as per (4)
14:                **if** $f(\mathbf{z}^{j,i}) < f(\mathbf{v}^j)$ **then**
15:                    $\mathbf{v}^j := \mathbf{z}^{j,i}$;
16:                **end if**
17:            **end for**
18:        **end for**
19:    **end if**
20:    return M into $P(t)$;
21: **end while**

Table 1. Types of functions comprising the CEC2014 benchmark suite

Functions	Type	Search space
$f_1 \sim f_3$	Unimodal	$S = [-100, 100]^d$
$f_4 \sim f_{16}$	Simple multimodal	
$f_{17} \sim f_{22}$	Hybrid	
$f_{23} \sim f_{30}$	Composition	

complementary/auxiliary search mechanisms. The second investigates the benefit of incorporating the proposal into already improved algorithm variants. The CEC2014 benchmark suite [13] for $d = 10$, and 30 was used in both parts. The 30 functions comprising the suite are categorised into four types as is concisely shown in Table 1.

4.1 Setup

The experiments were conducted as defined in [13]. Accordingly, 51 independent runs were performed, where each run was terminated after $d \cdot 10^4$ function evaluations (FEs) or earlier if an optimisation error below 10^{-8} was reached. In each run, the population was initialised uniformly at random inside the search space (S). The same parameter configuration i.e. $NP = 100$, $F = 0.5$, and $CR = 0.9$ was used for each algorithm. This configuration coincides with those used in the respective papers on the competing algorithms.

Table 2. Algorithms used with auxiliary search mechanisms

Algorithm	Mechanism params.	Note
DE	–	DE/rand/1/bin; Represents the standard/canonical DE algorithm
ODE [19]	$J_r = 0.3$	DE/rand/1/bin; OBL employed for population initialisation and the occasional creation of new solutions
QODE [20]	$J_r = 0.05$	DE/rand/1/bin; QOBL employed for population initialisation and the occasional creation of new solutions
DERW [24]	$RW = 0.1 \rightarrow 0.099$	DE/rand/1/bin; Modified binomial crossover operator
DEMM [15]	$p_c = 0.5 \rightarrow 0.005$, $p_{MM} = 0.05 \rightarrow 0.9$	DE/rand/1/bin; Macromutations are occasionally performed instead of common mutation and crossover
CDE [6]	$m = 10, k = \mathcal{U}[2, \sqrt{NP}]$	DE/rand/1/exp; Periodical clustering of the population using one-step of the k-means algorithm to create new solutions
DE$_{SOX}$	$p = 0.1$	DE/rand/1/bin; Proposed search mechanism

Table 3. Results of the comparison of DE algorithms with auxiliary search mechanisms on the CEC2014 functions for $d = 10$

Func.	DE	ODE	QODE	DERW	DEMM	CDE	DE$_{SOX}$
	$\downarrow/\circ/\uparrow$	$\downarrow/\circ/\uparrow$	$\downarrow/\circ/\uparrow$	$\downarrow/\circ/\uparrow$	$\downarrow/\circ/\uparrow$	$\downarrow/\circ/\uparrow$	¬
$f_1 \sim f_3$	0/3/0	1/2/0	2/1/0	0/3/0	2/1/0	0/3/0	–
$f_4 \sim f_{16}$	11/2/0	9/4/0	10/3/0	11/2/0	11/2/0	7/4/2	–
$f_{17} \sim f_{22}$	1/3/2	5/1/0	3/3/0	2/2/2	5/1/0	1/5/0	–
$f_{23} \sim f_{30}$	2/5/1	4/4/0	5/3/0	2/5/1	2/3/3	2/5/1	–
Total	14/13/3	19/11/0	20/10/0	15/12/3	20/7/3	10/17/3	–
$\overline{SR}$	0.15	0.11	0.13	0.15	0.05	0.20	0.19
$\overline{R}$	4.08	4.53	4.97	4.37	4.42	3.03	2.60

4.2   Comparison Against Other Mechanisms

A number of algorithms with various auxiliary search mechanisms have been proposed in the literature. The mechanisms they incorporate differ in both complexity and working principles. Hence, to put the benefit of the proposed mechanism into perspective, a comparison with some of these was performed. The competing algorithms are briefly summarised in Table 2, whilst all have been described earlier in the literature overview in more detail. The standard DE algorithm is included as well to serve as baseline.

The results obtained in the experiments are concisely reported in Table 3 and 4. A summary of the differences in attained average optimisation errors between the algorithm incorporating the proposal (DE$_{SOX}$) and the others is shown per function category. More precisely, a summary of the applications of the Wilcoxon signed-rank test [9] is shown, given a confidence interval of 95%. Accordingly, statistically significant differences in favour of DE$_{SOX}$ are denoted by $\downarrow$, absences of such differences are denoted by $\circ$, whilst those favouring the competition are denoted by $\uparrow$. Additionally, the average success-rate ($\overline{SR}$) in reaching the targeted optimisation error ($< 10^{-8}$) and the average ranks ($\overline{R}$) obtained from the Friedman test [9] are given at the bottom of each table.

Table 4. Results of the comparison of DE algorithms with auxiliary search mechanisms on the CEC2014 functions for $d = 30$

Func.	DE	ODE	QODE	DERW	DEMM	CDE	DE_{SOX}
	$\downarrow/\circ/\uparrow$	$\downarrow/\circ/\uparrow$	$\downarrow/\circ/\uparrow$	$\downarrow/\circ/\uparrow$	$\downarrow/\circ/\uparrow$	$\downarrow/\circ/\uparrow$	$\neg$
$f_1 \sim f_3$	0/2/1	3/0/0	3/0/0	0/3/0	2/1/0	1/2/0	–
$f_4 \sim f_{16}$	9/1/3	10/2/1	11/2/0	10/0/3	7/3/3	7/1/5	–
$f_{17} \sim f_{22}$	5/0/1	6/0/0	5/0/1	5/0/1	4/1/1	5/1/0	–
$f_{23} \sim f_{30}$	0/1/7	7/1/0	5/2/1	0/2/6	0/3/5	2/3/3	–
Total	14/4/12	26/3/1	24/4/2	15/5/10	13/8/9	15/7/8	–
$\overline{SR}$	0.12	0.00	0.02	0.10	0.07	0.15	0.10
$\overline{R}$	3.72	5.52	5.47	3.92	3.03	3.43	2.92

A brief glance at the average ranks reveals an overall advantage of DE_{SOX} (lowest $\overline{R}$). Surprisingly, the overall worst performance amongst the competing algorithms (indicated by the highest $\overline{R}$) was not attained by the standard algorithm, but by QODE which is closely followed by its conceptual predecessor ODE. Although both were shown to be highly competitive on classical benchmark functions (see [19, 20]), this is far from that case on the much more challenging functions considered here. This is also evident when viewed from a lower level, where statistically significant differences on the vast majority of functions are in favour of DE_{SOX} regardless of dimensionality. The closest competitor to DE_{SOX} on the problem instances for $d = 10$ was CDE, which however, lost its position to DEMM with the increase of problem dimensionality. Further, a more detailed inspection of the reported results reveals that proposed mechanism is especially suitable for simple multimodal functions ($f_4 \sim f_{16}$) and in most cases beneficial on hybrid functions ($f_{17} \sim f_{22}$). Although there are mostly no meaningful differences on unimodal functions ($f_1 \sim f_3$) amongst the competing algorithms, it is apparent that the proposal is not beneficial on composition functions ($f_{23} \sim f_{30}$), which is most notable on the problem instances for $d = 30$. Nevertheless, this does not reduce its viability and utility, given the favourable effect achieved on the simple multimodal and hybrid functions. The competitive performance of DE_{SOX} is also supported by the convergence graphs in Fig. 1, which provide insight into the behaviour of the various algorithms. Shown are the median optimisation errors in relationship to the performed FEs. As can be observed, virtually all algorithms demonstrate a similar convergence behaviour early into the search process. However, the algorithm incorporating the proposal is typically able to take and maintain a considerable lead in the second half of this process in many instances, again, most notably in the case of the simple multimodal and hybrid functions.

4.3   Incorporation into Improved Algorithm Variants

It should be noted that the considerations about the incorporation or design of auxiliary search mechanisms stated earlier do not take into account improvements that are possibly already incorporated into the algorithm. Presumably,

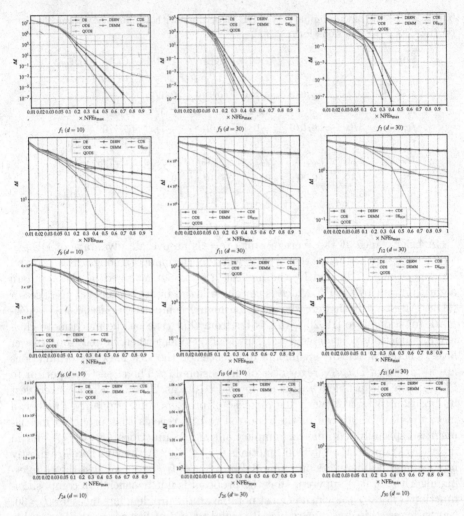

Fig. 1. Convergence graphs for DE algorithms with auxiliary search mechanisms on several CEC2014 functions of different dimensionality

Table 5. Improved DE variants used for incorporating the proposed mechanism

Algorithm	Improvement params.	Note
JADE [25]	$p = 0.05$	DE/current-to-$p$best/1/bin; Adaptation of F and CR; extended current-to-best/1 mutation strategy
jDE [5]	$\tau_1 = \tau_2 = 0.1$	DE/rand/1/bin; Self-adaptation of F and CR
RNDE [17]	$N_{lb} = 3$, $N_{ub} = 10$	Asynchronous DE/neighbour/1/bin; adaptive neighbour-based mutation scheme; Adaptation of CR
aDE [16]	–	DE/rand/1/exp; Self-adaptation of F and CR

Table 6. Results on the impact of amending improved DE variants with the proposed mechanism on the CEC2014 functions for $d = 10$

Func.	JADE	▷$_{SOX}$	jDE	▷$_{SOX}$	RNDE	▷$_{SOX}$	aDE	▷$_{SOX}$
	↓/○/↑	¬	↓/○/↑	¬	↓/○/↑	¬	↓/○/↑	¬
$f_1 \sim f_3$	0/3/0	–	0/3/0	–	0/3/0	–	0/3/0	–
$f_4 \sim f_{16}$	8/5/0	–	8/5/0	–	8/4/1	–	7/5/1	–
$f_{17} \sim f_{22}$	3/1/2	–	6/0/0	–	1/3/2	–	5/1/0	–
$f_{23} \sim f_{30}$	1/6/1	–	2/5/1	–	2/5/1	–	2/5/1	–
Total	12/15/3	–	16/13/1	–	11/15/4	–	14/14/2	–
$\overline{SR}$	0.18	0.18	0.17	0.21	0.21	0.22	0.19	0.22
R^+/R^-	193.0/272.0	¬	42.5/392.5	¬	158.5/276.5	¬	102.5/362.5	¬

Table 7. Results on the impact of amending improved DE variants with the proposed mechanism on the CEC2014 functions for $d = 30$

Func.	JADE	▷$_{SOX}$	jDE	▷$_{SOX}$	RNDE	▷$_{SOX}$	aDE	▷$_{SOX}$
	↓/○/↑	¬	↓/○/↑	¬	↓/○/↑	¬	↓/○/↑	¬
$f_1 \sim f_3$	0/2/1	–	0/3/0	–	1/2/0	–	0/2/1	–
$f_4 \sim f_{16}$	6/6/1	–	7/5/1	–	9/3/1	–	8/4/1	–
$f_{17} \sim f_{22}$	1/5/0	–	1/1/4	–	4/0/2	–	1/0/5	–
$f_{23} \sim f_{30}$	0/3/5	–	2/2/4	–	0/2/6	–	1/2/5	–
Total	7/16/7	–	10/11/9	–	14/7/9	–	10/8/12	–
$\overline{SR}$	0.19	0.17	0.14	0.13	0.14	0.14	0.16	0.15
R^+/R^-	203.0/262.0	¬	267.5/197.5	¬	194.0/241.0	¬	287.5/147.5	¬

this could represent an additional obstacle due to unpredictable interactions between the already present improvement(s) and the search mechanism to be introduced. Thus, with the aim of providing insight into possible issues related to this, a number of improved DE variants were used for the additional incorporation of the proposal. The selected DE variants, summarised in Table 5, include various improvements and thus represent a diverse testbed.

The results obtained in the pair-wise comparisons are reported in Table 6 and 7, where the algorithm variants amended with the proposed mechanisms are given to the right of the original and are denoted by ▷$_{SOX}$. The same presentation of the results is employed as previously. However, instead of the results of the Friedman test for multiple comparisons, the results of Wilcoxon test for pair-wise comparisons are shown at the bottom of these tables (R^+/R^-).

As may be observed, the presented results are not clear cut. Specifically, there is a discrepancy between the results obtained on problem instances for $d = 10$ and those obtained on the same for $d = 30$. Although the amended JADE and RNDE show favourable performance compared to their "unamended" counterparts ($R^+ < R^-$), this is not the case for jDE and aDE. The clear advantage of the amended versions of the latter on problems for $d = 10$ is turned into a disadvantage with the increase of problem dimensionality. The disadvantage on the problem instances for $d = 30$ is mainly reflected in a detrimental effect of the pro-

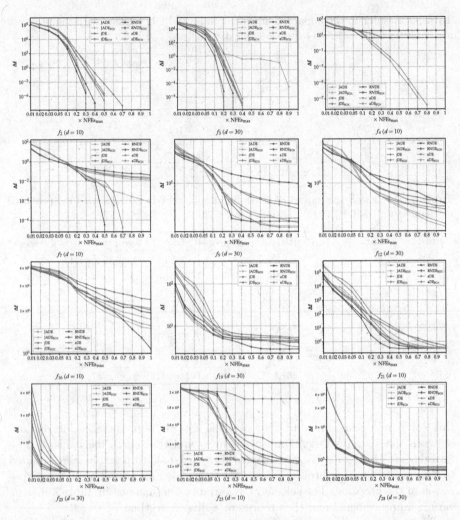

Fig. 2. Convergence graphs for improved DE variants and their amended counterparts on several CEC2014 functions of different dimensionality

posal on the hybrid and composition functions. On the upside, a positive impact on the simple multimodal functions is still maintained, which corroborates the utility of the proposal on this type of problems. The results support the presumption that the extension of already improved algorithm variants is not straightforward and shed some light on the unpredictability of interactions between algorithm extensions. Both jDE and aDE integrate only parameter control schemes. Yet, besides parameter control, JADE and RNDE, include enhanced mutation strategies, but the incorporation of the proposal was still beneficial. An answer as to why this is the case is difficult to obtain, to say the least. Nonetheless, this suggests that synergistic effects from such extensions/amendments should not be assumed a priori. Also, a similar observation was made in [1] with regard to the incorporation of an enhanced mutation scheme. Further insight into the effect of incorporating the proposal into the improved DE variants is provided by the convergence graphs in Fig. 2. In many cases, an impact on the behaviour of the algorithms in terms of increased convergence-rate is noticeable, mostly in the later phases of the optimisation process. This agrees with the observations made in the previous comparison.

5   Conclusion

Extending a well-established optimisation algorithm is by no means an easy task, as it requires careful consideration regarding any modification. Search mechanisms which do not alter the structure of the original algorithm but attempt to complement its search behaviour lend themselves reasonably well to achieving this task. One such mechanism for DE is proposed in this paper. The proposal utilises the principle of synthetic minority instance creation from the SMOTE algorithm. In this application context, a small number of population members acts as the minority and is oversampled to create new solutions competing for survival. The results showed its effectiveness when compared to various other auxiliary mechanisms from the literature on the CEC2014 benchmark functions of different dimensionality. It notably outperformed the competition on the simple multimodal and exhibited beneficial behaviour on the hybrid functions in most cases, whilst lacklustre behaviour, to some extent, could be observed on the composition functions. When incorporated into already improved DE variants, its beneficial effects on the simple multimodal problems did not subside. However, for some of the improved DE variants and types of problems a detrimental effect was observed with increasing problem dimensionality, which implies that extending already improved (DE) algorithm variants brings further challenges, since the introduction of any auxiliary search mechanisms could cause unpredictable interactions with already present enhancements.

In the end, the presented mechanism is only a single possible interpretation of SMOTE ideas in the context of optimisation algorithms. Different interpretations, e.g. considering which population members are oversampled (the meaning of minority), how new solutions are introduced into the population or even the meaning of the neighbourhood, could influence the behaviour of this mechanism considerably and are certainly worth exploring further. As stated earlier, the proposal is beneficial for certain types of functions/problems. It would be interesting to see whether the changes to the mechanism render it more suitable to other types of problems as well. On the other hand, applying the proposed algorithm on real-world problems stemming from machine learning, such as feature selection or RBFN model design could be interesting, as they represent various types of multimodal problems. Delving deeper into population diversity or estimating the usefulness of the mechanism in terms of the number of solutions that are introduced into the population could shed further light into its behaviour and, perhaps, provide some more explanations to its advantages and drawbacks.

References

1. Bajer, D.: Adaptive k-tournament mutation scheme for differential evolution. Appl. Soft Comput. **85**, 105776 (2019)
2. Bajer, D., Martinović, G., Brest, J.: A population initialization method for evolutionary algorithms based on clustering and Cauchy deviates. Expert Syst. Appl. **60**, 294–310 (2016)
3. Basu, M.: Quasi-oppositional differential evolution for optimal reactive power dispatch. Int. J. Electr. Power Energy Syst. **78**, 29–40 (2016)
4. Biedrzycki, R., Arabas, J., Jagodzinski, D.: Bound constraints handling in differential evolution: an experimental study. Swarm Evol. Comput. **50**, 100453 (2019)
5. Brest, J., Greiner, S., Boskovic, B., Mernik, M., Zumer, V.: Self-adapting control parameters in differential evolution: a comparative study on numerical benchmark problems. IEEE Trans. Evol. Comput. **10**(6), 646–657 (2006)
6. Cai, Z., Gong, W., Ling, C.X., Zhang, H.: A clustering-based differential evolution for global optimization. Appl. Soft Comput. **11**(1), 1363–1379 (2011)
7. Chawla, N.V., Bowyer, K.W., Hall, L.O., Kegelmeyer, W.P.: SMOTE: synthetic minority over-sampling technique. J. Artif. Intell. Res. **16**, 321–357 (2002)
8. Das, S., Mullick, S.S., Suganthan, P.N.: Recent advances in differential evolution - an updated survey. Swarm Evol. Comput. **27**, 1–30 (2016)
9. Derrac, J., García, S., Molina, D., Herrera, F.: A practical tutorial on the use of nonparametric statistical tests as a methodology for comparing evolutionary and swarm intelligence algorithms. Swarm Evol. Comput. **1**(1), 3–18 (2011)
10. Dudjak, M., Martinović, G.: An empirical study of data intrinsic characteristics that make learning from imbalanced data difficult. Expert Syst. Appl. **182**, 115297 (2021)
11. Eiben, A.E., Smith, J.E.: Evolutionary robotics. In: Introduction to Evolutionary Computing. NCS, pp. 245–258. Springer, Heidelberg (2015). https://doi.org/10.1007/978-3-662-44874-8_17
12. Fernández, A., García, S., Herrera, F., Chawla, N.V.: SMOTE for learning from imbalanced data: progress and challenges, marking the 15-year anniversary. J. Artif. Intell. Res. **61**, 863–905 (2018)
13. Liang, J.J., Qu, B.Y., Suganthan, P.N.: Problem definitions and evaluation criteria for the CEC 2014 special session and competition on single objective real-parameter numerical optimization, Technical report. Zhengzhou University and Nanyang Technological University (2013)
14. Mahdavi, S., Rahnamayan, S., Deb, K.: Opposition based learning: a literature review. Swarm Evol. Comput. **39**, 1–23 (2018)
15. Martinović, G., Bajer, D.: Data clustering with differential evolution incorporating macromutations. In: Panigrahi, B.K., Suganthan, P.N., Das, S., Dash, S.S. (eds.) SEMCCO 2013. LNCS, vol. 8297, pp. 158–169. Springer, Cham (2013). https://doi.org/10.1007/978-3-319-03753-0_15
16. Noman, N., Bollegala, D., Iba, H.: An adaptive differential evolution algorithm. In: IEEE Congress of Evolutionary Computation, pp. 2229–2236. IEEE (2011)
17. Peng, H., Guo, Z., Deng, C., Wu, Z.: Enhancing differential evolution with random neighbors based strategy. J. Comput. Sci. **26**, 501–511 (2018)
18. Price, K., Storn, R.M., Lampinen, J.A.: Differential Evolution: A Practical Approach to Global Optimization. Springer, New York (2005)
19. Rahnamayan, S., Tizhoosh, H.R., Salama, M.M.: Quasi-oppositional differential evolution. In: IEEE Congress of Evolutionary Computation, pp. 2229–2236. IEEE (2007)

20. Rahnamayan, S., Tizhoosh, H.R., Salama, M.M.: Opposition-based differential evolution. IEEE Trans. Evol. Comput. **12**(1), 64–79 (2008)
21. Rashid, M.A., Iqbal, S., Khatib, F., Hoque, M.T., Sattar, A.: Guided macromutation in a graded energy based genetic algorithm for protein structure prediction. Comput. Biol. Chem. **61**, 162–177 (2016)
22. Storn, R., Price, K.: Differential Evolution - a simple and efficient heuristic for global optimization over continuous spaces. J. Glob. Optim. **11**(4), 341–359 (1997)
23. Črepinšek, M., Liu, S.H., Mernik, M.: Exploration and exploitation in evolutionary algorithms: a survey. ACM Comput. Surv. **45**(3), 35:1–35:33 (2013)
24. Zhan, Z.H., Zhang, J.: Enhance differential evolution with random walk. In: 14th Annual Conference Companion on Genetic and Evolutionary Computation, pp. 1513–1514 (2012)
25. Zhang, J., Sanderson, A.C.: JADE: adaptive differential evolution with optional external archive. IEEE Trans. Evol. Comput. **13**(5), 945–958 (2009)

The Influence of Local Search on Genetic Algorithms with Balanced Representations

Luca Manzoni[1] (ID), Luca Mariot[2(✉)] (ID), and Eva Tuba[3,4] (ID)

[1] Dipartimento di Matematica e Geoscienze, Università degli Studi di Trieste, Via Valerio 12/1, 34127 Trieste, Italy
lmanzoni@units.it
[2] Digital Security Group, Radboud University, PO Bus 9010, 6500 GL Nijmegen, The Netherlands
luca.mariot@ru.nl
[3] Department of Computer Science, Trinity University, 1 Trinity Place, San Antonio, TX 78212, USA
etuba@ieee.org
[4] Faculty of Informatics and Computing, Singidunum University, Danijelova 32, 11000 Belgrade, Serbia

Abstract. Certain combinatorial optimization problems with binary representation require the candidate solutions to satisfy a balancedness constraint (e.g., being composed of the same number of 0s and 1s). A common strategy when using Genetic Algorithms (GA) to solve these problems is to use crossoveer and mutation operators that preserve balancedness in the offspring. However, it has been observed that the reduction of the search space size granted by such tailored variation operators does not usually translate to a substantial improvement of the GA performance. There is still no clear explanation of this phenomenon, although it is suspected that a balanced representation might yield a more irregular fitness landscape, where it could be more difficult for GA to converge to a global optimum. In this paper, we investigate this issue by adding a local search step to a GA with balanced operators, and use it to evolve highly nonlinear balanced Boolean functions. We organize our experiments around two research questions, namely if local search (1) improves the convergence speed of GA, and (2) decreases the population diversity. Surprisingly, while our results answer affirmatively the first question, they also show that adding local search actually *increases* the diversity among the individuals. We link these findings to some recent results on fitness landscape analysis for problems on Boolean functions.

Keywords: Genetic algorithms · Balanced crossover · Local search · Boolean functions · Nonlinearity

© The Author(s), under exclusive license to Springer Nature Switzerland AG 2022
M. Mernik et al. (Eds.): BIOMA 2022, LNCS 13627, pp. 232–246, 2022.
https://doi.org/10.1007/978-3-031-21094-5_17

1   Introduction

There exist three common approaches for constraint handling in the literature of Genetic Algorithms (GA): incorporate a penalty factor in the fitness function that punishes deviations from the desired constraints, use ad-hoc representations and variation operators, or employ repair operators. Penalty factors are fairly simple to implement and can be employed virtually in any optimization problem, once a suitable notion of distance from the required constraints has been defined. However, penalty factors can also be wasteful, since a GA may spend a great amount of fitness evaluations to satisfy them, driving the search effort away from the main optimization objective. The second approach requires designing suitable crossover and mutation operators, so that feasible parents produce feasible offspring. This makes the GA explore a smaller search space, which in principle should lead to better performance, since the fitness budget is entirely used to evolve feasible solutions only. Repair operators also make the GA to explore only the feasible space, although their approach is to transform invalid solutions into valid ones.

In this work, we focus on the second approach for handling *balancedness* constraints, namely when the binary representation of the candidate solutions must have a fixed number of ones. Such a constraint is relevant in several optimization problems related to cryptography, coding theory and combinatorial designs. To the best of our knowledge, Lucasius and Kateman [5] were the first to investigate balancedness-preserving crossover operators in GA, applying them to the subset selection problem. Millan et al. [12] used GA to evolve balanced Boolean functions with good cryptographic properties such as high nonlinearity and low deviation from correlation immunity. To this end, the authors devised a counter-based crossover operator that preserved the balancedness of the parent Boolean functions. Balanced crossover operators have also been designed for other optimization problems such as portfolio optimization [2,3] and multiobjective *k*-subset selection [11]. Further extensions of this approach include the design of balancedness-preserving operators for non-binary candidate solutions with non-binary representations [8,9] or for matrix-based representations where each column needs to be balanced [10].

More recently, we carried out in [6] a rigorous statistical investigation of three balanced crossover operators against different optimization problems related to cryptography and combinatorial designs. We found that balanced operators indeed give an advantage to GA over a classic one-point crossover coupled with a penalty factor. Hence, these results seem to confirm the aforementioned principle that reducing the search space by means of ad-hoc variation operators improves the GA performance. Nonetheless, the improvement is not substantial and does not scale well with respect to the problem size. This is especially evident when comparing a GA based on balanced crossover operators with other metaheuristics such as Genetic Programming (GP). In general, it has been observed that GP converges more easily to an optimal solution than GA on problems where balanced solutions are sought [9,10,14].

Clearly, the particular encoding of the candidate solutions used for ad-hoc operators can change the fitness landscape of a particular optimization problem. Indeed, one of the possible explanations for the meagre improvement of GA when using balanced crossover operators is that the resulting fitness landscape becomes more irregular. Hence, although searching a smaller space of feasible solutions, the GA could get stuck more easily on local optima. We started to investigate this hypothesis in [7] by considering an *adaptive bias* strategy where the counter-based crossover of [12] is allowed to produce partially unbalanced Boolean functions. The rationale is that, by slightly enlarging the search space, the GA might escape more easily from local optima, thus improving its explorability. Yet, the results showed that even this strategy provides only a marginal improvement in the GA performance.

In this paper, we further investigate the scarce improvement of GA with balanced crossover operators by augmenting them with a *local search step*. In particular, we consider the evolution of highly nonlinear balanced Boolean functions as an underlying optimization problem, for which an efficient local search move has already been developed in [13]. We perform an experimental evaluation of the three balanced crossover operators in [6] by combining them with three variants of local search. The first variant is the baseline GA where no local search is performed. The second variant applies only a single step of local search on a new offspring individual created through balanced crossover and mutation. The third variant, finally, is a steepest ascent strategy, which performs local search on an offspring individual until a local optimum is reached. The experiments are performed for Boolean functions of $6 \leq n \leq 9$ variables.

To assess the influence that local search has on the GA performance, we consider two research questions. The first one is whether local search improves the convergence speed of GA to a local optimum. As expected, the answer given by our experimental results is positive, especially for the third variant employing the steepest ascent strategy. On the other hand, the second research question is whether the use of local search decreases the diversity in the population, as measured by the pairwise Hamming distance. Indeed, a natural hypothesis for the scarce improvement of GA performance when using balanced crossover operators is that the solutions in the population become too similar, determining a premature convergence to a local optimum. Therefore, one would expect that such a phenomenon is magnified by augmenting the GA with a local search step. Surprisingly, our results indicate that the use of local search actually *increases* the population diversity. We discuss this interesting finding by linking it to a recent work on fitness landscape analysis for problems related to cryptographic Boolean functions [4]. In particular, the fact that the individuals in the population tend to be quite different among each other seem to indicate that the fitness landscape of balanced Boolean functions is characterized by many isolated local optima. This in turn suggests that a possible way to improve the GA performance is to use a different initialization strategy than the usual one where candidate solutions are generated uniformly at random.

The rest of this paper is organized as follows. Section 2 covers all background definitions related to balanced crossover operators and Boolean functions that the contributions of this paper are based upon. Section 3 defines the optimization problem of evolving highly nonlinear balanced Boolean function, and describes the local search algorithm used as a further optimization step after balanced crossover and mutation. Section 4 presents the experimental evaluation of our approach, discussing the experimental settings adopted and the obtained results. Finally, Sect. 5 concludes the paper by summarizing the main findings and pointing out directions for further research on the topic.

2   Background

In this section, we first describe the three balanced crossover operators introduced in [6], which we will use in our investigation. Next, we recall the basic notions related to Boolean functions and their cryptographic properties, that will be the basis of the underlying optimization problem for our experiments.

As a general notation, in what follows we denote by $\mathbb{F}_2 = \{0, 1\}$ the finite field with two elements, and $\mathbb{F}_2^n$ is the set of all n-bit strings, which is endowed with a vector space structure. In particular, the sum of two vectors $x, y \in \mathbb{F}_2^n$ corresponds to their bitwise XOR $x \oplus y$, while multiplication of $x \in \mathbb{F}_2^n$ by a scalar $a \in \mathbb{F}_2$ amounts to computing the logical AND of a with each coordinate of x. The *scalar product* of two vectors $x, y \in \mathbb{F}_2^n$ is defined as $\bigoplus_{i=1}^n x_i y_i$, i.e. the XOR of all bitwise AND of the two vectors. Given $[n] = \{1, \cdots, n\}$ for all $n \in \mathbb{N}$, the *Hamming distance* of $x, y \in \mathbb{F}_2^n$ is defined as $d_H(x, y) = |\{i \in [n] : x_i \neq y_i\}|$, i.e. the number of coordinates where x and y differ. The *Hamming weight* of a vector $x \in \mathbb{F}_2^n$, denoted by $w_H(x)$, is the Hamming distance of x from the null vector $\underline{0}$, or equivalently the number of ones in x. The number of binary strings with a fixed Hamming weight $k \in [n]$ is the binomial coefficient $\binom{n}{k}$, since it is equivalent to the number of k-subsets of $[n]$, when one interprets a vector $x \in \mathbb{F}_2^n$ as the characteristic function of a subset.

2.1   Balanced Crossover Operators

We start by giving a brief description of the three balanced crossover operators that we will use in our experiments. Further details about them and their pseudocode can be found in our previous paper [6]. In the remainder of this paper, we assume that the Hamming weight that we want to preserve is exactly half of the string length, i.e. the individuals in the population have an equal number of zeros and ones in their representation.

Counter-Based Crossover. The first operator employs two counters cnt_0 and cnt_1 to keep track respectively of how many zeros and ones the child individual has during the crossover process. Specifically, given two parent bitstrings $p_1, p_2 \in \mathbb{F}_2^{2m}$ such that $w_H(p_1) = w_H(p_2) = m$, a child chromosome $c \in \mathbb{F}_2^{2m}$ is obtained by randomly copying either the i-th bit of p_1 or p_2 with uniform probability, for each position $i \in [2m]$. Then, cnt_0 or cnt_1 is incremented depending on the value copied in the child. When one of the two counters reaches the threshold weight m, the remaining positions in the child are filled with the complementary value.

A natural question about this crossover operator is whether setting the last bits to a fixed value to preserve balancedness does not introduce a bias towards certain solutions in the search space. We considered this issue in our previous work [6], by comparing the basic "left-to-right" version of the operator described above with another one that randomly shuffles the order of the positions to be copied in the child chromosome. Results showed that in most cases there is no significant difference among the two variants, while in certain instances the shuffling strategy fares even worse than the basic "left-to-right" version. Hence, we used the latter for the experiments of this paper.

Zero-Length Crossover. The second crossover operator considered in our investigation is based on a different representation of the candidate solutions, namely their *zero-length* encoding. Formally, given a n-bit string x with $n = 2m$, the zero-length encoding of x is a vector r of length $m+1$ where each coordinate r_i represents the number of consecutive zeros (or equivalently, the run length of zeros) between two consecutive ones.

To correctly represent a balanced bitstring, the values in the zero-length encoding vector must sum to m. Sticking to our previous example, the zero-length encodings of $p_1 = (0,1,0,1,0,1,1,0)$ and $p_2 = (1,0,0,0,1,0,1,1)$ are respectively $r_1 = (1,1,1,0,1)$ and $r_2 = (0,3,1,0,0)$. At each position the zero-length crossover randomly copies the zero-length value of the first or second parent with uniform probability. An accumulator variable is used to represent the partial sums of the zeros' run lengths in the offspring chromosome. If the threshold value m is reached, the remaining positions of the offspring's zero-length vector are filled with zeros; thus, the bitstring representation will only contain ones in the last positions. Otherwise, the last coordinate of the zero-length vector is filled with the value that balances the sum to m; accordingly, the bitstring representation of the offspring will contain only zeros in the last positions.

Map-of-Ones Crossover. The third crossover considered in our experiments leverages on an integer-based representation of the candidate solutions. In particular, the map-of-ones is simply the vector that indicates the positions of the ones in a bitstring. Using our examples above, the map of ones for $p_1 = (0,1,0,1,0,1,1,0)$ and $p_2 = (1,0,0,0,1,0,1,1)$ are $b_1 = (2,4,6,7)$ and $b_2 = (1,5,7,8)$, respectively. Similarly to the previous two operators, the map-of-ones crossover works coordinate-wise by randomly copying either the value of the first or second parent's zero-length vector in the child chromosome. The only constraint that is enforced is that the map of ones of the child chromosome cannot have duplicate values, something that can occur if the bitstrings of the two parents have value one in the same position. For this reason, the crossover first computes a list of common positions between the two parents, and then checks whether the selected value has already been inserted before in the child or not. If this is the case, then the value from the other parent is copied instead.

2.2   Boolean Functions

We now describe the essential notions related to the optimization problem underlying our experiments on local search. A Boolean function of $f : \mathbb{F}_2^n \to \mathbb{F}_2$ is a mapping $f : \mathbb{F}_2^n \to \mathbb{F}_2$, i.e. a function that associates to each n-bit vector a single output bit, 0 or 1. The most common way to represent such a function is via its *truth table*: assuming that the vectors of $\mathbb{F}_2^n$ are lexicographically ordered, the truth table of f is the 2^n-bit vector

$$\Omega_f = (f(0, \cdots, 0), f(0, \cdots, 1), \cdots, f(1, \cdots, 1)) \ ,$$

i.e. the vector that specifies the output value $f(x)$ for each possible input vector $x \in \mathbb{F}_2^n$. A fundamental criterion for Boolean functions used in stream ciphers is that the truth table must be a balanced string, i.e. $w_H(f) = 2^{n-1}$, to resist basic statistical attacks.

Another way to uniquely represent a Boolean function commonly used in cryptography is the Walsh transform. Formally, the Walsh transform of $f : \mathbb{F}_2^n \to \mathbb{F}_2$ is the map $W_f : \mathbb{F}_2^n \to \mathbb{Z}$ defined as:

$$W_f(a) = \sum_{x \in \mathbb{F}_2^n} (-1)^{f(x) \oplus a \cdot x} = \sum_{x \in \mathbb{F}_2^n} (-1)^{f(x)} \cdot (-1)^{a \cdot x} \ , \tag{1}$$

for all $a \in \mathbb{F}_2^n$. The coefficient $W_f(a)$ measures the correlation between f and the linear function defined by the scalar product $a \cdot x$. A second important property for Boolean functions used in symmetric cryptography is their nonlinearity, which is defined as:

$$nl(f) = 2^{n-1} - \frac{1}{2} \max_{a \in \mathbb{F}_2^n} \{|W_f(a)|\} \ . \tag{2}$$

We refer the reader to [1] for further cryptographic implications and bounds related to the nonlinearity property. Here, we just limit ourselves to specify that the nonlinearity should be as high as possible. Taking into account also the balancedness property mentioned above, this gives rise to the following optimization problem:

Problem 1. Let $n \in \mathbb{N}$. Find a n-variable Boolean function $f : \mathbb{F}_2^n \to \mathbb{F}_2$ that is balanced and has maximum nonlinearity, as measured by the fitness function $fit(f) = nl(f)$.

Remark that it is still an open question to determine the maximum nonlinearity value attainable by a balanced Boolean function for $n > 7$ variables [1]. We will tackle Problem 1 in the experimental part of the paper using various combinations of balanced GA and local search.

3   Local Search of Boolean Functions

To perform local search, the first step is to define an *elementary move* between two candidate solutions. This further subsumes the notion of a topology over the search space, in order to give a precise meaning to the *neighborhood* of a solution. In our case, since we are dealing with fixed-length binary strings to represent the truth tables of Boolean functions, the most obvious choice is to adopt the topology induced by the Hamming distance. Therefore, the neighborhood of a candidate a solution $f : \mathbb{F}_2^n \to \mathbb{F}_2$ represented by its truth table $\Omega_f \in \mathbb{F}_2^{2^n}$ would be the set of all truth tables at Hamming distance 1 from Ω_f. Hence, the elementary move from f to a neighboring solution f' would be obtained by complementing a single bit in Ω_f. However, such a move would break the balancedness constraint, since the Hamming weight would change by ± 1. Hence, similarly to the mutation operator employed in our previous paper [6], we consider the *swap* between two different values in Ω_f as an elementary move for our local search procedure. In this way, the Hamming weight of the new candidate solution will still be 2^{n-1}.

Concerning the Walsh transform, a single swap in the truth table of f induces a change $\Delta(a) \in \{-4, 0, +4\}$ for each coefficient $a \in \mathbb{F}_2^n$, that can be computed with the following result proved in [13]:

Lemma 1. *Let $f : \mathbb{F}_2^n \to \mathbb{F}_2$ be a n-variable Boolean function, and assume that $y, z \in \mathbb{F}_2^n$ are such that $f(y) \neq f(z)$. Define $f^* : \mathbb{F}_2^n \to \mathbb{F}_2$ as the function obtained by swapping the values $f(y)$ and $f(z)$ in the truth table of f Then, for each $a \in \mathbb{F}_2^n$, the difference of the Walsh coefficients $W_f(a)$ and $W_{f^*}(a)$ equals:*

$$\Delta(a) = [(-1)^{f(y)} - (-1)^{f(z)}][(-1)^{a \cdot z} - (-1)^{a \cdot y}] . \tag{3}$$

Consequently, there is no need to recompute the Walsh transform from scratch when swapping two values in the truth table of f. Using Lemma 1, each coefficient can be updated from the old one as $W_{f'}(a) = W_f(a) + \Delta(a)$. This allows one to efficiently explore the neighborhood of a given function, since in this way the fitness of a single swap can be evaluated in linear time with respect to the length of the function's table. On the other hand, recomputation from scratch would entail a quadratic complexity by using the *fast Walsh transform* algorithm [1], which is the one employed by the GA to evaluate the fitness of a new individual created through crossover and mutation.

In summary, a single iteration of the GA combined with a local search step works as follows:

1. Select a pair of parents p_1, p_2 from the population.
2. Apply crossover and mutation to obtain a new balanced individual c.
3. Evaluate the fitness of c by computing the Walsh transform in Eq. 1 using the fast algorithm [1].
4. Apply one or more steps of local search to c as follows:
   (a) Generate the *2-Improvement set* of c, i.e. find all swaps in c such that the nonlinearity increases by 2. Use Eq. 3 to efficiently update the Walsh transform for each swap.

(b) Pick a swap in the 2-Improvement set and apply it to c, updating the fitness value as $W_{c'}(a) = W_c(a) + \Delta(a)$ for all $a \in \mathbb{F}_2^n$.

Since each swap in the improvement set increases the nonlinearity by 2, there is no ground to drive the selection. In our experiments, we pick the first generated swap. This is similar to the strategy adopted in [4] where local search was used to create the *Local Optima Network* of the search space of Boolean functions.

4   Experiments

As discussed in the Introduction, our aim is to assess the influence of local search as a further optimization step in the loop of a GA with balanced crossover. To this end, we consider the following two research questions:

- **RQ1**: does local search improve the convergence speed of GA, i.e. does it allow to reach a local optimum in less fitness evaluations?
- **RQ2**: does local search decrease the diversity of the GA population?

Remark that we deliberately excluded any research question pertaining the improvement of the best fitness. Indeed, it has already been remarked that balanced GA usually have a lower performance than other metaheuristics on combinatorial optimization problems such as Problem 1. Moreover, in [7] we observed that augmenting a balanced GA with a partially unbalanced crossover strategy does not improve significantly the best fitness. Considering also the evidence gathered in [12] where a balanced GA combined with hill climbing was used, our hypothesis is that local search step does not make a significant difference as well. As we will show in the next sections, this hypothesis was experimentally confirmed.

Nevertheless, it is reasonable to expect that adding local search in the loop may help GA to converge more quickly toward a local optimum, which motivates **RQ1**. Furthermore, crossover tends to exploit the genetic information of the current population, producing offspring individuals that resemble their parents, and thus decreasing the population diversity. Therefore, one may also expect that a local search step would magnify this effect, by tweaking the candidate solutions toward the nearest local optimum. This argument motivates **RQ2**.

In what follows, we describe the experimental settings used to investigate our research questions and the results obtained from our experiments.

4.1   Experimental Setting

For our experiments, we tested three variants of local search, namely:

- $LS0$: No local search, which corresponds to the basic balanced GA.
- $LS1$: Single-step local search, where only a single swap is performed on a new individual.
- $LS2$: Steepest ascent local search, with swaps performed until a local optimum is reached.

We considered counter-based $(CX1)$, zero-length $(CX2)$ and map-of-ones $(CX3)$ crossover. As for mutation, we adopted the simple swap-based operator used in [6]. Hence, we tested a total of 9 combinations of crossover operators and local search variants. Concerning the problem instances, we performed our experiments on Boolean functions of $6 \leq n \leq 9$ variables. Notice that the number of Boolean functions of n variables is 2^{2^n}, which means that $n = 6$ is the smallest problem instance from where it makes sense to apply metaheuristics, since it is not amenable to exhaustive search. The same holds even if we restrict our attention to the space of balanced Boolean functions, whose size is $\binom{2^n}{2^{n-1}}$: for $n = 6$ variables, the total amount of candidate solutions to search exhaustively would be approximately $1.83 \cdot 10^{19}$.

For the GA, we carried out a preliminary sensitivity analysis by performing small perturbations on the parameters that we adopted in our previous paper [6], to assess if significantly different results would arise. As this did not happen, we sticked to the same GA parameters. In particular, we used a population of 50 individuals, evolved for a budget of 500 000 evaluations, using a steady-state breeding policy with tournament selection of size $t = 3$: upon drawing 3 random individuals, the best two are crossed over, and the newly created offspring undergoes mutation with probability 0.7. After calculating the fitness, local search is performed according to the chosen variant, and then the obtained individual replaces the worst one in the tournament. Finally, each experiment (i.e. combination of problem instance, crossover operator and local search policy) was repeated for 30 independent runs to obtain statistically sound results. To compare two combinations of crossover operator and local search, we adopted the *Mann-Whitney-Wilcoxon* test, with the alternative hypothesis that the corresponding two distributions are not equal, with a significance value $\alpha = 0.05$.

4.2   Results

As expected, the use of local search did not improve significantly the performance of the GA, independently of the underlying combination of crossover and local search policy. The only significant differences arose with the largest instance of $n = 9$ variables, where the steepest ascent policy combined with the counter-based and the map-of-ones crossover consistently found functions with a slightly higher nonlinearity of 232 instead of 230 from the other combinations. Since the improvement is anyway too small, we avoid to report the distributions of the best fitness for this case as well.

Figure 1 depicts the boxplots for the distributions of the number of fitness evaluations required to reach the best fitness value obtained in each run. In general, it can be observed that the use of local search does have a substantial effect on the convergence speed of the GA towards a local optimum. This is particularly evident in the case of $n = 6$ variables for all three crossover operators. For $n = 7$ and $n = 8$, one can still see from the boxplot that the steepest ascent strategy gives the fastest convergence under all three crossovers, while the situation is less clear for the single-step variant. Looking at the p-values heatmaps in Fig. 2, one can indeed see that there are no significant differences between $LS1$

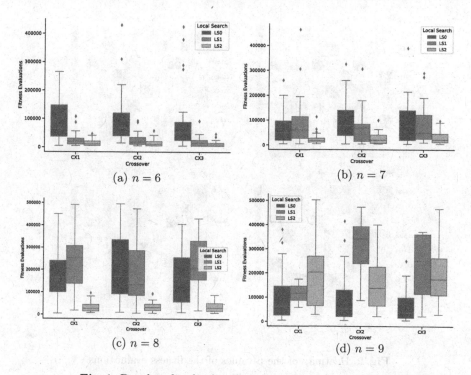

(a) $n = 6$

(b) $n = 7$

(c) $n = 8$

(d) $n = 9$

Fig. 1. Boxplots for the distributions of fitness evaluations.

and $LS0$ for all three crossover operators. The situation seems to be reversed for $n = 9$ variables, with the number of fitness evaluations required by the combinations that use the steepest ascent being higher than the variant where no local search is used. Although this finding seems odd at a first glance, it can be easily explained by the remark above on the best fitness. Since for $n = 9$ variables the steepest ascent strategy consistently finds Boolean functions with higher nonlinearity than in the basic case, it is reasonable to assume that more fitness evaluations are required to achieve them.

To investigate the solutions' diversity, at the end of each run we computed the Hamming distance of each pair of individuals in the population. Figure 3 reports the boxplots of the distributions for the median pairwise distance, while Fig. 4 gives the corresponding p-value heatmaps.

The conclusions that one can draw from these results seem counterintuitive: instead of decreasing the population diversity, *the use of local search either does not affect the diversity, or it even increases it in certain cases*. For example, one may see that for $n = 6$ there is no difference between the boxplots for each considered crossover, except maybe for $CX2$ where the diversity slightly drops with the steepest ascent policy. This is however not confirmed by the statistical tests, in that no significant differences were observed. By considering bigger instances, one can see that the local search actually starts to play a role

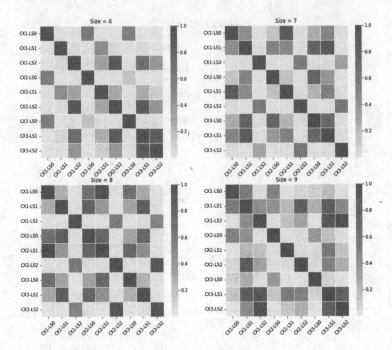

Fig. 2. Heatmap of the p-values of the fitness evaluations.

in increasing the median distance. This is particularly evident from the boxplots for $n = 8$ with the combination of counter-based crossover and steepest ascent, but also for the map-of-ones. The difference becomes even more pronounced for $n = 9$ variables, with the steepest ascent obtaining the boxplots with highest median and smallest interquartile range for all three crossovers. This is confirmed by significant differences in the corresponding heatmap. Moreover, in general one can also observe that the zero-length crossover achieves the highest median diversity for all problem instances, independently of the underlying local search policy. Indeed, one can see that the central 3×3 square in each heatmap reports non-significant differences in these cases.

4.3   Discussion

We now attempt to answer the two research questions formulated at the beginning of Sect. 4 in the light of the obtained results.

Concerning **RQ1**, the answer seems to be positive: as our initial intuition predicted, the use of local search in general increases the convergence speed of a balanced GA towards a local optimum, independently of the underlying crossover operator. Therefore, although there is no significant improvement in the best fitness (except a slight one for $n = 9$ variables), local search allows to reach the current best local optimum more quickly. This is somewhat expected, especially when using a local search step with steepest ascent policy: as each

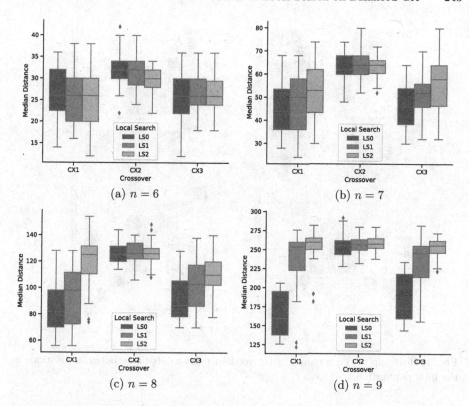

Fig. 3. Boxplots for the distributions of the median pairwise distance between solutions in the final population.

new individual created by GA undergoes local search until a local optimum is reached, the population is quickly filled by candidates that represent local optima, or candidate solutions close to them. Therefore, finding even better local optima by crossing over highly fit individuals in the population might become very unlikely already in the early stages of the optimization process. However, this finding could also indicate that by increasing substantially the fitness budget and the population size of the GA, maybe the best fitness could also improve by employing the steepest ascent local search variant. The rationale is that crossover and mutation could find something better in a large population composed of many local optima obtained through steepest ascent.

The most interesting finding concerns instead **RQ2**. Contrary to our expectations, the use of local search has either little influence on the population diversity, or it even contributes to increase the median Hamming distance among pairs of individuals. This is surprising, as the most natural explanation for the poor performance of balanced GA when compared to other metaheuristics was that the population would converge quickly around a single local optimum, therefore decreasing the population diversity. On the other hand, our experiments

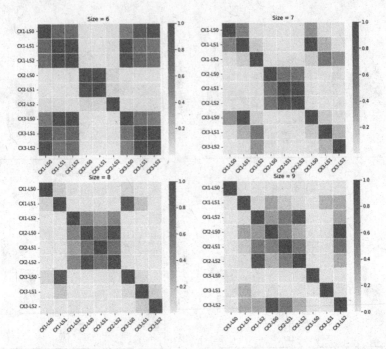

Fig. 4. Heatmap of the p-values of the median pairwise distance between solutions in the final population.

confirm that this is not the case, i.e. the final population is composed of many different local optima that are far apart from each other in the search space. A possible explanation of this phenomenon might be related to the shape of the fitness landscape for this particular problem. Indeed, Jakobovic et al. [4] already noticed that the Local Optima Networks (LONs) of generic Boolean functions (i.e., without balancedness constraints) are characterized by a huge number of isolated local optima. Although here we consider a restricted search space, it might still be the case that the resulting fitness landscape has a similar property, since it is a subset of the space of all Boolean functions. In particular, the authors in [4] explained that, to construct a meaningful LON, they had to change the initialization step of their hill climber, so that they could avoid ending up with many isolated local optima. Instead of starting each search trajectory from a completely random point, they employed a *lexicographic sampling*, where each subsequent starting point would be generated in lexicographic order from the first one, which was drawn at random.

Therefore, a possible insight from the discussion above is that the poor performance of GA in evolving highly nonlinear balanced Boolean functions is not only related to the underlying crossover operators, but also to the method used to initialize the population. Indeed, in our experiments we used a basic initialization step where each individual is generated at random with uniform probability. However, this is exactly what might contribute to cause a high median distance also in the final population, exacerbated by the use of local search, especially in its steepest-ascent version. In future experiments, it would be interesting to test different initialization method, such as the lexicographic sampling mentioned above of [4], or other methods where the population is created by small random tweaks from a single initial individual.

5   Conclusions

In this work, we investigated the effect of a local search step combined with balanced GA to evolve highly nonlinear balanced Boolean functions. The motivation was to analyze the possible causes of the poor performance of balanced GA on this particular optimization problem, when compared to other meta-heuristics such as GP. To this end, we set up our investigation by adding to the GA with balanced crossovers proposed in our previous paper [6] a local search strategy originally devised by Millan et al. [13]. We investigated three variants, namely no local search, single-step local search, and steepest-ascent local search, and applied it to the optimization of Boolean functions of $6 \leq n \leq 9$ variables. The investigation was centered around two main research questions: the first one concerned whether the use of local search increased the convergence speed of a balanced GA toward a local optimum. The second question asked if local search decreases the population diversity, as measured by the median pairwise Hamming distance between individuals. While our results answered affirmatively the first question as expected, the answer to the second question surprisingly turned out to be negative. In particular, local search either does not affect or even increases the median distance in the population. We discussed this finding by referring to a recent work on the fitness landscapes of Boolean functions [4], in the form of Local Optima Networks. In particular, the main insight gained from this discussion is that the poor performance of balanced GA might be connected to the initialization method of the population, which right now generates each individual independently with uniform probability.

Future experiments should consider other types of initialization, such as random walk from a single initial individual, or lexicographic generation. A more thorough tuning phase of the GA is also in order, to assess its sensitivity toward the population size and mutation rate. Beside this, several other directions for future research remained to be explored on the subject. Perhaps the most interesting one, after the finding of this paper, involves the analysis of the fitness landscape for the particular search space of balanced Boolean functions. Indeed, the analysis of Local Optima Networks in [4] considered the space of all Boolean functions, with no balancedness constraints. Therefore, it would be interesting to repeat the analysis for balanced functions, to see if similar properties like many isolated local optima still emerge. Further, we believe that it would be interesting to augment GA with local search also for other optimization problems that require balanced representations, such as the construction of bent functions and orthogonal arrays already considered in our previous paper [6].

References

1. Carlet, C.: Boolean Functions for Cryptography and Coding Theory. Cambridge University Press, Cambridge (2021)
2. Chen, J., Hou, J.: A combination genetic algorithm with applications on portfolio optimization. In: IEA/AIE 2006, Proceedings, pp. 197–206 (2006)
3. Chen, J., Hou, J., Wu, S., Chang-Chien, Y.: Constructing investment strategy portfolios by combination genetic algorithms. Expert Syst. Appl. **36**(2), 3824–3828 (2009)
4. Jakobovic, D., Picek, S., Martins, M.S.R., Wagner, M.: Toward more efficient heuristic construction of Boolean functions. Appl. Soft Comput. **107**, 107327 (2021)
5. Lucasius, C.B., Kateman, G.: Towards solving subset selection problems with the aid of the genetic algorithm. In: Männer, R., Manderick, B. (eds.) Parallel Problem Solving from Nature 2, PPSN-II, Brussels, Belgium, 28–30 September 1992, pp. 241–250. Elsevier (1992)
6. Manzoni, L., Mariot, L., Tuba, E.: Balanced crossover operators in genetic algorithms. Swarm Evol. Comput. **54**, 100646 (2020)
7. Manzoni, L., Mariot, L., Tuba, E.: Tip the balance: improving exploration of balanced crossover operators by adaptive bias. In: CANDAR 2021 - Workshops, Proceedings, pp. 234–240. IEEE (2021)
8. Mariot, L., Leporati, A.: A genetic algorithm for evolving plateaued cryptographic Boolean functions. In: Dediu, A.-H., Magdalena, L., Martín-Vide, C. (eds.) TPNC 2015. LNCS, vol. 9477, pp. 33–45. Springer, Cham (2015). https://doi.org/10.1007/978-3-319-26841-5_3
9. Mariot, L., Picek, S., Jakobovic, D., Leporati, A.: Evolutionary algorithms for the design of orthogonal Latin squares based on cellular automata. In: Bosman, P.A.N. (ed.) GECCO 2017, Proceedings, pp. 306–313. ACM (2017)
10. Mariot, L., Picek, S., Jakobovic, D., Leporati, A.: Evolutionary search of binary orthogonal arrays. In: Auger, A., Fonseca, C.M., Lourenço, N., Machado, P., Paquete, L., Whitley, D. (eds.) PPSN 2018. LNCS, vol. 11101, pp. 121–133. Springer, Cham (2018). https://doi.org/10.1007/978-3-319-99253-2_10
11. Meinl, T., Berthold, M.R.: Crossover operators for multiobjective k-subset selection. In: GECCO 2009, Proceedings, pp. 1809–1810 (2009)
12. Millan, W., Clark, A., Dawson, E.: Heuristic design of cryptographically strong balanced Boolean functions. In: Nyberg, K. (ed.) EUROCRYPT 1998. LNCS, vol. 1403, pp. 489–499. Springer, Heidelberg (1998). https://doi.org/10.1007/BFb0054148
13. Millan, W., Clark, A., Dawson, E.: Boolean function design using hill climbing methods. In: Pieprzyk, J., Safavi-Naini, R., Seberry, J. (eds.) ACISP 1999. LNCS, vol. 1587, pp. 1–11. Springer, Heidelberg (1999). https://doi.org/10.1007/3-540-48970-3_1
14. Picek, S., Jakobovic, D., Miller, J.F., Batina, L., Cupic, M.: Cryptographic Boolean functions: one output, many design criteria. Appl. Soft Comput. **40**, 635–653 (2016)

Trade-Off of Networks on Weighted Space Analyzed via a Method Mimicking Human Walking Track Superposition

Shota Tabata(✉) (iD)

Kajima Corporation, 3-1, Motoakasaka 1-chome, Minato-ku, Tokyo, Japan
tbtgoat.contact@gmail.com

Abstract. This study proposes a method for constructing networks with a small total weighted length and total detour rate by mimicking human walking track superposition. The present study aims to contribute to the scarce literature on multiple objectives, the total weighted length and the total detour rate, by allowing branching vertices on weighted space. The weight on space represents the spatial difference in the implementation cost, such as buildings, terrains, and land price. In modern society, we need to design a new transportation network while considering these constraints so that the network has a low total weighted length that enables a low implementation cost and a low total detour rate that leads to high efficiency. This study contributes to this requirement. The proposed method outputs solutions with various combinations of the total weighted length and the total detour rate. It approximates the Pareto frontier by connecting inherent non-dominated solutions. This approximation enables the analysis of the relationship between the weighted space and the limit of effective networks the space can generate quantitatively. Several experiments are carried out, and the result infers that the area with a huge weight significantly affects the trade-off relationship between the total weighted length and the total detour rate. Quantitatively revealing the trade-off relationship between the total weighted length and the total detour rate is helpful in managerial situations under certain constraints, including the budget, needed operational performance, and so on.

Keywords: Network design · Self-organize · Pareto frontier · Weighted space · Random Delaunay network

1 Introduction and Related Work

Transportation networks play a key role in modern society. Typical networks, such as highways, railways, and electric wires, make urban transportation efficient. Moreover, there is a possible requirement of new networks, including hydrogen fuel pipelines and quantum internet networks. These infrastructures should be designed from the viewpoint of construction feasibility and transportation sustainability. Transportation networks require an enormous implementation cost consisting of construction cost, land expropriation expenses, land compensation, and so on. The reduction of implementation

M. Mernik et al. (Eds.): BIOMA 2022, LNCS 13627, pp. 247–261, 2022.
https://doi.org/10.1007/978-3-031-21094-5_18

cost makes the design more feasible. Simultaneously, the high efficiency of networks makes transportation sustainable because of the decrease in the transportation cost based on time or energy for conveying people or goods. Hence, a network developed in the city is desired to have low implementation and transportation costs.

Designing a network in existing cities has to consider the spatial difference in implementation cost. Buildings are obstacles whereon links are forbidden to be made. Construction cost escalate on steep terrain because of the construction difficulty. Moreover, the influence on the implementation cost is not only physical. Land expropriation expenses and compensation differ due to land price, and the implementation cost changes depending on the place. This study aims to develop a network design method that can be used in this situation.

Herein, we measure two network indices. One is a total weighted length, which we consider proportional to the implementation cost. The other is a total detour rate proportional to the transportation cost. This study calls the implementation cost of the unit length of a link the weight at a certain location. We define the function $\mu : \mathbb{R}^2 \mapsto \mathbb{R}$ as weight of points $(x, y) \in \mathbb{R}^2$. We assume that the link construction cost is positive, that is, $\mu(x, y) \geq 0$. Moreover, normalized by the minimum weight, $\mu(x, y) \geq 1$. Considering a network $G(V, E)$ connecting D, where V is the set of vertices, E is the set of links, and $D \in V$ is the set of demand vertices that G must connect, we can define the total weighted length $L_{\text{total}}(G)$ and the total detour rate $D_{\text{total}}(G)$ as the following Eqs. (1) and (2):

$$L_{\text{total}}(G) = \sum_{e \in E} \int_e \mu(x, y) dl. \tag{1}$$

$$D_{\text{total}}(G) = \sum_{u, v \in D} \text{ND}(u, v|G) / \sum_{v, u \in D} \text{ED}(u, v). \tag{2}$$

Here, dl denotes the infinitesimal length of link e on (x, y), $\text{ND}(u, v|G)$ denotes the shortest distance between u and v on G, and $\text{ED}(u, v)$ denotes the Euclidean distance between u and v. We assume that moving demands occur uniformly between demand vertices, giving Eq. (2).

The network with the minimum total weighted length is the weighted Euclidean Steiner minimum tree (weighted ESMT) [1, 2], and that with the minimum total detour rate is the complete graph (CG), connecting each pair of D by a straight line. The weighted ESMT has a higher total detour rate, and the CG has a higher total weighted length. Hence, there is a trade-off relationship between $L_{\text{total}}(G)$ and $D_{\text{total}}(G)$. To build a feasible and sustainable transportation network, we need to obtain a non-dominated network of the two indices.

Some heuristic studies are tackling the problem without weight, based on inspiration from natural network generation mechanisms. Among them is the self-organized network on a green space denuded by human walking, referred to as the walking track superposition network (WTSN). Some studies pay attention to the effectiveness of the WTSN, and the simulation models are developed [3–6]. Notably, WTSNs are effective in a magnitude similar to or greater than proximity graphs [6], conventionally considered ideal networks in terms of the total length and the total detour rate [7]. Moreover, the network of amoeba connecting food sources is focused on in this context [8–10]. The mechanism of amoeba connecting food sources is modeled mathematically, and it can connect demand points by an effective network [11]. Another research analyzes the effectiveness of the network of strings loosely connecting demand points and gathering together when they contain water [12]. Considering the difficulty of obtaining the exact solution, the heuristic methods inspired by natural phenomena generating an effective network can be helpful.

The problem of obtaining the non-dominated network connecting demand points on a weighted plane space has not been considered. Mathematical studies have been conducted to prove the existence of a network with a certain total length and maximum detour rate [13–15]. Moreover, several methods to obtain a non-dominated network have been developed [16–18], and numerous studies have dealt with network optimization concepts, such as transit networks [19], electric power distribution networks [20], and supply chain networks [21]. These methods choose the given links. However, to the best of our knowledge, the method to construct the effective network of the total weighted length and the total detour rate forgiving additional vertices from zero links has not been developed. This study proposes a method inspired by WTSN extended to weighted space and discusses the influence of weighted space on network effectiveness.

This study aims to enable a quantitative analysis of the trade-off relationship between $L_{\text{total}}(G)$ and $D_{\text{total}}(G)$. The most relevant contributions are (i) the development of a method for constructing networks with small total weighted length and small total detour rate on weighted space mimicking WTSN, (ii) the approximation of the Pareto frontier by using the proposed method, and (iii) the quantitative analysis of the relationship between the weighted space and the limit of effective networks. In a managerial situation, the quantitatively revealed trade-off relationship between $L_{\text{total}}(G)$ and $D_{\text{total}}(G)$ helps the decision-makers select the network under certain constraints, including the budget, needed operational performance, and so on.

The remainder of this study is organized as follows: Sect. 2 presents the simulation method for WTSN on weighted planar space. Section 3 shows the result for four abstract spaces. Section 4 discusses the function of the method and the influence of the weighted space on network effectiveness. Finally, the conclusions are presented in Sect. 5.

2 Simulation Model of WTSN on Weighted Space

2.1 Generation Process of WTSN on a Mixture of Different Ground Conditions

Walking environment physically and psychologically creates walking resistance, and Pedestrians walk along the path with the shortest length weighted by walking resistance [26]. Walking resistance varies based on ground conditions, such as green space, muddy land, and forest, and depends on the small-detour preference of pedestrians. Pedestrians are less likely to walk where walking load is high; however, if pedestrians have a high small-detour preference, they walk even in areas that are difficult to walk. This way, pedestrians walk with less walking load and detours. We can expect that the track becomes Pareto-optimal for the weighted length and detour rate.

The proposed algorithm mimics the generation process of WTSN on the ground under a mixture of different conditions. Highly walked locations become more walkable as vegetation is eroded, the land trodden, and so on. So, the walking load decrease and other pedestrians will walk into those places. However, ground conditions recover, and walking load increases when pedestrians do not walk. Therefore, pedestrian tracks interact with each other through the ground condition, and the ground is divided into two parts: with low and high walking load. This form a WTSN. As we expect each track to be the Pareto-optimal path of the weighted length and detour rate, we expect a WTSN, as the result of the superposition and aggregation of pedestrian tracks between demand vertices, to achieve a good performance of the total weighted length and total detour rate.

2.2 Pareto-Optimal Path Between Two Demand Vertices

We obtain agents' path as the Pareto-optimal path of weighted length and detour rate. The Pareto-optimal path is evidently one curve connecting two demand vertices. Let s and t be the demand vertices and P be the curve connecting s and t. We define the total weighted length and the total detour rate of P as follows:

$$L_{\text{total}}(P) = \int_s^t \mu(x, y)dl = \int_s^t \mu(x, y)\sqrt{1 + \left(\frac{dy}{dx}\right)^2}\, dx. \tag{3}$$

$$D_{\text{total}}(P) = \int_s^t dl/\text{ED}(s, t) = \int_s^t \sqrt{1 + \left(\frac{dy}{dx}\right)^2}\, dx/\text{ED}(s, t). \tag{4}$$

From Eqs. (3) and (4), we obtain the Pareto-optimal path by solving below:

$$\text{Min.} f(P) = (1 - \kappa')L_{\text{total}}(P) + \kappa' D_{\text{total}}(P) \qquad (0 \le \kappa' \le 1)$$

$$= \int_s^t \left((1 - \kappa')\mu(x, y) + \frac{\kappa'}{\text{ED}(s,t)}\right)\sqrt{1 + \left(\frac{dy}{dx}\right)^2}\, dx \tag{5}$$

$$\Leftrightarrow \text{Min.} f(P) = \int_s^t ((1 - \kappa)\mu(x, y) + \kappa)\sqrt{1 + \left(\frac{dy}{dx}\right)^2}\, dx \qquad (0 \le \kappa \le 1)$$

Here, κ' and κ are parameters controlling the trade-off between $L_{\text{total}}(P)$ and $D_{\text{total}}(P)$. The equivalence deformation depends on the variable transformation $\kappa = \kappa'/(\text{ED}(s, t)(1 - \kappa') + \kappa')$. Equation (5) shows that the Pareto-optimal path is the shortest with walking resistance $(1 - \kappa)\mu(x, y) + \kappa$.

To approximate the Pareto-optimal path, we use a random Delaunay network, rDn for short (also known as Poisson Delaunay tessellation [22]). The rDn is generated by the Delaunay tessellation of numerous randomly distributed nodes (Fig. 1). The shortest path length on the rDn is approximately 1.04 times longer than the Euclidean distance isotropically [23, 24]. Because of this feature, the rDn can approximate the shortest path with uniformly weighted regions [25, 26]. We weight the edge e of the rDn with 100,000 nodes by $\max(\mu(x_1, y_1), \mu(x_2, y_2))$, where (x_1, y_1) and (x_2, y_2) are the end nodes of e, and conduct the following experiments to confirm the shortest path approximation performance of rDn on weighted space.

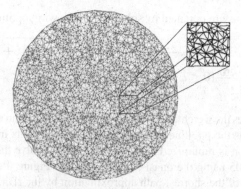

Fig. 1. Random Delaunay network.

On Space with a Uniformly Weighted Region

Figure 2 shows the setting of the space and the result of the shortest path approximation by rDn. We can determine the Pareto-optimal path giving the angle θ between the path and the gray area as the Pareto-optimal path is point-symmetric for $(x, y) = (90, 52.5)$ and does not go backward. Therefore, we can calculate κ corresponding to θ such that $f(P)$ is minimized. In this way, the Pareto frontier shown in Fig. 2(b) is obtained by changing θ by $1°$. The scatter plot in Fig. 2(b) shows $L_{\text{total}}/1.04$ and $D_{\text{total}}/1.04$ of the approximated shortest path by the rDn (because the length expands 1.04 times on the rDn to the Euclidean distance). The approximation was conducted by changing κ from 0.00 to 1.00 by 0.01 for five rDns. The locations and colors of the scatter plots are consistent with the exact Pareto frontier; an rDn can approximate the Pareto frontier of $L_{\text{total}}(P)$ and $D_{\text{total}}(P)$.

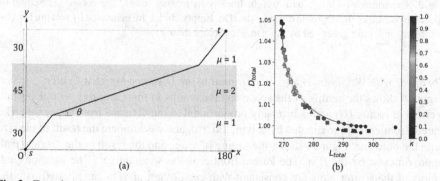

Fig. 2. Experiment on space with a uniformly weighted region. (a) Setting of space. (b) Result of the approximation by the rDn. Plot colors show κ, and plot shapes show the rDn types. The solid line is the exact Pareto frontier.

On Space with Weight Linearly Increasing Toward the Center

Figure 3 shows the setting of the space and the results in this case. $\mu(x, y)$ is conical, taking a value of 10 at the apex and 1 at the base and outside. We approximate the Pareto-optimal path as described below, and compare the rDn approximation with the outcome.

Approximate solutions are obtained by solving the following optimization problems.

$$\text{Min.} \quad \sum_{i=1}^{n-1} \left(\left(\frac{(1-\kappa)(\mu(x_i,y_i)+\mu(x_{i+1},y_{i+1}))}{2} + \kappa \right) \sqrt{(x_{i+1} - x_i)^2 + (y_{i+1} - y_i)^2} \right),$$
$$\text{s.t.} \quad x_i = -100 + \frac{200i}{n} \, (0 \le i \le n),$$
$$y_0 = 0, \, y_n = 0. \tag{6}$$

Equation (6) separates the weighted space into n parts with the same width in the direction of the x-axis and gets the polyline uniting straight-line segments in each part such that the objective function is minimized. We set $n = 100$ and solve the problem changing κ from 0 to 1 by 0.05 using the quasi-Newton method. Figure 3 shows the setting of space and the result of the shortest path approximation by the rDn. The Pareto frontier has a gentle s-shape. $L_{total}(P)$ and $D_{total}(P)$ become more sensitive to changes in κ and change their value significantly. The rDn approximation could not obtain the solutions in the middle of the frontier.

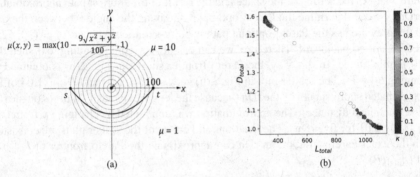

Fig. 3. Experiment on space with weight linearly increasing toward the center. (a) Setting of space. (b) Result of the approximation by the rDn. Empty circles are generated by solving Eq. (6), and filled circles are generated by the rDn. Plot colors show κ.

On Space with Weight Inversely Proportional to the Distance from the Center

Figure 4 shows the setting of the space and the results in this case. $\mu(x, y)$ is 1 outside a circle of radius 100 and is inversely proportional to the distance from the center. The Pareto frontier is approximated by solving Eq. (6), and we compare the result of the rDn with the outcome. Figure 4 shows the setting of space and the result of the shortest path approximation by the rDn. The Pareto frontier is the steep curve[1]. The locations and colors of the scatter plots are consistent with each other; an rDn can approximate the Pareto frontier of $L_{total}(P)$ and $D_{total}(P)$.

In summary, the approximation performance of the rDn for the Pareto frontier of $L_{total}(P)$ and $D_{total}(P)$ performed best on the space with a uniformly weighted region or the weight that is inversely proportional to the distance from a certain point. We can consider the space, similar to the former, as the construction difficulty affected by the terrain condition and the space, similar to the latter, as the land exploitation expenses or compensation expressed by a function inversely proportional to the distance from the city center. Thus, rDn can be useful when dealing with urban-scale issues.

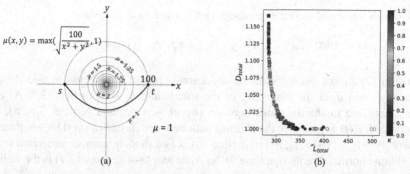

Fig. 4. Experiment on space with weight inversely proportional to the distance from the center. (a) Setting of space. (b) Result of the approximation by the rDn. Empty circles are generated by solving Eq. (6), and filled circles are generated by the rDn. Plot colors show κ.

2.3 Algorithm for WTSN on Weighted Space

Against the background of Sect. 2.1, we develop an algorithm for WTSN in weighted space (Algorithm 1). Algorithm 1 has five main parameters.

1. Decreasing speed of walking resistance N_-
2. Increasing speed of walking resistance N_+
3. Initial walking resistance w_{init}
4. Convergence walking resistance w_{conv}
5. Agent's small-detour preference degree κ

The weighted space $W \subset \mathbb{R}^2$ and set of demand vertices D are given. Table 1. shows the interpretation of the parameters and $\mu(x, y)$ in the context of pedestrian behavior.

Table 1. Interpretations of parameters in the context of pedestrian behavior.

	Interpretation in the context of a pedestrian
N_-	Speed of the ground becoming walkable by pedestrians trampling
N_+	Speed of the ground recovering while pedestrians do not walk
w_{init}	Walking load on a green space where the vegetation is complete
w_{conv}	Walking load on a space that pedestrians have trampled completely
κ	Pedestrian's small detour preference/inessential degree of the walking load on the walking resistance that agents feel
$\mu(x, y)$	Ratio of the walking load at (x, y) to that of green space

In Algorithm 1, W is discretized by an rDn (line 1), and we obtain D', the set of nearest nodes on rDn to D (line 2). Before starting the procedure, we provide the edges of rDn with the initial walking resistance (lines 3 and 4) as follows:

$$w_{\text{init}}(e) = \max(w_{\text{init}}((1 - \kappa)\mu(x_1, y_1) + \kappa), w_{\text{init}}((1 - \kappa)\mu(x_2, y_2) + \kappa)). \quad (7)$$

Equation (7) makes agents walk on the Pareto-optimal path between two demand vertices d_s and d_t at the beginning of the simulation, as seen in Sect. 2.2. At each step t, an agent randomly walks between two chosen demand vertices (line 8). The agent's path $P(t)$ is given by the shortest path between d_s and d_t on rDn, weighted by $(\lambda(w(e, t) - w_{\text{conv}}) + w_{\text{conv}})|e|$ on e (line 10). λ is a random number generated by the logarithmic normal distribution (line 9), $|e|$ is the length of e, and $w(e, t)$ is the walking resistance on e at step t.

λ represents the variation in the walking resistance perceived by pedestrians. We consider that λ relaxes the dependence of the outputs on the order of the two chosen vertices. $\lambda > 1$ indicates that pedestrians are sensitive to walking resistance and likely to walk along a path where the walking resistance is small. By contrast, $\lambda < 1$ implies that pedestrians are insensitive to walking resistance and tend to walk straight. Since $(\lambda(w(e, t) - w_{\text{conv}}) + w_{\text{conv}})|e|$ must exceed zero, we let λ follow the logarithmic normal distribution and empirically set the mean and variance of $\ln \lambda$ 0 and 0.5, respectively.

We then update the walking resistance for each edge on rDn (line 11). In line 14, we provide the walking resistance on e at step t as the following sigmoid function:

$$w(e, t) = \frac{w_{\text{init}}(e) - w_{\text{conv}}}{1 + \exp\left(\frac{1}{\alpha N_-}\left(c(e, t) - \frac{N_-}{2}\right)\right)} + w_{\text{conv}}. \quad (8)$$

Here, $c(e, t)$ is the parameter denoting how much e has been walked on until step t, and α is the parameter controlling the steepness of the sigmoid curve (empirically, $\alpha = 0.05$). We update $c(e, t)$ to $c(e, t + 1)$ as follow (line 13):

$$c(e, t + 1) = \begin{cases} 0 & \text{if } \eta < 0, \\ \eta & \text{if } 0 \leq \eta \leq N_-, \\ N_- & \text{if } \eta > N_-. \end{cases} \quad (9)$$

Here,

$$\eta = \begin{cases} c(e, t) - \frac{N_-/N_+}{|D|(|D|-1)} + 1 & \text{if } e \in P(t), \\ c(e, t) - \frac{N_-/N_+}{|D|(|D|-1)} & \text{if } e \notin P(t), \end{cases} \quad (10)$$

where $|D|$ is the number of demand vertices (line 12). In Eq. (10), the second term leads to an increase in the walking resistance, whereas the third term decreases it. The probability that $d_o, d_d \in D'$ are chosen is $1/|D|(|D| - 1)$. Therefore, we standardized the second term as $1/|D|(|D| - 1)$.

Based on Eqs. (7)–(10), the relationship between $w(e, t)$ and $c(e, t)$ depending on κ is shown in Fig. 5. Edges where $\mu = \max(\mu(x_1, y_1), \mu(x_2, y_2)) > 1$ have $w_{\text{init}}(e) = w(e, 0) > w_{\text{init}}$. The walking resistance decreases asymptotically to w_{conv} as an agent walks, and increases asymptotically to $w_{\text{init}}(e)$ when the agent does not. The speed of the decrease (increase) was controlled by N_- (N_+).

The output network consists of E_{acti}, a set of edges satisfying $w(e, t) \leq (w_{init}(e) - w_{conv})/2$ (lines 5 and 16). The convergence condition is that E_{acti} connects D' and $|E_{acti}|$ does not increase for $|D|(|D| - 1)$ steps (lines 5, 7, and 17–20). After the convergence, Algorithm 1 outputs the network by removing the nodes whose degrees are 1 from E_{acti} using the function Reconstitution(E_{acti}) (lines 21 and 22). The network dynamics of the simulation model are illustrated in Fig. 6.

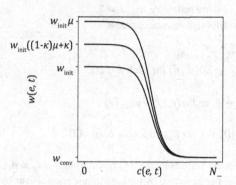

Fig. 5. Relationship between $w(e, t)$ and $c(e, t)$.

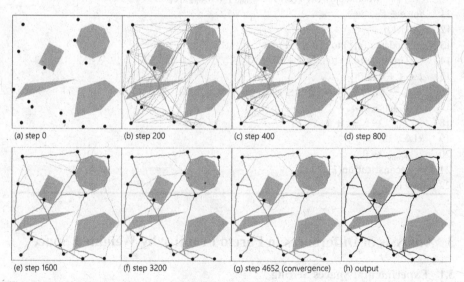

(a) step 0   (b) step 200   (c) step 400   (d) step 800

(e) step 1600   (f) step 3200   (g) step 4652 (convergence)   (h) output

Fig. 6. Network dynamics for the simulation model. Gray areas are the weighted regions (with weight equal to 2), the black dots are the demand vertices. Orange lines are rDn edges that agents have walked on, and their width depends on the walking resistance. The less the walking resistance, the thicker the width.

Algorithm 1. WTSN on weighted space.

Input: Weighted space W,

        set of demand vertices D,

        decreasing speed of walking resistance N_-,

        increasing speed of walking resistance N_+,

        initial walking resistance w_{init},

        convergence walking resistance $w_{conv} = 1$,

        agent's small-detour preference degree κ.

Output: Network G.

1 Generate rDn $\text{rDn}(V_D, E_D)$ on W.

2 $D' \leftarrow \{d' \mid \text{argmin}_{d' \in V_D} \text{ED}(d', d)$ for each $d \in D\}$.

3 $t \leftarrow 0$.

4 **for** $e \in E_D$, $c(e, t) \leftarrow 0$, and $w(e, t) \leftarrow w_{init}(e)$.

5 $E_{acti} \leftarrow \emptyset$, $t_{conv} = 0$.

6 **while** $t_{conv} \leq |D|(|D| - 1)$ or E_{acti} does not connect D',

7      $n_{prev} \leftarrow |E_{acti}|$.

8      $d_s, d_t \leftarrow$ two demand vertices randomly chosen from D' $(d_s \neq d_t)$.

9      $\lambda \leftarrow$ random number generated by the logarithmic normal distribution.

10    $P(t) \leftarrow$ the shortest path between d_s and d_t on rDn

           weighted by $(\lambda(w(e, t) - w_{conv}) + w_{conv})|e|$ on e.

11   **for** $e \in E_D$,

12   Get η.

13   Update $c(e, t)$ to $c(e, t + 1)$.

14   Update $w(e, t)$ to $w(e, t + 1)$.

15    $t \leftarrow t + 1$.

16    $E_{acti} \leftarrow \{e \mid e \in E_D, w(e, t) \leq (w_{init}(e) - w_{conv})/2\}$.

17   **if** $|E_{acti}| \leq n_{prev}$,

18      $t_{conv} \leftarrow t_{conv} + 1$.

19   **else**,

20      $t_{conv} \leftarrow 0$.

21 $G \leftarrow \text{Reconstitution}(E_{acti})$.

22 **return** G.

---

3 Analysis of Differences in Pareto Frontier by Weighted Space

3.1 Experimental Spaces Setting

We prepare several experimental spaces that assume real spaces, as shown in Fig. 7. Cases 1–1 and 1–2 have convex regions, considered as difficult construction areas, such as lakes, wetlands, forests, and other topographical barriers (Fig. 7(a)). Cases 2–1 and 2–2 have jagged regions (Fig. 7(b)). We can regard the concave parts as valleys and the sharp parts as mountains in these spaces. The weight of weighted regions is 2 in Cases 1–1 and 2–1, and 5 in Cases 1–2 and 2–2. The harder the construction is in such regions, the higher the unit implementation cost, as shown in the weight. Hence, in Cases

1–1, 1–2, 2–1, and 2–2, the weight shows the immaturity of the construction technique (the maturity of the construction technique leads the less difference in the weight). Cases 3 and 4 have a weight distribution inversely proportional to the distance from certain points, representing the land price in urban areas. In these cases, we consider the land expropriation or the land compensation proportional to the land price. Case 3 has one center (Fig. 7(c)), as defined by $\mu(x, y) = \max\left(\sqrt{100/(x^2 + y^2)}, 1\right)$, and Case 4 has two (Fig. 7(d)), as defined by $\mu(x, y) = \max(\left(\sqrt{75/((x - 25)^2 + (y - 25)^2)}\right)^{1/2}$, $\left(\sqrt{75/((x + 25)^2 + (y + 25)^2)}\right)^{1/4}, 1)$.

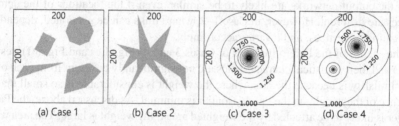

| (a) Case 1 | (b) Case 2 | (c) Case 3 | (d) Case 4 |

Fig. 7. Experimental weighted spaces.

3.2 Result of Pareto Frontier Approximation

We applied Algorithm 1 to the experimental spaces with each combination of $w_{init} = 1.1, 1.2, 1.5, 1.75, 2.0, 2.5, 3.0, 4.0, 5.0$, $N_+/N_- = 1.0, 1.5$ ($N_- = 15$) and $\kappa = 0.0, 0.1, 0.25, 0.5, 0.75, 0.999$. Using a large number of non-dominated solutions obtained through this procedure, we infer the form of the inherent Pareto frontier. We further approximated the Pareto frontier connecting many non-dominated solutions, the minimum spanning tree (MST), and the CG approximated by rDn. We ran the experiment 30 times by randomly changing the rDn of 100,000 nodes and the location of the demand vertices. Figure 8 shows the Pareto frontiers for the 30 runs. L_{total} is normalized by the L_{total} of MST^2.

We evaluate the effectiveness of the networks generable by the space using the approximated Pareto frontier. We cannot generate networks below and to the left of the Pareto frontier (with small L_{total} and D_{total}). Therefore, the upper and more right-sided the Pareto frontier is, the less effective networks the space can generate. The area on the L_{total}–D_{total} space where we cannot generate networks is the area surrounded by $L_{total} = 1$, $D_{total} = 1$, and the Pareto frontier. We call this value the complementary index c. The characteristics of c for each case are presented in Table 2..

We can infer that the space with regions with larger weights tends to generate effective networks with difficulty. The statistical characteristics of the c of Case 1–1 are smaller than those of Case 1–2. This trend can be observed in Cases 2–1 and Case 2–2. We can consider that this is because it takes a large L_{total} to make small detour links in the region with a large weight.

We can infer that the standard deviation of c describes the distance between the Pareto frontiers corresponding to the 30 locations of the demand vertices. The mean and standard deviation of c are denoted as $E(c)$ and $\sigma(c)$, respectively. In Cases 1–2 and 2–2, having weighted regions with a weight of 5, the $E(c)$ ratio is 0.99, whereas the $\sigma(c)$ ratio is 1.68. $\sigma(c)$ is significantly different in comparison to $E(c)$ between Cases 1–2 and 2–2. This difference is due to the positions of the weighted regions. In Case 2–2, the weighted region is centered on W, and links tend to be generated around the region. Thus, the output networks are likely to be similar, even if the locations of the demand vertices are different. However, in Case 1–2, where links can be generated, depends on the location of vertices, and c varies accordingly.

The shapes of the Pareto frontiers of Cases 3 and 4 (Fig. 8(e) and Fig. 8(f), respectively) and their statistical characteristics are not significantly different from each other. This similarity is because the area where the weight is considerable is so small that the influence of the weight on the Pareto frontier is minimal. As discussed above, the Pareto frontier is likely to be affected by the weighted area if the weight is large. The area with a weight of more than 2 is 4.9% in Case 3 and 2.9% in Case 4. The weight distributions of Cases 3 and 4 differ so slightly that there is little difference in the trade-off relationship between them.

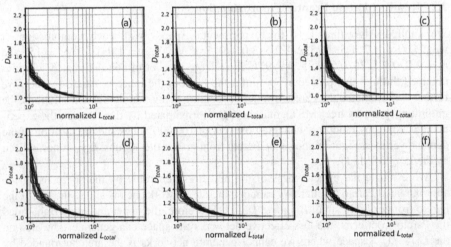

Fig. 8. Pareto frontier approximation by WTSN on weighted spaces. (a)–(f) correspond to the results of Cases 1–1, 1–2, 2–1, 2–2, 3, and 4, respectively.

Table 2. The characteristics of the complementary index.

	Case 1–1	Case 1–2	Case 2–1	Case 2–2	Case 3	Case 4
Count			30			
Mean	24.6	38.1	25.0	38.6	24.2	22.9
Std	2.2	4.2	1.5	2.5	2.0	2.0
Max	28.2	45.7	28.3	43.6	28.1	26.3
Median	25.1	37.4	24.8	38.5	24.2	22.7
Min	20.3	30.3	21.9	33.6	20.4	18.2

4 Discussion

In Algorithm 1, agents walk straight with no weighted region between the demand vertices at the beginning (Fig. 6(b)). As the process continues, the agents' paths superimpose each other and aggregate into a network (Fig. 6(c)–(g)). This behavior is consistent with previous studies [3–6]. Different from [3–6], the initial walking resistance is not homogeneous. At the start of the simulation, the agents walked on weighted regions (Fig. 6(b)). However, commonly used places are often passed on and become parts of the WTSN, whereas redundant links disappear (Fig. 6(g)). This agent's behavior is based on the setting of the initial walking resistance (Eq. (7)), and the agent's path is the Pareto-optimal path of the weighted length and detour rate, as mentioned in Sect. 2.2. The major difference from [6] is the applicability of the method to the weighted space, even though Algorithm 1 has similarities with [6].

The initial walking resistance includes space attribute $\mu(x, y)$ and personal attribute κ, although [26, 27] denotes the walking resistance as a single integrated value for predicting pedestrian paths. Equation (7) represents walking resistance in detail and can be helpful in the prediction of walking paths.

Our result indicates that technological maturity stabilizes the network design. As the technology matures, the weight, defined as the implementation cost of the unit link length on the place, decreases. Table 2. implies that with smaller weight, (i) we can design more functional networks (the complementary index is smaller) and (ii) the functional variance by the location of the demand vertices decreases (the standard deviation is small). However, an exact or approximate way to evaluate and discuss the accuracy of our proposed method still needs to be developed.

5 Conclusion and Further Work

In modern society, newly laid networks require to be designed in consideration of buildings, terrains, and land price, which lead to the spatial difference in implementation cost. This study's contributions to the literature are as follows: (i) we confirmed that an rDn could approximate the Pareto-optimal path of the weighted length and the detour rate between two demand vertices and have developed the method for constructing networks with small total weighted length and small total detour rate on weighted space mimicking human walking track superposition, (ii) we approximated the Pareto frontier by using the proposed method, and (iii) we quantitatively analyzed the relationship between the weighted space and the limit of effective networks. This study is novel vis-à-vis the conceptualization of a method for designing efficient networks by allowing branching vertices on weighted space.

As previously mentioned, a method to obtain the exact non-dominated network of the total weighted length and the total detour rate needs to be developed. Moreover, we assume that the moving demands are homogeneous. However, network design needs to consider the moving demand deviation. We have to extend the proposed method to output the effective networks of the total weighted length and the total detour rate weighted by moving demands for further developments.

Notes

1. The weight at the center is infinite; therefore, the base of $L_{total}(P)$ expands to infinity.
2. The network with the minimum L_{total} is the weighted ESMT. Therefore, L_{total} should be normalized by the weighted ESMT. However, MST is the network connecting the demand vertices with the minimum total weighted length using links between two demand vertices. In this study, we used MST instead of weighted ESMT because of the difficulty in obtaining weighted ESMT.

Acknowledgements. We would like to thank Editage (www.editage.com) for English language editing.

References

1. Garrote, L., Martins, L., Nunes, U.J., Zachariasen, M.: Weighted Euclidean Steiner trees for disaster-aware network design. In: 15th International Conference on the Design of Reliable Communication Networks (DRCN), pp. 138–145. IEEE, Coimbra, Portugal (2019). https://doi.org/10.1109/DRCN.2019.8713664
2. Tabata, S., Arai, T., Honma, K., Imai, K.: A heuristic for the weighted Steiner tree problem by using random Delaunay networks. J. City Plan Inst. Japan **55**(3), 459–466 (2020). https://doi.org/10.11361/journalcpij.55.459
3. Helbing, D., Molnár, P., Farkas, I., Bolay, K.: Active walker model for the formation of human and animal trail systems. Phys. Rev. E **56**, 2527–2539 (1991). https://doi.org/10.1002/bs.3830360405
4. Helbing, D., Keltsch, J., Molnár, P.: Modelling the evolution of human trail systems. Nature **388**, 47–50 (1997). https://doi.org/10.1038/40353
5. Helbing, D., Molnár, P., Farkas, I., Bolay, K.: Self-organizing pedestrian movement. Environ. Plan B: Plan. Des. **28**, 361–383 (2001). https://doi.org/10.1068/b2697
6. Tabata, S., Arai, T., Honma, K., Imai, K.: Method for constructing cost-effective networks by mimicking human walking track superposition. J. Asian Archit. Build Eng. (2022). https://doi.org/10.1080/13467581.2022.2047056
7. Watanabe, D.: Evaluating the configuration and the travel efficiency on proximity graphs as transportation networks. Forma **23**, 81–87 (2008)

8. Nakagaki, T., Kobayashi, R., Hara, M.: Smart network solutions in an amoeboid organism. Biophys. Chem. **107**(1), 1–5 (2004). https://doi.org/10.1016/S0301-4622(03)00189-3

9. Nakagaki, T., Kobayashi, R., Nishiura, Y., Ueda, T.: Obtaining multiple separate food sources: behavioural intelligence in the *Physarum* plasmodium. Biol. Sci. **271**(1554), 2305–2310 (2004). https://doi.org/10.1098/rspb.2004.2856

10. Tero, A., Yumiki, K., Kobayashi, R., Saigusa, T., Nakagaki, T.: Flow-network adaptation in *Physarum* amoebae. Theory Biosci. **127**, 89–94 (2008). https://doi.org/10.1007/s12064-008-0037-9

11. Tero, A., et al.: Rules for biologically inspired adaptive network design. Science **327**(5964), 439–442 (2010). https://doi.org/10.1126/science.1177894

12. Schaur, E.: IL39 Non-planned settlements characteristic features. Institute for Lightweight Structures, University of Stuttgart, Stuttgart, pp. 36–51 (1991)

13. Carmi, P., Chaitman-Yerushalmi, L.: Minimum weight Euclidean t-spanner is NP-hard. J. Discrete Algorithms **22**, 30–42 (2013). https://doi.org/10.1016/j.jda.2013.06.010

14. Le, H., Solomon, S.: Truly optimal Euclidean spanners. In: 2019 IEEE 60th Annual Symposium on Foundations of Computer Science (FOCS), pp. 1078–1100. IEEE, Baltimore, MD, USA (2019). https://doi.org/10.1109/FOCS.2019.00069

15. Bhore, S., Tóth, C.D.: On Euclidean Steiner (1+ε)-spanners. In: 38th International Symposium on Theoretical Aspects of Computer Science (STACS 2021), pp. 13:1–13:16. Schloss Dagstuhl, Dagstuhl, Germany (2021). https://doi.org/10.4230/LIPIcs.STACS.2021.13

16. Scott, A.J.: The optimal network problem: some computational procedures. Transp. Res. **3**(2), 201–210 (1969). https://doi.org/10.1016/0041-1647(69)90152-X

17. Ridley, T.M.: An investment policy to reduce the travel time in a transportation network. Transp. Res. **2**(4), 409–424 (1968). https://doi.org/10.1016/0041-1647(68)90105-6

18. Magnanti, T.L., Wong, R.T.: Network design and transportation planning: models and algorithms. Transp. Sci. **18**(1), 1–55 (1984). https://doi.org/10.1287/trsc.18.1.1

19. Lee, Y.-J., Vuchic, V.R.: Transit network design with variable demand. J. Transp. Eng. **131**, 1 (2005). https://doi.org/10.1061/(ASCE)0733-947X(2005)131:1(1)

20. Nowdeh, S.A., et al.: Fuzzy multi-objective placement of renewable energy sources in distribution system with objective of loss reduction and reliability improvement using a novel hybrid method. Appl. Soft. Comput. **77**, 761–779 (2019). https://doi.org/10.1016/j.asoc.2019.02.003

21. Sahebjamnia, N., Fathollahi-Fard, A.M., Hajiaghaei-Keshteli, M.: Sustainable tire closed-loop supply chain network design: hybrid metaheuristic algorithms for large-scale networks. J. Clean Prod. **196**, 273–296 (2018). https://doi.org/10.1016/j.jclepro.2018.05.245

22. Okabe, A., Boots, B., Sugihara, K., Chiu, S.N.: Spatial Tessellations Concepts and Applications of Voronoi Diagrams, 2nd edn., p. 391. Wiley, New York (2000)

23. Imai, K., Fujii, A.: A study on Voronoi diagrams with two-dimensional obstacles. J. City Plan Inst. Japan **42**(3), 457–462 (2007). https://doi.org/10.11361/journalcpij.42.3.457. (in Japanese)

24. Chenavier, N., Devilers, O.: Stretch factor in a planar Poisson-Delaunay triangulation with a large intensity. Adv. Appl. Probab. **50**(1), 35–56 (2018). https://doi.org/10.1017/apr.2018.3

25. Imai, K., Fujii, A.: An approximate solution of restricted Weber problems with weighted regions. J. City Plan. Inst. Japan **43**(3), 85–90 (2008). https://doi.org/10.11361/journalcpij.43.3.85. (in Japanese)

26. Tabata, S., Arai, T., Honma, K., Imai, K.: The influence of walking environments on walking tracks through reproduction of desire paths. J. City Plan Inst. Japan **54**(3), 1562–1569 (2020). https://doi.org/10.11361/journalcpij.54.1562. (in Japanese)

27. Al-Widyan, F., Al-Ani, A., Kirchner, N., Zeibots, M.: An effort-based evaluation of pedestrian route selection. Sci. Res. Essays **12**(4), 42–50 (2017)

Towards Interpretable Policies in Multi-agent Reinforcement Learning Tasks

Marco Crespi, Leonardo Lucio Custode⬛, and Giovanni Iacca⁽⊠⁾⬛

Department of Information Engineering and Computer Science, University of Trento,
Trento, Italy
marco.crespi@studenti.unitn.it,
{leonardo.custode,giovanni.iacca}@unitn.it

Abstract. Deep Learning (DL) allowed the field of Multi-Agent Reinforcement Learning (MARL) to make significant advances, speeding-up the progress in the field. However, agents trained by means of DL in MARL settings have an important drawback: their policies are extremely hard to interpret, not only at the individual agent level, but also (and especially) considering the fact that one has to take into account the interactions across the whole set of agents. In this work, we make a step towards achieving interpretability in MARL tasks. To do that, we present an approach that combines evolutionary computation (i.e., grammatical evolution) and reinforcement learning (Q-learning), which allows us to produce agents that are, at least to some extent, understandable. Moreover, differently from the typically centralized DL-based approaches (and because of the possibility to use a replay buffer), in our method we can easily employ Independent Q-learning to train a team of agents, which facilitates robustness and scalability. By evaluating our approach on the Battlefield task from the MAgent implementation in the PettingZoo library, we observe that the evolved team of agents is able to coordinate its actions in a distributed fashion, solving the task in an effective way.

Keywords: Reinforcement learning · Multi-agent systems ·
Grammatical evolution · Interpretability

1 Introduction

In recent years, the application of Deep Learning (DL) to the field of Multi-Agent Reinforcement Learning (MARL) led to the achievement of significant results in the field. While DL allows to train powerful multi-agent systems (MASs), it has some drawbacks. First of all, to exploit state-of-the-art deep reinforcement learning (RL) algorithms, one often has to employ centralized approaches for training [1], which limits the scalability of the system, i.e., no agents can be added after the MAS has been trained. Moreover, deep RL methods suffer from

M. Mernik et al. (Eds.): BIOMA 2022, LNCS 13627, pp. 262–276, 2022.
https://doi.org/10.1007/978-3-031-21094-5_19

an even worse drawback: the lack of *interpretability*[1]. In fact, in safety-critical or high-stakes contexts, DL approaches cannot be employed as they are not fully predictable [3–5] and, thus, they may exhibit unexpected behaviors in edge cases. While interpretability in RL is an important concern, in MARL it is even more important. In fact, in contrast to traditional RL setups where safety can be assessed by inspecting the trained agent, in MARL not only do we need to analyze each agent, but we also need to understand their *collective behavior*.

In this paper, we employ a recently proposed methodology [6] (originally designed for single-agent tasks) for training an interpretable MAS. More specifically, we extend the setup proposed in [6] by creating a *cooperative co-evolutionary algorithm* [7] in which each evolutionary process addresses the evolution of an agent of the MAS. As a baseline, we also provide the results obtained when a single policy is trained for all the agents in the MAS. We evaluate our approach on the Battlefield task from MAgent [8] (implemented in the PettingZoo library [9]). The teams evolved with our approach are able to obtain promising performances, eliminating the whole opponent team in up to 98% of the cases.

So, the main contribution of this paper are: 1) the extension of the approach presented in [6] to multi-agent reinforcement learning settings; 2) the introduction of two approaches, a co-evolutionary one and single-policy one; 3) a validation of the proposed methods on the Battlefield task.

The rest of the paper is structured as follows. In the next section, we briefly overview the related work. In Sect. 3, we describe the proposed method. Then, in Sect. 4 and 5 we present the experimental setup and the numerical results, respectively. Finally, we draw the conclusions in Sect. 6.

2   Related Work

In the following, we make a summary of the state-of-the-art in the field of MARL. For a more complete review, we refer the reader to [1, 2, 10–12].

In a preliminary work [11], the authors explained the advantages of adopting a multi-agent approach instead of a single, complex agent approach. Several approaches have then been proposed for MARL. In [13] the authors compared two function approximators in the iterated prisoner's dilemma: a table-based approach and a recurrent neural network (RNN). The experiments showed that the agents based on the tabular approach were more prone to cooperate than the ones trained using the RNN, indicating that the agents trained by using the tabular approach had learned a better approximation of the Q function. Littman [14] presented a novel algorithm based on Q-learning and minimax, named "minimax Q". This algorithm, in the experimental results, proved to be able to learn policies that were more robust than the policies learned by Q-learning. In [15] the authors made use of cooperative co-evolution with strongly-typed genetic programming (GP) to evolve agents for a predator-prey game. The evolved strategies were more effective than handcrafted policies.

---

[1] In the rest of this paper, we will define as an interpretable system one that can be *understood* and inspected by humans [2].

Independent Q-learning (IQL) [16] is another convenient approach to MARL, as it is scalable and decentralized. However, when using neural networks as function approximators for reinforcement learning, this method cannot be applied. In fact, the need for a replay buffer does not make this method suitable in settings with neural networks. To mitigate this issue, several approaches have been proposed [17–21]. Other approaches circumvent this problem by using instead the actor-critic model [22–26].

Recently, some approaches have been proposed to measure the interpretability of a machine learning model. For instance, Virgolin et al., [27], propose a metric of interpretability based on the elements contained in the mathematical formula described by the model. In [28], the authors suggest that the computational complexity of the model can be used as a measure of interpretability. In this paper we follow this approach, assuming that less complex models are easier to interpret, see Sect. 5.1.

3   Method

The goal of our work is to produce interpretable agents that are capable of cooperate to solve a given task. To do that, we evolve populations of interpretable agents in the form of decision trees. To evolve these decision trees, we use the same approach that was recently proposed in [6,29]. In particular, we use the Grammatical Evolution (GE) algorithm [30] to evolve a genotype made of integers that, by using a grammar translator, is converted into a decision tree. However, we do not build the full decision tree. Instead, we only build the inner structure (i.e., the tree without leaves). The reason behind this choice relies on the fact that we want to *exploit* at their best the rewards given by the environment, using them to train the state-action function embedded in the leaves. Moreover, using Q-learning allows the agents to refine (and modify) their behavior in real-time, without having to wait for the next generation to improve the performance, which is particularly useful in multi-agent settings.

Finally, it is important to note that our method employs a cooperative coevolutionary process [7], where each population optimizes the structure of the tree for a particular agent of the environment.

3.1   Creation of the Teams

To evaluate a genotype, we have to assess the quality of the corresponding phenotype when placed inside a team. Each agent (i.e., a member of the team) has its own evolutionary process (i.e., there is a separate population for each agent in the task). Thus, we assemble teams composed of one phenotype (i.e., a genotype transformed into a decision tree) taken from each agent-local population.

Each agent-local population has N_{ind} individuals, such that N_{ind} different teams are created. Each i-th team is formed by the corresponding i-th individuals (one per each agent-local population), where i is an index $\in [0, N_{ind} - 1]$. This approach guarantees that each individual from each agent-local population is evaluated exactly once. Note that the selection operator, when applied, shuffles the array of the individuals. This means that an individual from an agent-local population is generally not always evaluated with the same individuals taken from the other agent-local populations.

At the end of the evolutionary process, we form the final team by combining the best individuals from all the agent-local populations. Moreover, by using an adoption mechanism (described in Sect. 3.4), the structure of the best agents may be shared between different agent-local populations.

3.2 Fitness Evaluation

Once a team is created, it undergoes N_{ep} episodes of simulation of the task. In the simulation phase, the agents perform IQL (with a dynamic ε-greedy exploration approach) to learn the function that maps the leaves to actions. By using IQL, each agent does not have to take into account the choices made by the other agents, as these are modeled as part of the environment. Moreover, given a sufficient number of episodes for the evaluation, the continuous learning of all the agents results in a co-adaptation. After the simulation phase, the seventh decile of the returns (i.e., the cumulative reward for each episode) received by an agent is used as fitness. The choice of the seventh decile lies on the fact that our fitness function is meant to describe the quality of a genotype as the quality of the state-space decomposition function [6], which can only be measured when the performance of the agent converges. While also the mean, the maximum, and the median have been considered as aggregation functions to compute the fitness, they have been discarded for the following reasons. Since the agents initially use a high ε for the exploration, the initial returns have a significant impact on the mean, thus they do not reflect the true quality of the genotype. Using the median would also present problems: on the one hand, the median would discard all the episodes in which the co-operation between the agents was fruitful enough to receive high returns; on the other hand, since we expect the returns to grow towards the end of the simulation phase, using the median would mean that we take into account the performance of a not-fully-trained agent. Finally, if we used the maximum to aggregate the returns, we would give too much importance to spurious good performance that may occur in the simulation (e.g., returns obtained just by randomly effective behaviors), without taking into account the performance of the trained version of the agent. In a preliminary experimental phase, the seventh decile represented a good trade-off between the median and the maximum, reflecting more closely the performance of the agents. The fitness evaluation process is described in Fig. 1.

3.3 Individual Encoding

Each individual is represented as a list of integers, where each integer indicates the production to choose for the current rule (modulo the number of productions). Unlike the original version of GE, we do not use variable-length genotypes. Instead, the genotype is a list that has fixed length. The process used to create

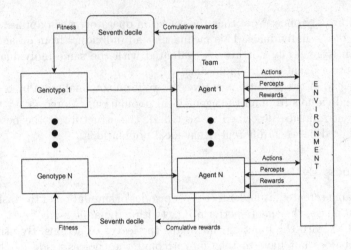

Fig. 1. Block diagram of the fitness evaluation process.

decision trees from genotypes (i.e., lists of integers) is the following. Starting from the first integer of the list, we apply the first (i.e., leftmost) non-expanded rule from the current phenotype by using the production rule indicated by the current integer (modulo the number of productions). The start symbol for the grammar is called "dt". The process can terminate in two different ways, depending on the case: (a) The phenotype does not contain any non-expanded rule: in this case, the phenotype is simply returned; or (b) all the parameters from the genotype have been converted into productions, but there are still non-expanded rules: in this case, the missing branches of the trees are linked to novel leaves.

3.4   Operators

Mutation. The mutation operator used in this work is the uniform mutation. This operator mutates each gene of the genotype with a probability p_{gene}. When a gene of the genotype is selected for mutation, its next value is selected uniformly $\in [0, M]$, where M is a number significantly bigger than the maximum number of choices in the grammar, to ensure that the productions are approximately uniformly distributed.

Crossover. The crossover operator used in this work is the one-point crossover operator. This operator simply chooses a random splitting point for the two fixed-length genotypes. Then, it produces two offspring by mixing the two substrings of the genotypes.

Selection. The individuals are selected by means of a tournament selection. This operator creates N_{ind} "tournaments" (i.e., random groups of s_t individuals taken from an agent-local population). Then, for each tournament, the best individual is selected to create the population for the next generation.

Replacement. Individuals in each population are replaced by their offspring, (obtained through mutation or crossover) when the new individuals perform better than their parents. If an individual is obtained through mutation, it will replace its parent only if it reaches a better fitness. In case of crossover (which involves two individuals and two parents) the individual of the offspring with the best fitness replaces the parent with the worst fitness. This mechanism also allows to systematically discard "adopted" individuals (see the next paragraph) that perform worse than their parents in the new population.

Adoption. The adoption of an individual happens at the end of each generation. An agent-local population is randomly chosen and its individual with the highest fitness is selected. At this point, the selected individual is copied into the other agent-local populations, replacing a randomly selected individual from the offspring. The adopted individual's parents are then assigned to the replaced individual's parents. The reason why we use this adoption mechanism lies in the reward system of the specific BattleField environment (see Sect. 4.1). As mentioned by the authors of PettingZoo: "Agents are rewarded for their individual performance, and not for the performance of their neighbors, so coordination is difficult"[2]. This means that only agents capable of hitting or killing enemies (this will become clearer in the next section) obtain a high fitness and the adoption mechanism allows sharing "knowledge" across agent-local populations.

4   Experimental Setup

4.1   Environment

We simulate a multi-agent environment by using the PettingZoo library [9]. More specifically, we use the Battlefield environment from the MAgent [8] suite. A screenshot of the environment is shown in Fig. 2.

In this task, there are two teams: the red team and the blue team. As the name of the environment suggests, the goal of each team is to defeat the other team by killing all of its members. The environment is an 80×80 grid. To win the battle, the agents have to learn to collaborate with their team in order to eliminate the enemies, and to move through the map to overcome walls and obstacles.

Each agent has a perceptive field of 13×13 squares and can either move or attack at each turn. The agents' perception is composed of: local presence/absence of an obstacle in a square; local presence/absence of a teammate/enemy in a square; health points (hp) of the teammate/enemy in a square; global density of teammates/enemies. A square represents a 7×7 quadrant of the environment. Note that each agent's local perception area corresponds to a circle with a radius of six squares around that agent. Moreover, to simplify the learning phase (and the interpretability of the agents evolved), we perform a pre-processing of these features, based on domain knowledge, in order to

---

[2] https://www.pettingzoo.ml/magent/battlefield (accessed on 02/02/2022).

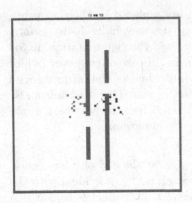

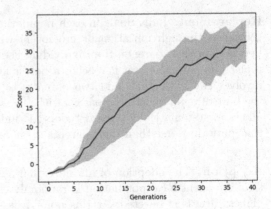

Fig. 2. A screenshot from the Battlefield environment.

Fig. 3. Maximum return (average ± std. dev. across 10 runs of the proposed co-evolutionary approach) at each generation.

obtain higher-level features that are then fed as inputs to the decision tree. The selected features, extracted from the raw observations, are reported in Table 1. The "Abbreviation" column shows the abbreviation that we will use throughout the text to refer to a specific feature.

Both local and global density are calculated based on the active agents in the environment, i.e., killed agents are not taken into account.

Each agent initially has 10 hp. When an agent attacks another agent (called target), the target's hp are decreased by 2 hp. Moreover, each turn increases the agents' health points by 0.1 hp (unless the agent already has already 10 hp).

An agent, at each step, can perform 21 discrete actions: no action; move to any of the 8 adjacent squares; move to two squares on either left, right, up, down; attack any of the 8 adjacent squares.

Table 2 shows the action that can be performed by the agent. As in Table 1, the "Abbreviation" column shows how we refer to the actions in the remainder of the text. The rewards obtained by the environment are the following: 5 points if the agent kills an opponent; -0.005 points for each timestep (a time penalty, thus the quicker the team wins, the higher the reward); −0.1 for attacking (to make the agent attack only when necessary); 0.9 when the agent hits an opponent (to give a quicker feedback to the agent, without having to wait for killing an agent to obtain a positive reward that encourages hitting enemies); −0.1 if the agent dies. At each timestep, the agent receives a combination of these rewards based on the events that happened in the last timestep. For instance, if an agent attacks and hits an enemy, it obtains a total reward of $r = 0.9 - -0.1 - -0.005$.

While there is no reward for collaboration, we decided to not alter the reward function to encourage it, to preserve the original configuration of the environment. Note that we evolve only one of the two teams (the blue one), while the other team (the red one) uses a random behavior for all the agents. This choice has been made in order to provide a non-biased baseline policy, i.e., to prevent

Table 1. Extracted features, their abbreviation and their domain.

Feature	Abbreviation	Domain
Obstacle 2 squares above	o_{2a}	$\{0, 1\}$
Obstacle 2 squares left	o_{2l}	$\{0, 1\}$
Obstacle 2 squares right	o_{2r}	$\{0, 1\}$
Obstacle 2 squares below	o_{2b}	$\{0, 1\}$
Obstacle 1 square above-left	o_{1al}	$\{0, 1\}$
Obstacle 1 square above	o_{1a}	$\{0, 1\}$
Obstacle 1 square above-right	o_{1ar}	$\{0, 1\}$
Obstacle 1 square left	o_{1l}	$\{0, 1\}$
Obstacle 1 squares right	o_{1r}	$\{0, 1\}$
Obstacle 1 square below-left	o_{1bl}	$\{0, 1\}$
Obstacle 1 squares below	o_{1b}	$\{0, 1\}$
Obstacle 1 squares below-right	o_{1br}	$\{0, 1\}$
Allied global density above	ag_a	$[0, 1]$
Allied global density left	ag_l	$[0, 1]$
Allied global density same quadrant	ag_s	$[0, 1]$
Allied global density right	ag_r	$[0, 1]$
Allied global density below	ag_b	$[0, 1]$
Enemies global density above	eg_a	$[0, 1]$
Enemies global density left	eg_l	$[0, 1]$
Enemies global density same quadrant	eg_s	$[0, 1]$
Enemies global density right	eg_r	$[0, 1]$
Enemies global density below	eg_b	$[0, 1]$
Enemies local density above	el_a	$[0, 1]$
Enemies local density left	el_l	$[0, 1]$
Enemies local density right	el_r	$[0, 1]$
Enemies local density below	el_b	$[0, 1]$
Enemy presence above-left	e_{al}	$\{0, 1\}$
Enemy presence above	e_a	$\{0, 1\}$
Enemy presence above-right	e_{ar}	$\{0, 1\}$
Enemy presence left	e_l	$\{0, 1\}$
Enemy presence right	e_r	$\{0, 1\}$
Enemy presence below-left	e_{bl}	$\{0, 1\}$
Enemy presence below	e_b	$\{0, 1\}$
Enemy presence below-right	e_{br}	$\{0, 1\}$

Table 2. Actions that the agent can perform.

Action	Abbreviation
Move 2 squares above	m_{2a}
Move 1 square above-left	m_{1al}
Move 1 square above	m_{1a}
Move 1 square above-right	m_{1ar}
Move 2 squares left	m_{2l}
Move 1 square left	m_{1l}
No action	m_n
Move 1 squares right	m_{1r}
Move 2 squares right	m_{2r}
Move 1 square below-left	m_{1bl}
Move 1 squares below	m_{1b}
Move 1 squares below-right	m_{1br}
Move 2 squares below	m_{2b}
Attack above-left	a_{al}
Attack above	a_a
Attack above-right	a_{ar}
Attack left	a_l
Attack right	a_r
Attack below-left	a_{bl}
Attack below	a_b
Attack below-right	a_{br}

the evolved policies from overfitting to a specific handmade policy for the red team. Furthermore, we decided not to competitively co-evolve the policies for both teams (blue and red) to reduce the complexity of the evolutionary process, and focus on the interpretability of the evolved policy for the blue team. We reserve this kind of investigations for future works. For each fitness evaluation, N_{ep} episodes are simulated, each of 500 timesteps.

4.2 Parameters

The parameters used for GE and Q-learning are shown in Table 3. To ensure that the Q function tends to the optimal one, we employ a learning rate of $\alpha = \frac{1}{v}$, where v is the number of visits made to the state-action pair [31]. The grammar for the GE algorithm is shown in Table 4. Note that we constrain the grammar to evolve orthogonal decision trees, i.e., decision trees whose conditions are in the form $x < c$, where x is a variable and c is a constant.

Table 3. Parameters used for the two algorithms (Grammatical Evolution and Q-learning) used in the experimentation.

Algorithm	Parameter	Value
Grammatical Evolution	N_{ind}	60
	N_{gen}	40
	p_{xover}	0.4
	p_{mut}	0.8
	p_{gene}	0.05
	Genotype length	500
	Selection	Tournament
	s_t	3
Q-learning	α	$1/v$
	ε	1
	N_{ep}	400
	$decay_\varepsilon$	0.99

Table 4. Grammar used to evolve the decision trees. "|" denotes the possibility to choose between different productions; "dt" indicates the start symbol.

Rule	Production
Dt	$\langle root \rangle$
Root	$\langle condition \rangle$	leaf
Condition	if $\langle input\_index \rangle < \langle float \rangle$
	then $\langle root \rangle$ else $\langle root \rangle$
Input_index	$[0, 33]$, step 1
Float	$[0.1, 0.9]$, step 0.1

5   Experimental Results

We perform 10 independent evolutionary runs to evolve the policy of each agent in the blue team. Figure 3 shows the average maximum return (across the 10 runs) during the evolutionary process generation. The shaded area indicates the standard deviation across runs. We should note that while the average trend did not reach yet a plateau after the considered number of generations, we had to limit the total duration of our runs due to constraints on the available computational resources. On average, one full run of our approach takes approximately 30 h on a 16-core machine with parallelization at the level of the individual evaluation.

Since the goal of the task is the elimination of the opponent team, we use two metrics to analyze the results in a post-hoc test phase (i.e., after the evolutionary process): the number of opponents killed, and the agents' returns over 100 unseen episodes. Table 5 shows the results of this test phase. For each of the 10 evolutionary runs, we report the statistics obtained with a team composed of the best agents (one for each population) evolved in that run over unseen episodes. The "Team kills" row shows the descriptive statistics of the number of enemies killed in each episode. Note that a team is formed by 12 agents therefore in a single episode the number of enemies killed is limited between 0 and 12. The "Agents' returns" row shows the descriptive statistics of the average returns of all the agents in the team. The "Completed" column shows the percentage of episodes in which the team was able to eliminate the entire opponent team.

We observe that, for most runs, the obtained teams are able to complete the task (i.e., kill all the enemies) in most cases. In fact, the average number of kills is very close to the maximum achievable value and the standard deviation confirms that the behaviour of the teams is quite consistent.

Table 5. Summary of the test results (co-evolutionary approach).

Run	Type	Mean	Std	Best	Worst	Completed
1	Team kills	11.96	0.20	12	11	96.0%
	Agents' returns	7.92	0.93	9.08	4.01	
2	Team kills	11.85	0.62	12	8	94.0%
	Agents' returns	8.06	0.93	8.96	3.33	
3	Team kills	11.98	0.14	12	11	98.0%
	Agents' returns	8.20	0.56	9.09	5.10	
4	Team kills	11.97	0.22	12	10	98.0%
	Agents' returns	7.85	0.94	9.00	2.16	
5	Team kills	11.81	0.73	12	7	91.0%
	Agents' returns	8.09	1.09	9.21	3.10	
6	Team kills	11.91	0.71	12	5	97.0%
	Agents' returns	8.17	0.82	8.94	1.92	
7	Team kills	8.98	1.60	12	4	1.0%
	Agents' returns	4.43	1.19	8.22	1.02	
8	Team kills	11.65	0.77	12	9	79.0%
	Agents' returns	6.83	2.13	9.20	0.07	
9	Team kills	11.15	1.46	12	5	63.0%
	Agents' returns	7.00	1.60	9.38	2.29	
10	Team kills	11.9	0.46	12	8	93.0%
	Agents' returns	8.22	0.95	8.95	3.14	

5.1 Interpretation

In this section we practically demonstrate the interpretability of the obtained agents.

Figure 4 shows the decision tree of one of the agents evolved in one of the evolutionary runs presented before. For space reasons, we cannot present all the evolved agents from each run. However, similar considerations apply also to the other evolved agents in the various runs.

By reading the decision tree in the figure, we can describe how the agent moves in the environment. In the following, please remember that we evolve only the blue agents' behavior, and that these agents always start on the right side of the environment (see Fig. 2). To facilitate the description of the evolved policy, we added an id to each node in the decision tree.

The selected agent moves up to the left (id 24) until the local density of enemies below (id 6) or to the right (id 18) reaches a certain threshold. In both cases the agent changes the direction and moves towards the enemies (ids 11 and 21). It also moves to the right (id 25) if there is a high global density of enemies on its right (id 22). This means that this agent moves to the top left of the map and intercepts the enemies it finds in that area. Another interesting behaviour of this agent relies in the fact that it tries not to be on the front lines, in fact if there is a high density of allies in the same quadrant (id 14) it tends to move to the right (id 17), therefore from the direction from which its team started. This agent appears to behave like a "wing": it moves above the enemies and tries to eliminate the ones that try to move in the space between it and the allies below.

The attack actions are easy to understand: if an enemy is located in a certain square, the agent simply attacks that square. There are two particular cases. One is caused by the few visits of the leaf (id 9). The other one happens when there is an enemy above the agent (id 16): in this case, the agent tries to escape to the right (id 26), unless there is an obstacle in the above right square (id 19), in which case it attacks the enemy (id 27). Since an obstacle can be either a wall or an ally, this particular condition leads to two different behaviors. If the obstacle is an ally, the agent helps to kill the opponent, otherwise it tries to escape on the right. If the obstacle is present and is a wall, this means that the agent is located on the left side of that wall, since there is no possibility to have an opponent above while the agent is located next to a wall. This means that if there is a wall on the right the agent cannot escape and has to fight. According to the role that this agent appears to have, this behavior tells us that the agent tries to support other allies in the area, while it retreats if enemies are trying to surround them.

This behavior is quite common, in fact in every run agents can be seen that move to the top left of the environment and capture the enemies in that area.

Other interesting behaviours emerge from the observation of the teams in the environment. A common behavior of the agents starting closer to the opponent team is to go through the gap in the walls to reach the enemies. There are also more complex behaviors. In some runs it is possible to see some agents moving to the top of the environment, passing the walls from above and then descending to hit the enemies they encounter. Much rarer is the reverse behavior, where agents pass the walls from below and then move up.

5.2   Comparison with a Non Co-Evolutionary Approach

To provide a baseline for the proposed co-evolutionary approach, we also performed experiments in the same environment using a single phenotype for the entire team, i.e., by cloning the phenotype and assigning it to each member of the team. The parameters used in these experiments are the same shown in Table 3. Note that in this case each agent in the team shares the same decision tree structure, but each one develops its own leaf by using IQL.

In this case, the fitness evaluation is realized using the average of the seventh decile of the returns obtained by each agent over the training episodes. This choice is motivated by the following rationale. Since the structure of the agents is shared, we must favor the phenotype that, besides guaranteeing a high number of kills, also gives high importance to agents that do not kill any enemy.

Table 6 shows the test results obtained by the agents evolved in each of 10 runs over 100 unseen episodes. We can compare these results with the ones obtained in the co-evolutionary setup (shown in Table 5) by using the number

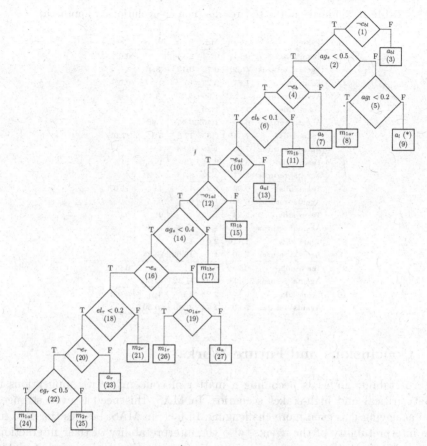

Fig. 4. Decision tree of the selected agent. The "(*)" notation indicates that the leaf has been visited a number of times that is not sufficient to train it, thus it can be seen as a random action. The numbers in parentheses are the identifiers of the nodes.

of kills at test time. In this regard, we observe that there is a large difference in performance between the two setups, being the co-evolutionary clearly superior. This indicates that, even though the agents have similar goals in both setups, the co-evolutionary setup can indeed find much better solutions. This may be due to the fact that the adoption mechanism used in the co-evolutionary approach allows for a quicker spreading of high-performing genotypes in the populations.

Another observation concerns the completion percentage: by looking at it, it appears that the performance of the non co-evolutionary is much less robust across runs. This suggests that the performance of this setup is heavily impacted by the initialization, with only a few occasional runs achieving a satisfactory completion percentage. A possible improvement of the non co-evolutionary setup would be to include an ad hoc method, e.g. based on domain knowledge, to provide a smarter initialization. We will consider this possibility in future works.

Table 6. Summary of the test results (non co-evolutionary approach).

Run	Type	Mean	Std	Best	Worst	Completed
1	Team kills	0.51	0.74	3	0	0.0%
	Agents' returns	-2.60	0.72	0.13	-3.47	
2	Team kills	1.87	2.23	10	0	0.0%
	Agents' returns	-1.15	1.66	4.27	-3.42	
3	Team kills	11.03	1.20	12	6	45.0%
	Agents' returns	6.59	1.45	9.45	2.60	
4	Team kills	11.55	1.33	12	5	87.0%
	Agents' returns	7.90	1.48	9.88	1.96	
5	Team kills	8.08	2.81	12	1	14%
	Agents' returns	3.71	2.27	8.13	-2.23	
6	Team kills	9.99	2.27	12	4	41.0%
	Agents' returns	6.00	1.95	9.22	0.81	
7	Team kills	4.24	1.93	10	0	0.0%
	Agents' returns	1.50	1.44	5.46	-1.99	
8	Team kills	11.16	2.01	12	2	81.0%
	Agents' returns	7.30	190	10.02	-0.25	
9	Team kills	0.20	0.57	2	0	0.0%
	Agents' returns	-3.08	0.48	-1.32	-3.56	
10	Team kills	6.8	2.33	12	3	3.0%
	Agents' returns	3.00	1.74	7.29	-0.50	

6   Conclusions and Future Works

Interpretability in AI is becoming a matter of concern for its applications in safety-critical and high-stakes scenarios. In MAS, this need is even stronger, and achieving it is even more challenging. In fact, in MAS, besides the need for the interpretability of the agents, also the interpretability of their interactions is important. In this paper, we proposed a co-evolutionary approach to interpretable RL in MARL settings. We evaluated our approach on the Battlefield environment from MAgent, obtaining promising results in most of the runs. In contrast, a non co-evolutionary approach obtained poorer performance.

Future work includes: 1) evaluating the proposed approach on different tasks; 2) introducing the possibility of communication between agents (both symbolic [32] and sub-symbolic [33]); 3) designing more efficient methodologies for training interpretable MARL systems, including for instance using other RL algorithms (different from Q-learning), and comparing them with existing methods, as well as handmade problem-specific policies; and 4) performing a sensitivity analysis for the proposed method.

References

1. OroojlooyJadid, A., Hajinezhad, D.: A review of cooperative multi-agent deep reinforcement learning (2020) . arXiv:1908.03963
2. Barredo Arrieta, A., et al.: Explainable Artificial Intelligence (XAI): Concepts, taxonomies, opportunities and challenges toward responsible AI. Inf. Fus. **58**, 82–115 (2020)

3. Rudin, C.: Stop explaining black box machine learning models for high stakes decisions and use interpretable models instead. Nat. Mach. Intell. **1**(5), 206–215 (2019)
4. Rudin, C., Radin, J.: Why are we using black box models in AI when we don't need To? A lesson from an explainable ai competition. Harvard Data Sci. Rev .**1**(2) (November 2019)
5. Rudin, C., Chen, C., Chen, Z., Huang, H., Semenova, L., Zhong, C.: Interpretable machine learning: fundamental principles and 10 grand challenges, July 2021. arXiv:2103.11251
6. Custode, L.L., Iacca, G.: Evolutionary learning of interpretable decision trees (2020)
7. Potter, M.A., De Jong, K.A.: A cooperative coevolutionary approach to function optimization. In: Davidor, Y., Schwefel, H.-P., Männer, R. (eds.) PPSN 1994. LNCS, vol. 866, pp. 249–257. Springer, Heidelberg (1994). https://doi.org/10.1007/3-540-58484-6_269
8. Zheng, L., et al.: MAgent: a many-agent reinforcement learning platform for artificial collective intelligence. In: Proceedings of the AAAI Conference on Artificial Intelligence, vol. 32, pp. 8222–8223 (2018)
9. Terry, J.K., et al.: Pettingzoo: gym for multi-agent reinforcement learning (2020). arXiv:2009.14471
10. Busoniu, L., Babuska, R., De Schutter, B.: A comprehensive survey of multiagent reinforcement learning. IEEE Trans. Syst. Man Cybernet. Part C (Applications and Reviews) **38**(2) 156–172 (2008)
11. Stone, P., Veloso, M.: Multiagent Systems: A Survey from a Machine Learning Perspective: Technical report. Defense Technical Information Center, Fort Belvoir, VA, December 1997
12. Yu, C., Liu, J., Nemati, S.: Reinforcement Learning in Healthcare: a survey, April 2020. arXiv:1908.08796
13. Sandholm, T.W., Crites, R.H.: On multiagent Q-learning in a semi-competitive domain. In: Weiß, G., Sen, S. (eds.) IJCAI 1995. LNCS, vol. 1042, pp. 191–205. Springer, Heidelberg (1996). https://doi.org/10.1007/3-540-60923-7_28
14. Littman, M.L.: Markov games as a framework for multi-agent reinforcement learning. In: Machine Learning Proceedings 1994. Morgan Kaufmann, San Francisco (CA), pp. 157–163 (1994)
15. Haynes, T., Wainwright, R.L., Sen, S., Schoenefeld, D.A.: Strongly typed genetic programming in evolving cooperation strategies. In: International Conference on Genetic Algorithms, San Francisco, CA, USA, pp. 271–278. Morgan Kaufmann Publishers Inc. (July 1995)
16. Tan, M.: In: Multi-agent Reinforcement Learning: Independent vs, pp. 487–494. Cooperative Agents. Morgan Kaufmann Publishers Inc., San Francisco (1997)
17. Lauer, M., Riedmiller, M.A.: An algorithm for distributed reinforcement learning in cooperative multi-agent systems. In: International Conference on Machine Learning, San Francisco, CA, USA, pp. 535–542. Morgan Kaufmann Publishers Inc. (2000)
18. Fuji, T., Ito, K., Matsumoto, K., Yano, K.: Deep multi-agent reinforcement learning using DNN-weight evolution to optimize supply chain performance. In: Hawaii International Conference on System Sciences, pp. 1278–1287. Honolulu, HI, USA, HICSS, (2018)

19. Omidshafiei, S., Pazis, J., Amato, C., How, J.P., Vian, J.: Deep decentralized multitask multi-agent reinforcement learning under partial observability. In: International Conference on Machine Learning, pp. 2681–2690. Sydney, NSW, Australia, JMLR.org, August 2017

20. Matignon, L., Laurent, G.J., Le Fort-Piat, N.: Hysteretic q-learning: an algorithm for decentralized reinforcement learning in cooperative multi-agent teams. In: International Conference on Intelligent Robots and Systems, pp. 64–69. New York, NY, USA, IEEE/RSJ (2007)

21. Tampuu, A., et al.: Multiagent cooperation and competition with deep reinforcement learning, November 2015. arXiv:1511.08779

22. Chu, X., Ye, H.: Parameter sharing deep deterministic policy gradient for cooperative multi-agent reinforcement learning, October 2017. arXiv:1710.00336

23. Singh, A., Jain, T., Sukhbaatar, S.: Learning when to communicate at scale in multiagent cooperative and competitive tasks (2018). arXiv:1812.09755

24. Macua, S.V., et al.: Diff-DAC: distributed actor-critic for average multitask deep reinforcement learning (2019). arXiv:1710.10363

25. Sunehag, P., et al.: Value-decomposition networks for cooperativae multi-agent learning based on team reward. In: International Conference on Autonomous Agents and MultiAgent Systems, Stockholm, Sweden, International Foundation for Autonomous Agents and Multiagent Systems, pp. 2085–2087, July 2018

26. Yang, J., Nakhaei, A., Isele, D., Fujimura, K., Zha, H.: CM3: cooperative multi-goal multi-stage multi-agent reinforcement learning, January 2020. arXiv:1809.05188

27. Virgolin, M., De Lorenzo, A., Medvet, E., Randone, F.: Learning a formula of interpretability to learn interpretable formulas. In: Bäck, T., et al. (eds.) Parallel Problem Solving from Nature, pp. 79–93. Springer International Publishing, Cham (2020)

28. Barceló, P., Monet, M., Pérez, J., Subercaseaux, B.: Model interpretability through the lens of computational complexity. In: Proceedings of 33rd conference on Advances in Neural Information Processing Systems (2020)

29. Custode, L.L., Iacca, G.: A co-evolutionary approach to interpretable reinforcement learning in environments with continuous action spaces. In: 2021 IEEE Symposium Series on Computational Intelligence (SSCI), pp. 1–8, December 2021

30. Ryan, C., Collins, J.J., Neill, M.O.: Grammatical evolution: evolving programs for an arbitrary language. In: Banzhaf, W., Poli, R., Schoenauer, M., Fogarty, T.C. (eds.) EuroGP 1998. LNCS, vol. 1391, pp. 83–96. Springer, Heidelberg (1998). https://doi.org/10.1007/BFb0055930

31. Sutton, R.S., Barto, A.G.: Reinforcement Learning: An Introduction. A Bradford Book, Cambridge (2018)

32. Foerster, J., Assael, I.A., de Freitas, N., Whiteson, S.: Learning to communicate with deep multi-agent reinforcement learning. In Lee, D., Sugiyama, M., Luxburg, U., Guyon, I., Garnett, R., eds.: Advances in Neural Information Processing Systems, vol. 29, Curran Associates, Inc. Red Hook (2016)

33. Lotito, Q.F., Custode, L.L., Iacca, G.: A signal-centric perspective on the evolution of symbolic communication. In: Proceedings of the Genetic and Evolutionary Computation Conference. Association for Computing Machinery, pp. 120–128. New York, NY, USA, June (2021)

Author Index

Printed in the United States
by Baker & Taylor Publisher Services